The Minkowski and Conformal Superspaces

The Classical and Quantum Descriptions

The Minkowski and Conformal Superspaces

The Classical and Quantum Descriptions

Rita Fioresi
Università di Bologna, Italy

María Antonia Lledó
Instituto de Física Corpuscular (Universitat de València
and Consejo Superior de Investigaciones Científicas), Spain

World Scientific

NEW JERSEY • LONDON • SINGAPORE • BEIJING • SHANGHAI • HONG KONG • TAIPEI • CHENNAI

Published by

World Scientific Publishing Co. Pte. Ltd.

5 Toh Tuck Link, Singapore 596224

USA office: 27 Warren Street, Suite 401-402, Hackensack, NJ 07601

UK office: 57 Shelton Street, Covent Garden, London WC2H 9HE

Library of Congress Cataloging-in-Publication Data

Fioresi, Rita, 1966–

The Minkowski and conformal superspaces : the classical and quantum descriptions / by Rita Fioresi (Università di Bologna, Italy), María Antonia Lledó (Universitat de València, Spain).

pages cm

Includes bibliographical references and index.

ISBN 978-9814566339 (hardcover : alk. paper)

1. Supersymmetry. 2. Generalized spaces. 3. Quantum groups. 4. Minkowski geometry. I. Lledó, María Antonia. II. Title.

QC174.17.S9F56 2015

539.7'25--dc23

2015000733

British Library Cataloguing-in-Publication Data

A catalogue record for this book is available from the British Library.

Printed in Singapore by B & Jo Enterprise Pte Ltd

Preface

This monograph originated from the desire to describe, in a mathematical rigorous way, the construction of the Minkowski and conformal superspaces as homogeneous spaces for the Poincaré and conformal supergroups, respectively. Many authors, at different times, have contributed to the development of the modern theory of the Minkowski superspace since the pioneering works in supersymmetry [77, 151, 158].

Very little is novel in these notes. Our work has grown out from our papers [26, 27, 66] on this subject and we have drawn heavily, especially in the treatment of the infinitesimal setting, from the comprehensive and beautiful monograph by V. S. Varadarajan [147].

Ideally these notes are aimed at graduate students in mathematics and physics. We have made an effort to keep our treatment as elementary as possible and self contained, providing the reader with a terse, but intuitive introduction to those advanced topics (like the theory of sheaves or the functor of points approach) which are necessary for our purposes.

As for the physics part, we try to build the physical intuition from the beginning, with introductions to special relativity, quantum mechanics and even quantum field theory. This aims to describe the fundamentals, so one cannot expect a treatment of supersymmetric field theories here. Even when our presentation greatly differs from the way in which supersymmetry is usually introduced by physicists, we try to make contact with their language.

We are very grateful to Prof. V. S. Varadarajan for his constant help and encouragement while this book was written. We have learnt this subject following his 2000 seminar at UCLA and we are deeply indebted to him.

We also wish to thank Prof. A. Waldron, Prof. A. Schwarz and Dr. E. Latini for their enthusiasm for this work; without them this book would not have been written.

We finally wish to thank Prof. L. Andrianopoli, Prof. A. Brini, Prof. C. Carmeli, Prof. A. Cattaneo, Prof. N. Ciccoli, Prof. F. Gavarini, Dr. S. D. Kwok, Prof. L. Migliorini and Prof. J. Navarro-Salas for helpful discussions and remarks. We also want to thank the UCLA Department of Mathematics, the Dipartimento di Matematica, Università di Bologna and the Departament de Física Teòrica, Universitat de València for support and hospitality during the realization of this work.

Introduction

In the sixties and seventies particle physics underwent a very fast development with many discoveries both, at the theoretical and the experimental level. The Standard Model, the model for electromagnetic and nuclear interactions, was taking its first steps.

There are two different nuclear interactions: the weak interaction and the strong interaction.

The weak interaction is responsible for the radioactive decays of nuclei of unstable isotopes of atoms by emission of particles and for the nuclear fusion, in which two nuclei colliding at high speed form a new nucleus. Nuclear fusion of two hydrogen nuclei into one helium nucleus is the process that fuels the Sun. With the works of Glashow, Weinberg and Salam, the electromagnetic and the weak nuclear interactions were unified into a single theory called the theory of the electroweak interactions.

The strong nuclear interaction is the force that keeps together the protons and neutrons (nucleons) in the nucleus in spite of the electric repulsion of the positively charged protons. Not only nucleons, but other heavy particles, generically called hadrons, were being discovered, and it seemed that they could not all be fundamental particles. Instead, one could assume the existence of only three fundamental particles, called quarks, that would combine to produce all the different hadrons. The quarks are not observed directly and their interactions are described by a theory that, in its current form, is called Quantum Cromodynamics (QCD). QCD has not been unified with the electroweak theory, at least not in the way in which electromagnetic and weak interactions are unified, but they are both very similar at the mathematical level. They are non abelian Yang-Mills theories with gauge symmetry groups $SU(2) \times U(1)$ (electroweak) and $SU(3)$ (QCD). The word 'non abelian' simply means that the gauge symmetry groups are non abelian. Electroweak theory and QCD form what is known as the Standard Model (SM) of particle interactions. It is worth to point out that quarks also undergo electroweak interactions.

This leaves out the fourth interaction, gravity, described by a completely different theory: Einstein's General Relativity (GR). This is a theory in which the geometry of the spacetime becomes the dynamical variable, and its mathematical treatment greatly differs from the SM. Nevertheless, since it incorporates the special theory of relativity (one says at a 'local level') one can think on the Poincaré algebra (the Lorentz algebra plus translations) as an algebra of infinitesimal symmetries.

In this state of things, physicists were looking for new constructions using bigger symmetry groups to help with unresolved problems. Nowadays we know six different types of quarks: we say that quarks come in six different 'flavors', up (u), down (d), strange (s), charm (c), bottom (b) and top (t), together with their antiparticles, $\bar{u}, \bar{d}, \bar{s}, \bar{c}, \bar{b}, \bar{t}$. The first quarks discovered were the up and down quarks. They have similar masses and there is an approximate SU(2) symmetry of the Standard Model in which (u, d) conform a doublet of SU(2), while $(\bar{u}, \bar{d})$ are in the dual or contragradient representation. This symmetry is called isospin: it is in fact an old acquaintance, since it is manifested in the similar masses and behaviour of the neutron (ddu) and the proton (uud), which also form a doublet of SU(2). Wigner was the first to propose, back in 1936, a theory that mixed the SU(2) isospin symmetry of nucleons together with the non relativistic SU(2) spin, a classic remnant of the spin of the Poincaré group. Wigner proposed an SU(4) symmetry group that would contain both SU(2)'s. Perhaps wishfully, or ahead of his time, Wigner called the SU(4) representations or multiplets of particles, supermultiplets.

The isospin group SU(2) was enhanced to SU(3) with the prediction of the strange quark, s. Then, (u, d, s) would be a triplet of SU(3). Hadrons then could be organized in other representations of SU(3) which are obtained by tensoring the fundamental representation of SU(3) and its dual several times. This successful classification of the hadrons was known as *the eightfold way*, in reference to the 8-dimensional adjoint representation of SU(3). It was proposed independently by Gell-Mann and Ne'eman in 1961. Soon after, Wigner's supermultiplets were upgraded to representations of an SU(6) symmetry group which would contain the SU(2) spin group and the SU(3) flavor group [76, 130].

The quarks u, d, s are the light quarks, being the c, b and t increasingly heavier and behaving too differently as to suppose that there is an even approximate symmetry among them.

Although this theory explained some facts, it presented problems of interpretation regarding the spin-statistics connection. Nevertheless, this line of research came to an end in 1967, with the no-go theorem of Coleman and Mandula [33], which stated that, under reasonable assumptions, the spacetime symmetries (the Poincaré group in this case) and the 'internal' symmetries such as flavor or gauge symmetries, could only appear together combined as a direct product. This was a very strong result that also seemed to put a theoretical barrier for the unification of General Relativity with the Standard Model.

Gol'fand and Likhtman [77] were the first to use a new object to try to circumvent the Coleman-Mandula theorem. They proposed a Lie superalgebra, as opposed to a Lie algebra, as the algebra of infinitesimal symmetries of a physical theory. Lie superalgebras are vector spaces with a bilinear bracket that is symmetric or antisymmetric depending on the elements on which it acts. The Jacobi identity needs only a modification to work also on Lie superalgebras.

Pioneering works on superalgebras applied to physical theories are Volkov and Akulov [151], Wess and Zumino [158, 159], Ferrara and Zumino [49] and Salam and Strathdee [131, 132] to mention just a few.

This prompted a development of physics in a different direction and ultimately led to the discovery of supersymmetric theories and supergravity. In supergravity, all of the interactions stem out of a geometric principle so they are all treated in the same way.

The most obvious restriction that physics imposes on the suitable Lie superalgebras is that the elements or generators that have a symmetric bracket (or anticommutator) among them must be in a spinorial representation of the Lorentz algebra, $\mathfrak{so}(1,3)$. These are representations that do not lift to representations of the Lorentz group SO(1,3) but of its spin group, $\mathrm{SL}(2,\mathbb{C})_\mathbb{R}$, which is its double cover and, consequently, has the same Lie algebra. A simpler example of this phenomenon is found with the group SO(3). It has the same Lie algebra as SU(2), its double (universal) cover, but the doublet of SU(2) is not a representation of SO(3). Then, the doublet is a spinorial representation of the orthogonal Lie algebra $\mathfrak{so}(3)$. The restriction comes from the spin-statistics connection: ordinary (non spinorial) representations are associated to generators with antisymmetric bracket and spinorial representations to generators with symmetric bracket generators.

The superalgebra of Gol'fand and Likhtman and Volkov and Akulov was the first Lie superalgebra introduced in physics. Nowadays it is known as the Poincaré superalgebra. Before, all of the physical symmetries belonged to a Lie algebra and transformed in ordinary representations of the Lorentz group. From the knowledge that we have today, it is perhaps difficult to conceive the blindness that the physics community had in the past to this possibility. It is even harder to understand that this natural generalization is still today seen with diffidence in some circles.

On the mathematical side, supersymmetry has raised a lot of interest. In Refs. [90–92] Kac studied sistematically Lie superalgebras and obtained a thorough classification of the complex simple ones, together with their real forms. He also provided a discussion of their representations, which are still nowadays a very active area of research.

A beautiful new field of mathematics called *supergeometry* has emerged from this seed. The pioneering works of Berezin [14], Kostant [97], Leites [101] and Manin [108] introduced, with mathematical rigour, the concept of supermanifold in the real differentiable and complex analytic categories. This idea was somehow implicitly present in the work of the physicists, with the notion of the so called 'odd variables', but it is only with Berezin in Ref. [14] that we see the first organic treatment. It was followed by the foundational papers [97], [101]. Later on, the work of Manin [108] and then of Bernstein [37] (notes by Deligne and Morgan) began to make an extensive use of the functor of points, a technique borrowed from algebraic geometry and especially fruitful in this context.

The idea of supermanifold and the one of supervariety, which is the algebraic counterpart, are a generalization of the notion of scheme introduced by Grothendieck in algebraic geometry. An affine variety in ordinary algebraic geometry is associated, in this framework, to a finitely generated associative commutative algebra. Such an algebra appears as a quotient of the polynomial algebra in n variables (for a suitable n):

$$k[x^1, \dots, x^n]/(f_1, \dots, f_r)\,,$$

with

$$k[x^1, \dots, x^n] := F/I, \qquad F = k\langle x^1, \dots, x^n\rangle\,, \tag{0.1}$$

where k is the ground field (typically $k = \mathbb{C}$ or $k = \mathbb{R}$ [1]), F is the free algebra in the variables $x^1, \dots, x^n$, and I is the ideal of the standard commutation

[1]Whenever dealing with $k = \mathbb{R}$, special care must be exerted.

relations

$$x^i x^j - x^j x^i = 0, \qquad i, j = 1, \dots, n.$$

The algebraic relations $f_1, \dots, f_r$ in the indeterminates $x^1, \dots, x^n$ describe the algebraic variety in the affine space of dimension n: the points of the algebraic variety correspond to the common zeroes of $f_1, \dots, f_r$, but the algebraic structure encoded in the algebra is of course much richer than just the topological points.

In supergeometry, one generalizes this model in the following way. The algebra associated to an algebraic supervariety is now a quotient of the polynomial superalgebra

$$k[x^1, \dots, x^m, \theta^1, \dots, \theta^n]/(f_1, \dots, f_r, \phi_1, \dots, \phi_s),$$

with

$$k[x^1, \dots, x^m, \theta^1, \dots, \theta^n] := F'/I',$$

where

$$F' = k\langle x^1, \dots, x^m, \theta^1, \dots, \theta^n \rangle$$

is the free algebra in the variables $x^1, \dots, x^m, \theta^1, \dots, \theta^n$ and I' is the ideal of (anti)commutation relations

$$\begin{gathered} x^i x^j - x^j x^i = 0, \qquad \theta^\alpha \theta^\beta + \theta^\beta \theta^\alpha = 0, \qquad x^i \theta^\alpha - \theta^\alpha x^i = 0, \\ i, j = 1, \dots, m, \quad \alpha, \beta = 1, \dots, n. \end{gathered} \tag{0.2}$$

The variables $x^1, \dots, x^m$ and $\theta^1, \dots, \theta^n$ are called, respectively, even and odd coordinates. An affine algebraic supervariety is then described by the homogeneous[2] algebraic even $f_1, \dots, f_r$ and odd $\phi_1, \dots, \phi_s$ relations among the variables $x^1, \dots, x^m$ and $\theta^1, \dots, \theta^n$. This amounts to take the quotient of F'/I' by the homogeneous ideal generated by such relations.

Supervarieties are then modeled locally by affine supervarieties, in the same way as ordinary varieties are modeled locally by affine ones.

The differential picture, historically developed earlier than the algebraic one, requires the machinery of sheaves in order to be properly described. Intuitively, a supermanifold consists of a topological space which locally

[2]'Homogeneity' is understood with respect to the natural $\mathbb{Z}_2$-grading coming from the definition of even (degree zero) and odd (degree one) variables.

admits even coordinates $x^1, \dots, x^m$ and odd coordinates $\theta^1, \dots, \theta^n$ obeying the commutation rules (0.2). The topological space itself has an ordinary manifold structure (the underlying manifold) with local coordinates $x^1, \dots, x^m$.

Notice that both in the algebraic and in the differential setting, we have that $(\theta^\alpha)^2 = 0$, hence, the odd coordinates are nilpotent. This property has many consequences in supergeometry. The differentiable functions on a supermanifold are locally expressed in terms of the even and odd coordinates as follows:

$$f(x,\theta) = f_0(x) + \sum_I f_I(x)\theta^I, \qquad \theta^I = \theta^{i_1} \dots \theta^{i_r}, \quad I = (i_1 < \cdots < i_r),$$

where f_0 and f_I are genuine differentiable functions on the underlying manifold. Notice that the sum is finite, thanks to the nilpotency of the θ's.

Both, in the differentiable and algebraic setting, it is possible to define the notion of supergroup. Lie superalgebras are then naturally associated to Lie supergroups (algebraic and differentiable), which play a key role as symmetries of supermanifolds and supervarieties.

Notice that the supergeometric generalization that we have described is a very particular one, much more limited, in a way, than another generalization of algebraic geometry known as non commutative geometry (see [34, 109]). In non commutative geometry, the algebras of non commutative spaces are still associative but have arbitrary commutation relations. For example, one can consider a generalization of (0.1) where the ideal of commutation relations is

$$x^i x^j - q_{ij} x^j x^i = 0, \qquad q_{ij} \in k, \qquad i,j = 1, \dots, n. \tag{0.3}$$

(No sum over repeated indices is understood.) This algebra is a higher dimensional version of the quantum plane, a noncommutative space that carries an action of a quantum group.

There are more general non commutative spaces than just the quantum plane: the subject of non commutative geometry is growing in both the physics and the mathematical fronts. From the physics point of view it is very natural, since in quantum mechanics the observables form a non commutative algebra that generally can be seen as a deformation of the commutative algebra of classical observables. The biggest departure from the ordinary algebraic geometry framework is that non commutative spaces do not have, generically, local models. The concept of functor of points cannot be defined generally in non commutative geometry. This is the mathematical rephrasing of the well known uncertainty principle of quantum physics:

a non commutative space is a 'fuzzy' space in which points cannot be resolved due to the non vanishing commutation relations of the variables. So many characteristics of commutative spaces are lost in this generalization.

This is not the case of supergeometry, which preserves all of the bounties of ordinary algebraic (and differential) geometry and even gives new meaning and prominence to concepts as the functor of points.

In fact, non commutative superspaces also can be considered. This is the case of a generalization of (0.3) [110]

$$x^i x^j - q_{ij} x^j x^i = 0, \qquad \theta^\alpha \theta^\beta + p_{\alpha\beta}\theta^\beta\theta^\alpha = 0, \qquad x^i\theta^\alpha - r_{i\alpha}\theta^\alpha x^i = 0,$$
$$q_{ij},\ p_{\alpha\beta} \in k, \qquad i,j = 1,\dots,n, \quad \alpha,\beta = 1,\dots,m\,.$$

Supergeometry and non commutative geometry are then frameworks that can be superposed, as the levels at which they generalize ordinary geometry are very different. They are both needed in physics for different reasons.

The philosophy underlying the present work is to study relativistic spacetime, its supergeometric generalization and the kinematical, relativistic groups or supergroups acting on them.

At the non super level, we find two relevant groups: the Poincaré group of special relativity $\mathcal{P}$ and the conformal group $\mathcal{C}$, which contains $\mathcal{P}$, and is not a universal symmetry as the Poincaré group is. It is, for example, a symmetry of electromagnetism in absence of sources. Nevertheless, the conformal group plays a very important conceptual role, and it has revealed itself as a far reaching tool in modern developments as the gauge/gravity correspondence.

Minkowski space, the spacetime of special relativity, is just $\mathbf{M} \cong \mathbb{R}^4$, and we will denote its standard coordinates as x^μ, $\mu = 0,1,2,3$, the notation generally used in physics. What is crucial about Minkowski space is the quadratic form

$$(x^0)^2 - (x^1)^2 - (x^2)^2 - (x^3)^2\,,$$

and the group of linear transformations preserving it, the Lorentz group $\mathcal{L} = \mathrm{O}(1,3)$. Together with spacetime translations, it forms the Poincaré group $\mathcal{P} = \mathcal{L} \ltimes \mathbb{R}^4$, the group of transformations among inertial systems.

When the Poincaré symmetry is enlarged to the conformal group, $\mathcal{C} = \mathrm{O}(2,4)$, spacetime has to be enlarged too by adding some points at infinity that make the vector fields of the action of the conformal group complete. The result is the conformal space, $\mathbf{C}$, a smooth, analytic and algebraic, compact variety that contains densely the Minkowski space. One says that the conformal space $\mathbf{C}$ is a compactification of the Minkowski space. More is true: **C** is covered by open dense affine spaces and the Minkowski space is identified with one of such spaces called the big cell.

Schematically, we can represent the group actions and the relations between the conformal and Minkowski space and their respective groups of symmetries in the following commutative diagram

$$\begin{array}{ccc} \mathcal{C} \times \mathbf{C} & \longrightarrow & \mathbf{C} \\ \uparrow & & \uparrow \\ \mathcal{P} \times \mathbf{M} & \longrightarrow & \mathbf{M} \end{array} \tag{0.4}$$

where the vertical arrows are the inclusions.

Physicists work often in local coordinates, which is enough for many purposes. Then, conformal and Minkowski spaces look alike. But the importance of the global picture cannot be underestimated. We just mention two situations where the conformal space is of vital importance: the *twistor* approach of Penrose [124] to spacetime and the gauge/gravity conjecture, where conformal symmetry plays a pivotal role.

We describe a generalization of the picture represented in (0.4) to the super case. We will introduce the conformal and Minkowski superspaces together with the Poincaré and conformal supergroups. We will demonstrate how these supergroups act naturally on the respective superspaces so that the diagram (0.4) holds unchanged. Furthermore, the Minkowski superspace is realized exactly as in the ordinary setting, as the big cell inside the conformal superspace.

The Minkowski superspace has been used by physicists almost since the advent of supersymmetry. As in the non super case, conformal superspace is perhaps not so widely known, but of fundamental importance.

Our main purpose is to describe these superspaces and the corresponding supergroups from a modern, supergeometric point of view. As we will see through the book, the generalization is far from trivial and presents many aspects that perhaps have not been treated comprehensively before.

Manin [108] was the first who developed a systematical treatment. However, he did not describe the realization of the Minkowski and conformal superspaces as homogeneous superspaces, essentially because the theory of homogeneous superspaces was not yet fully understood. For us, inspired by physics, this aspect will be of utmost importance.

Mathematically, the development of the theory of Minkowski and conformal superspaces is essentially an application of supergeometry techniques, which have found their correct layout only fairly recently (see the seminal works by Berezin [14], Kostant [97], Leites [101] and the more modern treatments by Manin [108], Bernstein [37], Varadarajan [147], and one of the authors and collaborators [25]).

Also from the mathematical viewpoint, the problem was faced originally in its infinitesimal version only. However, the non linearity of the construction plays a key role, since it encodes the information, not only of the action of the Poincaré supergroup, but also of the conformal supergroup. The conformal superspace is indeed a supermanifold with no global coordinates on it: it was so in the non super version too. It is our purpose to reach a complete understanding of the global properties of the conformal superspace.

The organization of our work is as follows.

In Chapter 1 we give a quick introduction to supergeometry and to the mathematical tools that are needed in the subsequent sections. Such tools include a careful construction of quotients of Lie supergroups in the analytic and differentiable setting, some notions on algebraic supergeometry and functors and a brief treatment of Hopf algebras and superalgebras.

In Chapter 2 we first give an introduction to special relativity, its physical motivation and the questions that gave rise to the notions of Minkowski space, Lorentz group and Poincaré group. We then introduce conformal symmetry as it appears in physics. These are issues that could be suppressed in a strictly axiomatic presentation, but one would lose all the physical insights. We intend to explain to the reader with, perhaps, less formation in physics than in mathematics, the compelling reasons that forced physicists to introduce the new concepts. We then review the mathematical construction of the complex conformal space in terms of a Grassmannian. We will see it as an analytic manifold, as a projective variety and as an homogeneous space. We will use the language which will suit best the extension to supergeometry. In particular, we give also the functor of points of the ordinary conformal space. We then proceed to obtain the Minkowski space, defined as the big cell inside the conformal space. At this point, the connection with the physical picture will become evident. The last section is devoted to the real form, a delicate issue that requires special care and that we have preferred to treat separately.

In Chapter 3 we introduce the concepts that are needed to understand what the hypothesis of supersymmetry means in physics. Although mathematically independent from the quantum theory, the role of supersymmetry in physics could not be understood from a purely classical perspective. We start with an introduction to quantum physics. It is of crucial importance to understand what are the precise concepts of *particle* and *spin* in the relativistic setting. These are intimately related to the unitary representations of the Poincaré group, so we start by describing them. As we have mentioned, the spin-statistics connection imposes restrictions on the form of the Lie superalgebras suitable for physics, so we give also a brief description of this theorem of Quantum Field Theory. We then proceed to explain with some detail the historical developments and the philosophy that led to the introduction of supersymmetry in physics. As before, this is not strictly needed for the mathematical approach that follows, although it should be useful for a mathematician interested in supersymmetry.

Chapter 4 is the heart of our treatment. After a technical interlude on Clifford algebras that are needed for the subsequent developments, we obtain the conformal superalgebra (or Wess-Zumino superalgebra) from some basic premises. We will be able to show that it is unique, as it is the Poincaré superalgebra sitting inside the conformal one, given some reasonable requirements. We then introduce the conformal superspace, first the *chiral* one, which is a simpler version, in terms of a super Grassmannian and then the full one, given as a superflag. We describe carefully the properties of these superspaces, as we did in the non super case. Then we consider the Minkowski superspaces as big cells inside the super Grassmannian and the superflag. As before, the complex description comes first and then we deal with the real form. We end the chapter with a careful translation to the language used in physics of the results.

In Chapter 5 we take a further step and we give a non commutative deformation of the conformal and Minkowski spaces and superspaces, based on the quantum group deformation of their symmetry groups. We give a motivation to the idea of considering a non commutative spacetime. Then we give a very brief account on quantum groups that suffices for the deformations that we present. The main point is that the diagram (0.4) can be transported to the quantum version for both the non super and the super cases, with some appropriate changes. These changes amount to the fact that in the quantum world we do not speak of spaces (or superspaces) anymore, but only of algebras (or superalgebras) generalizing the algebras of functions. The diagram (0.4) will then be replaced by its corresponding 'dual' diagram, that is a diagram with spaces replaced by the algebras of functions on them and the arrows reversed. The algebraic setting is most fruitful for this translation and this is the language that we shall use.

In the end we provide five appendices, with somehow different purposes. The first appendix is on category theory and has the intention to get the reader acquainted with some of the prerequisites and to establish our terminology. We urge the unexperienced reader to spend time to get confidence on this matter. The appendix on the Representability Theorem is a further step into the abstraction of categories. However, it provides an equivalent way, in some sense more natural, to obtain the Grassmannian and the flag manifolds and their super counterparts. One could approach the Minkowski and conformal spaces and superspaces construction through the Representability Theorem, however we felt that such path was unnecessarily adding abstraction and complications, so we decided not to take it. The appendices on super Harish-Chandra pairs and simple Lie superalgebras are not strictly needed in the text. They are fundamental issues that one cannot ignore and, furthermore, they add insight to the theory that we have developed for a very special case. Finally, the last appendix collects a few known facts on quantum groups and supergroups which are scattered in the literature: to have them collected in one place may help a non expert reading the last chapter.

The book has been written having in mind two types of readers: a mathematician interested in supersymmetry and a theoretical or mathematical physicist interested in learning the techniques of supergeometry. In both cases, the book can be considered an introduction to the subject through a specific example which, nonetheless, has a great importance in physics. For example, we have tried to give thorough proofs and explanations on the functor of points for projective space, Grassmannians, flag manifolds and their super versions. In this way the reader can learn how the concept works in practice. We have also introduced all of the physical concepts needed to understand the motivation behind supersymmetry and tried to give a translation from the mathematical language to the language used in physics. The study of supersymmetric field theories is much beyond the scope of this book, but we hope to have contributed to the mutual understanding of the two communities.

Contents

Chapter 1

Introduction to Supergeometry

Supergeometry represents a generalization of the theory of manifolds and algebraic varieties. In classical geometry, geometric objects such as manifolds are best understood as a pair consisting of a topological space and an assignment of the appropriate functions on each open set of the topological space. We consider differentiable functions when we speak of differentiable manifolds, algebraic functions when we speak of algebraic varieties and so on. In algebraic geometry we allow 'functions' with a nilpotent part, so strictly speaking, they are not functions on the topological space, but sections of a certain sheaf called the structural sheaf of the manifold, variety or scheme. In supergeometry, a further generalization is made by making the local rings $\mathbb{Z}_2 = \mathbb{Z}/2\mathbb{Z}$-graded and supercommutative. This generalization is thus ultimately based on the category of super vector spaces.

We therefore will develop the theory starting from super vector spaces and linear superalgebra, moving on then to the notion of supermanifolds and supervarieties. Supergroups, both in the differential and algebraic setting, together with the notion of action and homogeneous space, are also treated extensively, given their importance in our applications.

1.1 Linear superalgebra

In this section we introduce the basic definitions of linear superalgebra including super vector spaces, superalgebras, super Lie algebras and their morphisms, together with their main properties. Our treatment is self contained, however for more details we refer the reader to Refs. [25], [147], [37] or the classical references [108], [14].

Let k be our ground field, $\mathrm{char}(k) = 0$. In all our applications $k = \mathbb{R}$ or $k = \mathbb{C}$.

Definition 1.1.1. A *super vector space* is a $\mathbb{Z}_2$-graded vector space

$$V = V_0 \oplus V_1$$

where V_0 and V_1 are subspaces of V. The elements of V_0 are called *even*, while those in V_1 are called *odd*. The *parity* of $v \in V$, denoted $p(v)$ or $|v|$, is defined only on non zero *homogeneous* elements, that is, elements of either V_0 or V_1

$$p(v) = |v| = \begin{cases} 0 \text{ if } v \in V_0 \setminus \{0\}, \\ \\ 1 \text{ if } v \in V_1 \setminus \{0\}. \end{cases}$$

Since any element may be expressed as the sum of homogeneous elements, it suffices to consider only homogeneous elements in the statement of definitions, theorems, and proofs.

We say that $p|q$ is the *super dimension* of a finite dimensional super vector space $V = V_0 \oplus V_1$ if $\dim(V_0) = p$ and $\dim(V_1) = q$ as ordinary vector spaces.

The *tensor product* $V \otimes W$ of two super vector spaces V and W is again a super vector space with:

$$(V \otimes W)_0 = (V_0 \otimes W_0) \oplus (V_1 \otimes W_1)\,,$$

$$(V \otimes W)_1 = (V_0 \otimes W_1) \oplus (V_1 \otimes W_0)\,.$$

■

The most important example of super vector space is $k^{p|q} := k^p \oplus k^q$. If $(e_1, \dots, e_p)$ is the standard basis of the even part k^p and $(\epsilon_1, \dots, \epsilon_q)$ is the standard basis of the odd part k^q, we call $(e_1, \dots, e_p, \epsilon_1, \dots, \epsilon_q)$ the *canonical basis* of $k^{p|q}$.

More generally if V is any finite dimensional super vector space, with $\dim(V) = p|q$, fixing a basis $(v_1, \dots, v_p)$ of V_0 and a basis $(\nu_1, \dots, \nu_q)$ of V_1, V becomes identified with $k^{p|q}$.

Notice that we use roman letters for the even elements and Greek letters for the odd ones. We shall conform to this convention as much as possible.

Definition 1.1.2. A *superalgebra* A is a $\mathbb{Z}_2$-graded algebra, $A = A_0 \oplus A_1$, with multiplication morphism respecting the parity, that is:

$$p(ab) = p(a) + p(b), \qquad \mod (2)\,.$$

The subspace A_0 is an algebra, while A_1 is an A_0-module. We denote with J_A the (two-sided) ideal of A generated by the nilpotent elements and we call $A_r = A/J_A$ the *reduced algebra* associated with A.

A *morphism* of superalgebras is a parity preserving morphism of algebras.

■

The tensor product $A \otimes B$ of two superalgebras A and B is again a superalgebra, with multiplication defined as

$$(a \otimes b)(c \otimes d) = (-1)^{|b||c|}(ac \otimes bd)\,. \tag{1.1}$$

The basic principle in linear superalgebra is that the interchange of two odd elements introduces a minus sign. This is the *rule of signs* and is behind the above definition of product in $A \otimes B$. The fundamental reason why the rule of signs works in linear superalgebra lies in the following theorem.

Theorem 1.1.3. *There is a unique action L of the permutation group S_n on $V \otimes \cdots \otimes V$ (n factors) such that for any $i < n$, the element s_i of S_n that sends i to $i+1$ and vice-versa and fixes all the others goes over to the map*

$$L(s_i) : v_1 \otimes \cdots \otimes v_n \longmapsto (-1)^{p(v_i)p(v_{i+1})} v_1 \otimes \cdots \otimes v_{i+1} \otimes v_i \otimes \cdots \otimes v_n\,.$$

For an arbitrary $s \in S_n$:

$$L(s)(v_1 \otimes \cdots \otimes v_n) = (-1)^{p(s)} v_{s^{-1}(1)} \otimes \cdots \otimes v_{s^{-1}(n)}\,,$$

where

$$p(s) = \#\{(k,l) \,|\, k < l,\, v_k,\, v_l \,\text{both odd},\, s(k) > s(l)\}\,.$$

Proof. (Sketch). For $n = 3$ one can check that for any $i < n$ we have

$$L(s_i)L(s_{i+1})L(s_i) = L(s_{i+1})L(s_i)L(s_{i+1})\,.$$

Since S_n is generated by the s_i subject to the relations

$$s_i^2 = 1, \qquad s_i s_{i+1} s_i = s_{i+1} s_i s_{i+1}, \qquad 1 \leq i \leq n-1\,,$$

it is immediate that there is an action of S_n on $V \otimes \cdots \otimes V$ that sends s_i to $L(s_i)$ for all i. If we disregard the sign factors this is the action that we want. We shall prove the formula for the sign factor by induction on the length of s. The length of s, denoted as $\ell(s)$, is the minimal number of transpositions (i.e., permutations of only two elements) in which the permutation s can be decomposed. First notice that if $\ell(s) = 1$ then s can be decomposed in an odd number of $s = s_i$ and then the result is obvious. Suppose that $\ell(s) > 1$ and assume the result for permutations of smaller length. We can find i such that $(i, i+1)$, $s(i) > s(i+1)$. We define $t = ss_i$. It is then easily verified

$$\ell(t) = \ell(s) - 1 .$$

By induction the result is true for t. Now

$$\begin{aligned} L(s)(v_1 \otimes \cdots \otimes v_n) &= (-1)^{p(v_i)p(v_{i+1})} L(t) v_{s_i(1)} \otimes \cdots \otimes v_{s_i(n)} \\ &= (-1)^{p(s)} v_{s^{-1}(1)} \otimes \cdots \otimes v_{s^{-1}(n)} . \end{aligned}$$

■

The following corollary is an immediate and important consequence of this theorem. It tells us that applying a permutation through successive exchanges at the level of tensors is independent of the way in which the permutation is realized as product of adjacent transpositions, ultimately justifying the consistency of the sign rule that we have introduced.

Corollary 1.1.4. *Consider the super vector spaces V_i, $i = 1, \dots, n$. For each $s \in S_n$ let*

$$\begin{aligned} V_1 \otimes \cdots \otimes V_n &\xrightarrow{L(s)} \quad V_{s^{-1}(1)} \otimes \cdots \otimes V_{s^{-1}(n)} \\ v_1 \otimes \cdots \otimes v_n &\longrightarrow (-1)^{p(s)} v_{s^{-1}(1)} \otimes \cdots \otimes v_{s^{-1}(n)} . \end{aligned}$$

If $s = s_{i_1} \cdots s_{i_r}$ then $L(s) = L(s_{i_1}) \cdots L(s_{i_r})$. In particular, $L(st) = L(s)L(t)$.

■

We now continue our brief treatment of the foundations of supergeometry and introduce the fundamental notion of *commutativity*.

Definition 1.1.5. A superalgebra A is said to be *commutative* if for any two homogeneous elements x, y

$$xy = (-1)^{p(x)p(y)} yx .$$

The category of commutative superalgebras will be denoted by (salg), while the set of morphisms between the commutative algebras A and B with $\mathrm{Hom}_{(\mathrm{salg})}(A, B)$. ■

The most important example of commutative superalgebra is the *polinomial superalgebra*; any finitely generated commutative superalgebra is a quotient of it.

Definition 1.1.6. We define the *polynomial superalgebra* in the indeterminates $x_1, \dots, x_m$, $\xi_1, \dots, \xi_n$ as the super vector space:

$$k[x_1, \dots, x_m, \xi_1, \dots, \xi_n] := k[x_1, \dots, x_m] \otimes \wedge(\xi_1, \dots, \xi_n)\,,$$

where $k[x_1, \dots, x_m]$ is the usual polynomial algebra on the even indeterminates $x_1, \dots, x_m$, while $\wedge(\xi_1, \dots, \xi_n)$ is the exterior algebra on the odd indeterminates $\xi_1, \dots, \xi_n$. The parity is as follows:

$$k[x_1, \dots, x_m, \xi_1, \dots, \xi_n]_0 := \left\{ f_0 + \sum_{|I| \text{ even}} f_I \xi_I \quad \Big| \quad I = \{i_1 < \dots < i_r\} \right\},$$

with $\xi_I = \xi_{i_1} \xi_{i_2} \dots \xi_{i_r}$, $|I| = r$ and $f_0, f_I \in k[x_1, \dots, x_m]$, and

$$k[x_1, \dots, x_m, \xi_1, \dots, \xi_n]_1 := \left\{ \sum_{|J| \text{ odd}} f_J \xi_J \quad \Big| \quad J = \{j_1 < \dots < j_s\} \right\},$$

where

$$x_i x_j = x_j x_i, \qquad x_i \xi_k = \xi_k x_i, \qquad \xi_k \xi_l = -\xi_l \xi_k\,.$$

■

1.2 Lie superalgebras

Lie superalgebras and their representations are very important in our treatment. Once the sign rule is established (see Theorem 1.1.3) their definition is very natural.

Definition 1.2.1. Let $\mathfrak{g} = \mathfrak{g}_0 \oplus \mathfrak{g}_1$ be a super vector space. We say that $\mathfrak{g}$ is a *Lie superalgebra*, if we have a *bracket* $[\ ,\ ] : \mathfrak{g} \times \mathfrak{g} \longrightarrow \mathfrak{g}$ which satisfies the following properties (as usual, x, $y \in \mathfrak{g}$ are homogeneous):

- *Antisymmetry:*

$$[x, y] + (-1)^{p(x)\,p(y)} [y, x] = 0\,,$$

- *Jacobi identity:*

$$(-1)^{p(x)\,p(z)} [x, [y, z]] + (-1)^{p(y)\,p(x)} [y, [z, x]] + (-1)^{p(z)\,p(y)} [z, [x, y]] = 0\,.$$

■

The basic example of a Lie superalgebra is $\mathrm{End}(V)$ the endomorphisms of the super vector space V, with $\mathrm{End}(V)_0$ the endomorphisms preserving parity and $\mathrm{End}(V)_1$ the endomorphisms reversing parity. The bracket is defined as:

$$[X, Y] := XY - (-1)^{|X||Y|} YX \,.$$

If $V = k^{p|q}$ we write $\mathrm{End}\left(k^{p|q}\right) := \mathrm{End}(V)$ or $\mathfrak{gl}(p\,|q) := \mathrm{End}(V)$. In the canonical basis, $\mathrm{End}(V)_0$ consists of block diagonal matrices, while $\mathrm{End}(V)_1$ consists of block off diagonal matrices, all with entries in k. Notice that $\mathrm{End}(V)$ is also an associative (non commutative) superalgebra with the composition.

More in generally, if A is an associative superalgebra we can always define a Lie superalgebra structure on A by setting:

$$[x, y] = xy - (-1)^{p(x)p(y)} yx \,.$$

Given a super vector space, we can ask what are the requirements that a bilinear map must satisfy in order to define a Lie superalgebra structure. For practical purposes the following proposition is very useful.

Proposition 1.2.2. *Let $\mathfrak{g}$ be a super vector space and $[\, , \,] : \mathfrak{g} \times \mathfrak{g} \longrightarrow \mathfrak{g}$ a bilinear map. Then, $[\, , \,]$ is a super Lie bracket if and only if*

(1) $\mathfrak{g}_0$ is an ordinary Lie algebra with $[\, , \,]|_{\mathfrak{g}_0 \times \mathfrak{g}_0}$.
(2) $\mathfrak{g}_1$ is a $\mathfrak{g}_0$-module for the action $y \mapsto [x, y]$, $x \in \mathfrak{g}_0$, $y \in \mathfrak{g}_1$.
(3) $x \otimes y \mapsto [x, y]$ is a symmetric $\mathfrak{g}_0$-module map from $\mathfrak{g}_1 \otimes \mathfrak{g}_1$ to $\mathfrak{g}_0$.
(4) (Cubic Condition). For all $x \in \mathfrak{g}_1$, we have $[x, [x, x]] = 0$.

Proof. (Sketch). If $\mathfrak{g}$ is a Lie superalgebra, then the conditions $(1)-(4)$ are verified. Now assume the $(1)-(4)$ are true. Condition (1) implies the antisymmetry and the Jacobi identity for even elements. Condition (2) for two even elements is equivalent to the Jacobi identity for two even and one odd elements. Condition (3) is equivalent to the Jacobi identity for one even and two odd elements and the antisymmetry condition for two odd elements. The condition (4) is equivalent to the Jacobi identity for three odd elements which reads

$$[x, [y, z]] + [y, [z, x]] + [z, [x, y]] = 0, \qquad x, y, z \in \mathfrak{g}_1 \,.$$

If we assume that $[a, [a, a]] = 0$, with $a = x + y + z$, we obtain the Jacobi identity. ■

The condition (4) in Proposition 1.2.2, called the *cubic condition*, is non linear and it is the most difficult condition to verify in applications, when Lie superalgebras are constructed by putting together an ordinary Lie algebra and a module.

Definition 1.2.3. A *morphism* of Lie superalgebras is a linear map respecting the bracket (and of course the parity). A morphism of a Lie superalgebra $\mathfrak{g}$ into $\mathrm{End}(V)$ is called a *representation* of $\mathfrak{g}$ in V. ∎

Remark 1.2.4. In the definition of Lie superalgebra, the Jacobi identity looks bizarre, but it can be seen to be natural as follows. For $X \in \mathfrak{g}$ let $\mathrm{ad}X$ be the linear map $Y \to [X, Y]$ of $\mathfrak{g}$ into itself. Then the Jacobi identity is equivalent to the statement that $X \to \mathrm{ad}X$ is a morphism of $\mathfrak{g}$ into $\mathrm{End}(\mathfrak{g})$ (or equivalently, a representation of $\mathfrak{g}$ into itself). ∎

1.3 Interlude on sheaf theory

The concept of sheaf allows us to treat in a unified conceptual way very different kinds of geometric objects like differentiable manifolds and algebraic varieties. In this section we shall give the definition of sheaf and briefly outline the essential results that we are going to employ later in our treatment. The reader who is familiar with sheaf theory can very well skip this section and proceed to the next one.

Definition 1.3.1. Let M be a set with a topology and let $|M|$ denote the corresponding topological space. A *presheaf* of commutative algebras (respectively superalgebras, rings, groups, sets ...) $\mathcal{F}$ on M is an assignment

$$U \longrightarrow \mathcal{F}(U),$$

where U is open in $|M|$ and $\mathcal{F}(U)$ is a commutative algebra, such that:

(1) If $U \subset V$ are two open sets in $|M|$, there exists a morphism

$$\mathcal{F}(V) \xrightarrow{\mathrm{r}_{V,U}} \mathcal{F}(U),$$

called the *restriction morphism* and often denoted with $\mathrm{r}_{V,U}(f) = f|_U$, such that:

i) $\mathrm{r}_{U,U} = \mathrm{id}$,

ii) $\mathrm{r}_{W,U} = \mathrm{r}_{V,U} \circ \mathrm{r}_{W,V}$.

The presheaf $\mathcal{F}$ is called a *sheaf* if:

(2) Given $\{U_i\}_{i\in I}$, an open covering of U and a family $\{f_i\}_{i\in I}$, $f_i \in \mathcal{F}(U_i)$, such that $f_i|_{U_i\cap U_j} = f_j|_{U_i\cap U_j}$ for all $i,j \in I$, there exists a unique $f \in \mathcal{F}(U)$ such that $f|_{U_i} = f_i$.

The elements in $\mathcal{F}(U)$ are called *sections* over U; when $U = |M|$ we call such elements *global sections* and denote them with $\mathcal{F}(M)$.

As we shall see in the examples that follow, when each $\mathcal{F}(U)$ is a ring of numerical functions, $r_{V,U}$ is just the usual restriction map. This is the reason why we call the $r_{V,U}$ restriction morphisms.

Let $\mathcal{F}$ and $\mathcal{G}$ be presheaves on $|M|$. A *morphism of presheaves* $\phi : \mathcal{F} \longrightarrow \mathcal{G}$ is a collection of morphisms $\phi_U : \mathcal{F}(U) \longrightarrow \mathcal{G}(U)$ for each open set U in $|M|$ such that for all $V \subset U$ the following diagram commutes:

$$\begin{array}{ccc} \mathcal{F}(U) & \xrightarrow{\phi_U} & \mathcal{G}(U) \\ {\scriptstyle r_{UV}}\downarrow & & \downarrow{\scriptstyle r_{UV}} \\ \mathcal{F}(V) & \xrightarrow{\phi_V} & \mathcal{G}(V) \end{array}$$

A *morphisms of sheaves* is just a morphism of the underlying presheaves. ■

Here are some examples of particular relevance to us.

Example 1.3.2. Consider the assignment which associates to an open set U in the differentiable manifold M the commutative algebra $C^\infty(U)$, consisting of the differentiable functions on U. $r_{V,U}$ is taken to be the usual restriction map. This assignment is a sheaf and it is called the *sheaf of differentiable functions on* M. The verification of the properties (1) and (2) of the previous definition is straightforward. Similarly, one can define the *sheaf of holomorphic functions* on the complex manifold M by the assignment: $U \to \mathcal{H}(U)$, where $\mathcal{H}(U)$ are the holomorphic functions on the open set U. Finally, if X is an algebraic variety, the assignment $U \to \mathcal{O}_X(U)$ with U an open set in the algebraic variety X and $\mathcal{O}_X(U)$ the ring of polynomial functions on X, defines the *sheaf of algebraic functions* on X.

These are called *structural sheaves* since, together with the topological space structure, they identify uniquely the geometric structure. ■

A most important object associated to a presheaf is the *stalk* at a point.

Definition 1.3.3. Let $\mathcal{F}$ be a presheaf on the topological space $|M|$ and let x be a point in $|M|$. We define the *stalk* $\mathcal{F}_x$ of $\mathcal{F}$ at the point x as follows: $\mathcal{F}_x$ consists of the disjoint union of all pairs (U, s) with U open in $|M|$, $x \in U$, and $s \in \mathcal{F}(U)$, modulo the equivalence relation: $(U, s) \cong (V, t)$ if and only if there exists an open neighbourhood W of x, $W \subset U \cap V$, such that $s|_W = t|_W$.

The elements in $\mathcal{F}_x$ are called *germs of sections* at x. ∎

Clearly, any morphism of presheaves induces a morphism on the stalks: $\phi_x : \mathcal{F}_x \longrightarrow \mathcal{G}_x$. The sheaf property, i.e. property (2) in the Definition 1.3.1, ensures that if we have two morphisms of sheaves ϕ and ψ such that $\phi_x = \psi_x$ for all x, then $\phi = \psi$.

We say that the morphism of sheaves ϕ is *injective (resp. surjective)* if ϕ_x is injective (resp. surjective).

The next two propositions are very useful when one wants to define a sheaf on a given topological space starting from local data. The proofs of such statements are an easy exercise and we refer the reader to Ref. [45], Chapter 1, for more details.

Definition 1.3.4. Let $\mathcal{B}$ be a base for the topology in a topological space $|M|$. A *$\mathcal{B}$-sheaf* on $|M|$, is an assignment that assigns to each open set in $\mathcal{B}$, a commutative algebra (or superalgebra, ring, group, set etc.), and satisfies property (1) as in 1.3.1 and property (2) in 1.3.1, where we replace the intersection $U_i \cap U_j$ with $V \in \mathcal{B}$, $V \subset U_i \cap U_j$. ∎

Proposition 1.3.5. *Let $\mathcal{B}$ be a base for the open sets in the topological space $|M|$.*

(1) Every $\mathcal{B}$-sheaf extends uniquely to a sheaf on $|M|$.
(2) If $\mathcal{G}$ and $\mathcal{H}$ are two sheaves on $|M|$ and for all U in $\mathcal{B}$ we have a collection of morphisms

$$\psi_U : \mathcal{G}(U) \longrightarrow \mathcal{H}(U)$$

commuting with restrictions, then there is a unique sheaf morphism $\phi : \mathcal{G} \longrightarrow \mathcal{H}$ such that $\phi_U = \psi_U$, for all $U \in \mathcal{B}$.

∎

Proposition 1.3.6. *Let $\{U_i\}_{i \in I}$ be an open covering of the topological space $|M|$. Assume the following:*

(1) We have defined sheaves $\mathcal{F}_{U_i}$ for all i.

(2) $\phi_{U_iU_j} : \mathcal{F}_{U_i}|_{U_i \cap U_j} \longrightarrow \mathcal{F}_{U_j}|_{U_i \cap U_j}$ *are isomorphisms satisfying the compatibility conditions:*

$$\phi_{U_iU_j}\phi_{U_kU_i} = \phi_{U_kU_j}, \qquad on \quad U_i \cap U_j \cap U_k, \quad \forall i,j,k \in I\,.$$

Then there exists a unique sheaf $\mathcal{F}$ on $|M|$ such that $\mathcal{F}|_{U_i} = \mathcal{F}_{U_i}$. ■

We are ready for the definition of *ringed space.*

Definition 1.3.7. A *ringed space* is a pair $M = (|M|, \mathcal{F})$ consisting of a topological space $|M|$ and a sheaf of commutative rings $\mathcal{F}$ on $|M|$. We say that the ringed space $(|M|, \mathcal{F})$ is a *locally ringed space* if the stalk $\mathcal{F}_x$ is a local ring for all $x \in |M|$ (i.e. it has a unique maximal ideal).

A *morphism* of ringed spaces $\phi : M = (|M|, \mathcal{F}) \longrightarrow N = (|N|, \mathcal{G})$ consists of a morphism $|\phi| : |M| \longrightarrow |N|$ of the topological spaces (in other words, $|\phi|$ is a continuous map) and a sheaf morphism $\phi^* : \mathcal{O}_N \longrightarrow \phi_* \mathcal{O}_M$ where $\phi_*\mathcal{O}_M$ is the sheaf on $|N|$ defined as

$$(\phi_*\mathcal{O}_M)(U) = \mathcal{O}_M(\phi^{-1}(U))$$

for all U open in $|N|$. A morphism of ringed spaces induces a morphism on the stalks for each $x \in |M|$ $\phi_x : \mathcal{O}_{N,|\phi|(x)} \longrightarrow \mathcal{O}_{M,x}$. If M and N are locally ringed spaces, we say that the morphism of ringed spaces ϕ is a *morphism of locally ringed spaces* if ϕ_x is local, i.e. if

$$\phi_x^{-1}(m_{M,x}) = m_{N,|\phi|(x)}\,,$$

where $m_{N,|\phi|(x)}$ and $m_{M,x}$ are the maximal ideals in the local rings $\mathcal{O}_{N,|\phi|(x)}$ and $\mathcal{O}_{M,x}$ respectively. ■

Observation 1.3.8. If M is a differentiable manifold and $\mathcal{O}_M$ is its structural sheaf, the pair $(|M|, \mathcal{O}_M)$ is a locally ringed space. The same is true if M is a complex manifold or an algebraic variety. Moreover, the morphisms in the differential or in the algebraic setting respectively, correspond precisely to morphisms of locally ringed spaces.

At this point it is not hard to convince ourselves that we can take a different point of view on the definition of differentiable manifold. Namely, we can equivalently define a differentiable manifold as a locally ringed space $M = (|M|, \mathcal{O}_M)$, locally isomorphic to the locally ringed space $(\mathbb{R}^p, C^\infty_{\mathbb{R}^p})$. Similarly, we can define a complex manifold M as a locally ringed space $M = (|M|, \mathcal{O}_M)$ locally isomorphic to the locally ringed space $(\mathbb{C}^p, \mathcal{H}_{\mathbb{C}^p})$. ■

It is important to keep this point of view in mind, since this is the path to the definition of *supermanifolds* in the differential and analytic settings and of *supervarieties* in the analytic setting.

1.4 Supermanifolds

In this section we shall introduce real differentiable and complex analytic supermanifolds. As it happens in the ordinary setting, these are very different categories, however, they share many features in common. In this preliminary section we shall state only definitions and results holding for both. For this reason we shall write k to mean our ground field to be $\mathbb{R}$ or $\mathbb{C}$, with the understanding that we are either in the real differentiable or in the complex analytic category respectively.

Definition 1.4.1. A *superspace* $S = (|S|, \mathcal{O}_S)$ is a topological space $|S|$ endowed with a sheaf of superalgebras $\mathcal{O}_S$ such that the stalk at a point $x \in |S|$, denoted by $\mathcal{O}_{S,x}$, is a local superalgebra[1] for all $x \in |S|$. More generally, we speak also of *ringed superspace* whenever we have a topological space and a sheaf of superrings.

A *morphism* $\phi : S \longrightarrow T$ of superspaces is given by $\phi = (|\phi|, \phi^*)$, where $|\phi| : |S| \longrightarrow |T|$ is a map of topological spaces and $\phi^* : \mathcal{O}_T \longrightarrow \phi_* \mathcal{O}_S$ is such that

$$\phi_x^{-1}(\mathbf{m}_x) = \mathbf{m}_{|\phi|(x)} ,$$

where $\mathbf{m}_{|\phi|(x)}$ and $\mathbf{m}_x$ are the maximal ideals in the stalks $\mathcal{O}_{T,|\phi|(x)}$ and $\mathcal{O}_{S,x}$ respectively. ∎

Example 1.4.2. The superspace $k^{p|q}$ is the topological space k^p endowed with the following sheaf of superalgebras. For any open subset $U \subset k^p$

$$\mathcal{O}_{k^{p|q}}(U) = \mathcal{O}_{k^p}(U) \otimes \wedge(\xi^1 \dots \xi^q) ,$$

where $\mathcal{O}_{k^p}$ denotes the C^∞ sheaf on k^p when $k = \mathbb{R}$ and the complex analytic sheaf on k^p when $k = \mathbb{C}$. ∎

Definition 1.4.3. A *supermanifold* of dimension $p|q$ is a superspace $M = (|M|, \mathcal{O}_M)$ which is locally isomorphic to $k^{p|q}$, i.e. for all $x \in |M|$ there exist an open set $V_x \subset |M|$ and $U \subset k^p$ such that V_x and U are homeomorphic and there is an isomorphism of sheaves

$$\mathcal{O}_M|_{V_x} \cong \mathcal{O}_{k^{p|q}}|_U .$$

A *morphism* of supermanifolds is simply a morphism of superspaces.

The classical manifold $|M|$ underlying the supermanifold M is called the *reduced space* of M. Its sheaf at any open set U is $\mathcal{O}_M(U)$ modulo

[1] A local superalgebra is a superalgebra with a unique (homogeneous) maximal ideal.

the ideal generated by the nilpotent elements $J_{\mathcal{O}_M(U)}$. Notice that, with a slight but common abuse of notation, we are using $|M|$ to denote both the classical manifold and its underlying topological space. ■

From now on, with an abuse of notation, $k^{p|q}$ denotes both the super vector space $k^p \oplus k^q$ and the superspace $(k^p, \mathcal{O}_{k^{p|q}})$, the context making clear which one we mean. $k^{p|q}$ plays a key role in the definition of supermanifold, since it is the local model and of course it is itself a supermanifold. If $t^1, \dots, t^p$ arc global coordinates for k^p and $\theta^1, \dots, \theta^q$ for k^q, we shall speak of $t^1, \dots, t^p, \theta^1, \dots, \theta^q$ as a set of *global coordinates* for the superspace $k^{p|q}$.

Definition 1.4.4. Let $S = (|S|, \mathcal{O}_S)$ be a supermanifold. Given an open subset $|U| \subset |S|$, the pair $(|U|, \mathcal{O}_S|_{|U|})$ is a supermanifold, called the *open subsupermanifold* or *submanifold* for short, associated to $|U|$. ■

The next example is very important for our applications.

Example 1.4.5. *Supermatrices* $\mathrm{M}(p|q)$ *and the General Linear Supergroup* $\mathrm{GL}_{p|q}$. Let $\mathrm{M}(p|q) = k^{p^2+q^2|2pq}$. This is the superspace corresponding to the super vector space of $p|q \times p|q$ matrices. The underlying topological space is $\mathrm{M}_p \times \mathrm{M}_q$, the direct product of $p \times p$ and $q \times q$ matrices. As a super vector space

$$\mathrm{M}(p|q) = \left\{ \begin{pmatrix} A & B \\ C & D \end{pmatrix} \right\}, \ (\mathrm{M}(p|q))_0 = \left\{ \begin{pmatrix} A & 0 \\ 0 & D \end{pmatrix} \right\}, \ (\mathrm{M}(p|q))_1 = \left\{ \begin{pmatrix} 0 & B \\ C & 0 \end{pmatrix} \right\},$$

where A, B, C, D are respectively $p \times p$, $p \times q$, $q \times p$ and $q \times q$ matrices with entries in k. Hence as a superspace $\mathrm{M}(p|q)$ has $p^2 + q^2$ even global coordinates

$$t^{ij}, \qquad 1 \le i, j \le p \ \text{or} \ p+1 \le i, j \le p+q$$

and $2pq$ odd ones

$$\theta^{kl}, \qquad 1 \le k \le p \ \text{and} \ p+1 \le l \le p+q, \ \text{or} \ p+1 \le k \le p+q, 1 \le l \le p\,.$$

The structural sheaf of $\mathrm{M}(p|q)$ is the assignment

$$V \longrightarrow \mathcal{O}_{\mathrm{M}(p|q)}(V) = \mathcal{O}_{\mathrm{M}_p \times \mathrm{M}_q}(V) \otimes \wedge(\theta^{kl}), \qquad \text{for all } V \text{ open in } \mathrm{M}_p \times \mathrm{M}_q\,.$$

As one can readily check, the superspace $\mathrm{M}(p|q) = (\mathrm{M}_p \times \mathrm{M}_q, \mathcal{O}_{\mathrm{M}(p|q)})$ is a supermanifold called the supermanifold of *supermatrices*.

Now, let us consider in the topological space $\mathrm{M}_p \times \mathrm{M}_q = k^{p^2+q^2}$, the open set U consisting of the points for which $\det(t^{ij})_{1 \le i,j \le p} \ne 0$ and $\det(t^{ij})_{p+1 \le i,j \le p+q} \ne 0$. We define the superspace $\mathrm{GL}(p|q) := (U, \mathcal{O}_{\mathrm{M}(p|q)}|_U)$ to be the open subsuperspace of $\mathrm{M}(p|q)$ associated with the open set U. Again, this is a supermanifold and as we shall see, it has a *Lie supergroup* structure and it is called the *general linear supergroup*. ■

Notation. If $f \in \mathcal{O}_M(U)$ is a section of the structural sheaf of M, we define the *value* of f at $x \in |M|$, and we write $f(x)$, the image of f under the natural morphisms $\mathcal{O}_M(U) \longrightarrow \mathcal{O}_{M,x}/\mathbf{m}_{M,x} \cong k$. $f(x)$ coincides with $f_r(x)$, the value at x of the ordinary function f_r, which is the *reduced function* underlying f and it is defined as the image of f in the quotient $\mathcal{O}_M(U)/J_{\mathcal{O}_M(U)}$ of $\mathcal{O}_M(U)$ by the ideal generated by all the nilpotents $J_{\mathcal{O}_M(U)}$. ■

We now state the Chart Theorem, which prescribes how to obtain a morphism between *superdomains*, that is, supermanifolds $(U, \mathcal{O}_{k^{m|n}}|_U)$, where U is an open subset of $k^{m|n}$. For the proof and more details see Ref. [25], Chapter 4.

Theorem 1.4.6. *Chart theorem.*

- *Local version. Let $U \subset k^{p|q}$ and $V \subset k^{m|n}$ be open superdomains. There is a bijection between*

  (i) the set of morphisms $\phi : V \longrightarrow U$ and

  (ii) the set of systems of p even functions t^i and q odd functions θ^j in $\mathcal{O}_{k^{m|n}}(V)$ such that $(t^1(m), \ldots, t^p(m)) \in |U|$ for all $m \in |V|$.
- *Global version. Let $U \subset k^{p|q}$ be a superdomain and M a supermanifold. There is a one-to-one correspondence between the morphisms $M \longrightarrow U$ and the set of $(p+q)$-uples of p even functions t^i and q odd functions θ^j on M, such that $(t^1(x), \ldots, t^p(x)) \in |U|$ for all $x \in |M|$.*

Proof. See Ref. [25] Chapter 4 and Ref. [108] Chapter 4. ■

Let $M = (|M|, \mathcal{O}_M)$ be a supermanifold. By its very definition, at each point $x \in |M|$, there exists $V_x \ni x$ open such that $\mathcal{O}_M|_{V_x} \cong \mathcal{O}_{k^{p|q}}|_U$, hence, by the Chart Theorem, we can find $t^1, \ldots, t^p$ even and $\theta^1, \ldots, \theta^q$ odd elements in $\mathcal{O}_M(V_x)$, realizing such isomorphism. $(t^1, \ldots, t^p, \theta^1, \ldots, \theta^q)$ are called *local coordinates* on M at x while $m|n$ is called the *superdimension* of M. The pair consisting of the open set V_x and the local coordinates at x is called a *superchart* or *chart* for short. $k^{m|n}$ or any of its open subsets comes naturally equipped with a chart called the *canonical chart.*

We end this section with a very important example of a complex analytic supermanifold: the *complex projective superspace.*

Example 1.4.7. Let $\mathbb{P}^m = \mathbb{C}^{m+1} \setminus \{0\}/\sim$ be the ordinary complex projective space of dimension m. A point in $\mathbb{P}^m$ is an equivalence class of elements $(x_0, \dots, x_m) \in \mathbb{C}^{m+1}$ with the equivalence relation

$$(x_0, \dots, x_m) \sim (x'_0, \dots, x'_m), \quad \text{iff} \quad x'_i = \lambda x_i, \quad i = 0, \dots, m, \quad \lambda \in \mathbb{C}^\times .$$

We denote, as usual, an equivalence class in $\mathbb{P}^m$ as the $m+1$-uple $[x_0, \dots, x_m]$. The complex numbers $x_0, \dots, x_m$ are called *homogeneous coordinates.*

The projective space $\mathbb{P}^m$ is a complex manifold (see for example Ref. [79] Chapter 1). We now want to show that we can define a complex analytic supermanifold whose reduced part is the projective space; we will call it the *projective superspace.*

Let $\{U_i\}_{i=1,\dots,m}$ be the affine cover with

$$U_i = \{[x_0, \dots, x_m] \mid x_i \neq 0\}, \quad U_i \cong \mathbb{C}^m .$$

On each U_i we take the global ordinary coordinates

$$u_0^i, \dots, \hat{u}_i^i, \dots, u_m^i, \qquad u_k := x_k / x_i$$

($\hat{u}_i^i$ means, as usual, that we omit that variable). We now want to define the sheaf of superalgebras $\mathcal{O}_{\mathcal{U}_i}$ on the topological space U_i

$$\mathcal{O}_{\mathcal{U}_i}(V) = \mathcal{H}_{U_i}(V) \otimes \wedge(\xi_1, \dots, \xi_n), \qquad V \text{ open in } U_i ,$$

where $\mathcal{H}_{U_i}$ is the sheaf of holomorphic functions on U_i and $\xi_1, \dots, \xi_n$ are odd generators.

As one can readily check $\mathcal{U}_i = (U_i, \mathcal{O}_{\mathcal{U}_i})$ is a supermanifold isomorphic to $\mathbb{C}^{m|n}$. We now define the morphisms $\phi_{ij} : \mathcal{U}_j \cap \mathcal{U}_j \mapsto \mathcal{U}_j \cap \mathcal{U}_j$, where the domain is thought as an open submanifold of $\mathcal{U}_i$, while the codomain as an open submanifold of $\mathcal{U}_j$. By the Chart Theorem, the ϕ_{ij} are determined by the ordinary morphisms together with the choice of m even and n odd sections in $\mathcal{O}_{\mathcal{U}_i}(U_i \cap U_j)$. We write:

$$\phi_{ij}(u_0^i, \dots, \hat{u}_i^i, \dots, u_m^i, \xi_1, \dots, \xi_n) = \left(\frac{u_0^i}{u_j^i}, \dots, \frac{1}{u_j^i}, \dots, \frac{u_m^i}{u_j^i}, \frac{\xi_1}{u_j^i}, \dots, \frac{\xi_n}{u_j^i} \right) ,$$

where on the right hand side the $1/u_j^i$ appears in the ith position and the jth position is omitted. One can easily check that the ϕ_{ij}'s satisfy the compatibility conditions:

$$\phi_{ij}\phi_{ki} = \phi_{jk}, \qquad \text{on} \quad \mathcal{U}_i \cap \mathcal{U}_j \cap \mathcal{U}_k ,$$

hence they allow us to define uniquely a sheaf by 1.3.6 denoted with $\mathcal{O}_{\mathbb{P}^{m|n}}$. This is a supermanifold structure on the topological space $\mathbb{P}^m$, $(\mathbb{P}^m, \mathcal{O}_{\mathbb{P}^{m|n}})$, called the *projective superspace of dimension $m|n$.*

1.5 Morphisms of supermanifolds

The theory of supermanifolds resembles very closely the ordinary manifold theory. One can, for example, define tangent bundles, vector fields and the differential of a morphism in a similar way than in the classical case. Local models for morphisms of supermanifolds are very much the same as in the ordinary theory: we have the *inverse function*, the *immersion* and the *submersion theorems* both in the differential and the analytic setting. In this section we are going to briefly recall all the definitions and statements concerning such fundamental results sending the reader to Refs. [25], [101], [147] for more details.

We start with the notion of *derivation*.

Definition 1.5.1. A *left derivation* of degree $m = 0, 1$ of a superalgebra $\mathcal{A}$ is a linear map $D^L : \mathcal{A} \to \mathcal{A}$ such that

$$D^L(a \cdot b) = D^L(a) \cdot b + (-1)^{m|a|} a \cdot D^L(b), \qquad \forall a, b \in \mathcal{A} .$$

Graded left derivations or *left super derivations* span a super vector space.

In the same way one can define *right derivations* (also *right super derivations*),

$$D^R(a \cdot b) = (-1)^{m|b|} D^R(a) \cdot b + a \cdot D^R(b) .$$

Notice that derivations of degree zero are both, right and left derivations. Moreover, given a left derivation D^L of degree m one can define a right derivation D^R also of degree m in the following way

$$D^R a = (-1)^{m(|a|+1)} D^L a . \tag{1.2}$$

■

From now on, when we say 'derivation' or 'super derivation' without further specification, we mean *left* derivation.

Definition 1.5.2. Let $M = (|M|, \mathcal{O}_M)$ be a supermanifold. A *tangent vector* X_m at $m \in |M|$ is a (super) derivation $X_m : \mathcal{O}_{M,m} \to k$. The *tangent space* $T_m M$ of M at m is the super vector space of all derivations.

A *super vector bundle* E on M is a vector bundle $\pi : E \to M$ whose fiber is a super vector space and whose transition functions respect the parity. As it happens in the ordinary setting, we have that super vector bundles are in one-to-one correspondence with locally free sheaves, that is, sheaves whose stalk at x is a free $\mathcal{O}_{M,x}$ module. In such correspondence, if E is a super vector bundle, we build a sheaf $\mathcal{E}$ by taking $\mathcal{E}(U)$ as the sections of the bundle E, i.e. those functions $s : U \to E$ such that $\pi \circ s = \mathrm{id}$. ■

The most important vector bundle on a supermanifold M is its *tangent bundle.*

Definition 1.5.3. A *super vector field* V on a supermanifold M is a family of super derivations $V_U : \mathcal{O}_M(U) \longrightarrow \mathcal{O}_M(U)$ compatible with restrictions and defined for all open subsets $U \subset |M|$. The assignment associating to $U \subset |M|$ the super vector fields on U is a locally free sheaf with rank $\dim(M)$, hence it is a super vector bundle, which we call the *tangent bundle* TM, of the supermanifold M.

The fiber of TM at a point m of $|M|$ is the super vector space T_mM. Hence, the super vector fields on a supermanifold M are the sections of the tangent bundle of M. The composition of two super vector fields is not in general a vector field (it is a differential operator of higher order), however the Lie super bracket of two super vector fields is again a super vector field

$$[X,Y] = X \circ Y - (-1)^{|X||Y|} Y \circ X\,.$$

Then, super vector fields form a Lie superalgebra. ■

Definition 1.5.4. If $f : M \longrightarrow N$ is a morphism, we define its *differential* $(\mathrm{d}f)_m : T_mM \longrightarrow T_{f(m)}N$ as

$$(\mathrm{d}f)_m(X_m)\alpha = X_m(f^*(\alpha)) \qquad \forall\, X_m \in T_mM\,,$$

where $f^*(\alpha) = \alpha \circ f$. ■

Definition 1.5.5. Let $f : M \longrightarrow N$ be a supermanifold morphism. We say that f is a *submersion* at $m \in |M|$, if $(\mathrm{d}f)_m$ is surjective. Likewise, f is an *immersion* at m, if $(\mathrm{d}f)_m$ is injective. Finally f is a *diffeomorphism* at m if it is a submersion and an immersion.

When we say that f is a submersion (respectively immersion or diffeomorphism) we mean that f a submersion at all points of $|M|$. ■

As in the classical setting, submersions and immersions have the usual local models. We give the following propositions without proof, which can be found explained in Section 5.2 of Ref. [25].

Proposition 1.5.6. *The Inverse Function Theorem. Let $\phi : M \to N$ be a supermanifold morphism and let $m \in |M|$ be such that $(\mathrm{d}\phi)_m$ is bijective. Then, there exist charts U and V around m and $|\phi|(m)$ respectively such* *that* $|\phi|(U) \subseteq V$, $\phi_{|_U} : U \to V$ *is invertible and its inverse is differentiable (i.e. it is a supermanifold morphism).* ■

Proposition 1.5.7. *Immersion Theorem. Let* $\phi : M \longrightarrow N$ *be a supermanifold morphism, with* $\dim(M) = m|n \leq \dim(N) = m+p|n+q$. *The following facts are equivalent:*

(1) $\phi : M \to N$ *is an immersion at* x*;*
(2) $(\mathrm{d}\phi)_x$ *has rank* $(m|n)$*;*
(3) There exist charts U, $\{t^i\}_{i=1}^m$, $\{\theta^j\}_{j=1}^n$ *around* x *and,* $V = V_1 \times V_2$, $\{t^i\}_{i=1}^m$, $\{s^a\}_{a=1}^p$, $\{\theta^j\}_{j=1}^n$, $\{\eta^b\}_{b=1}^q$ *around* $|\phi|(x)$ *such that the restriction of the map to* U *has the form*

$$\begin{array}{ccc} U & \xrightarrow{\phi|_U} & V \\ t^i & \longrightarrow & t^i \\ \theta^j & \longrightarrow & \theta^j \\ s^a,\, \eta^b & \longrightarrow & 0\,. \end{array}$$

■

Proposition 1.5.8. *Submersion Theorem. Let* $\phi : M \longrightarrow N$ *be a supermanifold morphism, with* $\dim(M) = m+p|n+q \geq \dim(N) = m|n$. *The following facts are equivalent*

(1) $\phi : M \to N$ *is a submersion at* x*;*
(2) $(\mathrm{d}\phi)_x$ *has rank* $(m|n)$*;*
(3) There exist charts $U = U_1 \times U_2$, $\{t^i\}_{i=1}^m$, $\{s^a\}_{a=1}^p$, $\{\theta^j\}_{j=1}^n$, $\{\eta^b\}_{b=1}^q$ *around* x *and* V, $\{t^i\}_{i=1}^m$, $\{\theta^j\}_{j=1}^n$ *around* $|\phi|(x)$ *such that the restriction of the map to* U *has the form*

$$\begin{array}{ccc} U & \xrightarrow{\phi|_U} & V \\ t^i & \longrightarrow & t^i \\ \theta^j & \longrightarrow & \theta^j\,. \end{array}$$

■

We now want to discuss briefly subsupermanifolds.

Definition 1.5.9. We say that N is an *open subsupermanifold* or simply an *open submanifold* of M if $|N|$ is an open submanifold of $|M|$ and $\mathcal{O}_N$ is the restriction of $\mathcal{O}_M$ to N.

We say that N is a *closed subsupermanifold* or simply a *closed submanifod* of M if $|N|$ is a closed submanifold of $|M|$ and there exists a map $f : N \to M$ which is an immersion and such that $|f|$ is the embedding $|N| \subset |M|$. ■

Closed subsupermanifolds can be determined by using the super version of the Submersion Theorem.

Theorem 1.5.10. *Let $f : M \longrightarrow N$ be a submersion and let $|P| = |f|^{-1}(n)$ for $n \in |N|$. Then $|P|$ admits a supermanifold structure. Locally we have that for $p \in |P|$, $\mathcal{O}_{P,p} = \mathcal{O}_{M,p}/f^*(I_n)$, where I_n is the ideal in $\mathcal{O}_{N,n}$ of elements vanishing at n. Moreover,*

$$T_x P = \ker((\mathrm{d}f)_x), \qquad \dim P = \dim M - \dim N \,.$$

Frobenius theorem plays a fundamental role in constructing subsupermanifolds of a given manifold. For its proof and more details see Section 4.7 of Ref. [147] (pg. 157) and Chapter 7 of Ref. [25].

Definition 1.5.11. We say that $\mathcal{D}$ is a *super distribution* of dimension $r|s$ on the supermanifold M if it is a subbundle of the tangent bundle of M, which is locally a direct factor of dimension $r|s$. We say that $\mathcal{D}$ is *involutive* if given vector fields X, Y in $\mathcal{D}$, we have that $[X, Y]$ is also in $\mathcal{D}$. ■

Theorem 1.5.12. *Let M be a supermanifold and let $\mathcal{D}$ be an involutive super distribution on M of dimension $r|s$. Then,*

- *Local Frobenius Theorem. At each point there exists a coordinate system (x, ξ) such that the distribution at that point is spanned by ∂_{x_i}, ∂_{ξ_α}, with $1 \le i \le r$ and $1 \le \alpha \le s$.*
- *Global Frobenius Theorem. At each point there exists a unique maximal subsupermanifold N such that its tangent bundle is identified with the distribution $\mathcal{D}$.*

■

1.6 Supervarieties

In ordinary geometry, an affine algebraic variety X over an algebraically closed field k consists of the points in a certain affine space of k^n, which form the set of zeros of a finite set F of polynomials. Such set is assumed to be *irreducible*, i.e. it is not the union of two subsets, each of which consists of the zeros of a finite set of polynomials. The coordinate ring of X is defined as $\mathcal{O}(X) := k[x_1 \dots x_n]/I$, where I is the ideal of all polynomials vanishing on X and it is generated by the elements in F. The irreducibility of X amounts to the condition for I to be a prime ideal or equivalently for $\mathcal{O}(X)$ to be an integral domain (i.e. $\mathcal{O}(X)$ has no zero divisors). The ring $\mathcal{O}(X)$ allows us to fully recover the algebraic variety X, its maximal ideals being in one-to-one correspondence with the points in X. Consequently a classical affine variety X can be effectively identified with the algebra $\mathcal{O}(X)$, which is a finitely generated integral domain. Such algebras are called *affine algebras*. In this correspondence, which is an equivalence of categories, the morphisms of affine varieties correspond contravariantly to morphisms of the coordinate rings. An algebraic variety, which is not necessarily affine, is then obtained by gluing affine varieties.

There is an equivalent approach to algebraic varieties through sheaf theory: we understand a variety as a locally ringed space which is locally isomorphic to an affine algebraic variety (see Refs. [45], [85] for more details on the ordinary setting). We shall take this point of view in defining a supervariety.

We now turn to examine the supergeometric setting. Let k be our ground field. In most of our applications, $k = \mathbb{C}$.

Definition 1.6.1. We say that a commutative superalgebra A is an *affine superalgebra* if:

- A_0 is a finitely generated superalgebra such that its reduced associated algebra $A_{0,r} = A_0/J_{A_0}$ is an affine algebra (as always, J_{A_0} denotes the ideal generated by the odd nilpotents);
- A_1 is a finitely generated A_0-module.

■

Let A be an affine superalgebra. Then A_0 is an ordinary algebra and its reduced algebra $A_{0,r}$ corresponds to an affine variety X_0 with underlying topological space $|X| \subset k^n$. The topology on $|X|$ is the *Zariski topology* inherited from the affine ambient space and it is independent of the imbedding $X_0 \hookrightarrow k^n$. In this topology, a set is closed if and only if it consists of the zeros of an ideal in $k[x_1, \dots, x_n]$, the ring of regular functions (i.e. polynomials) on k^n.

Let us consider a non zero $f \in A_0$. We can define the localization of the A_0-module A at the element f as

$$A_f = \left\{ \frac{g}{f} \mid g \in A \right\} .$$

By the very definition of Zariski topology,

$$U_f = \{x \in |X| \mid f(x) \neq 0\}$$

is open in $|X|$ and in fact, $\{U_f\}_{f \in A_0}$ form a basis for the topology of $|X|$. One can check that the assignment $U_f \to A_f$ defines a $\mathcal{B}$-sheaf on $|X|$. By Proposition 1.3.5, there exists a unique sheaf of algebras $\mathcal{O}_X$ on $|X|$ such that $\mathcal{O}_X|_{U_{f_i}} = \mathcal{O}_X$.

It is very important to remark that A_0 is *not* an affine algebra, i.e. it does not correspond to an ordinary algebraic variety. This is because A_0 contains many nilpotents, coming from products of elements in A_1 (see Example 1.6.4). Consequently, the even part $\mathcal{O}_{X,0}$ of the sheaf $\mathcal{O}_X$, is not the sheaf of algebraic functions on the topological space $|X|$. Such a sheaf, that we denote with $\mathcal{O}_{X,r}$, is obtained by taking the quotient of $\mathcal{O}_X$ with respect to the ideal sheaf of nilpotents:

$$\mathcal{O}_{X,r} = \mathcal{O}_X / \mathcal{J}_X \ (= \mathcal{O}_{X,0} / \mathcal{J}_{X,0}) ,$$

where $\mathcal{J}_X(U)$ is the ideal generated by the nilpotents in superring $\mathcal{O}_X(U)$ and similarly $\mathcal{J}_{X,0}(U)$ is the ideal generated by the nilpotents in the ring $\mathcal{O}_{X,0}(U)$ the even part of $\mathcal{O}_X(U)$.

This prompts our definition of affine supervariety.

Definition 1.6.2. Let $\mathcal{O}(X)$ be an affine superalgebra. We define the *affine supervariety* X associated with $\mathcal{O}(X)$ as the superspace $X = (|X|, \mathcal{O}_X)$, where $|X|$ is the topological space of the ordinary affine variety defined by $\mathcal{O}(X)_r := \mathcal{O}(X)/J_{\mathcal{O}(X)}$, while $\mathcal{O}_X$ is the sheaf of superalgebras on $|X|$ described above. Notice that the set of global sections of $\mathcal{O}_X$ is $\mathcal{O}(X)$.

We define the *reduced variety* underlying X as the locally ringed space $(|X|, \mathcal{O}_{X,r})$ which corresponds to an ordinary variety and that we denote with X_r.

A *morphism* of affine supervarieties is a morphism of the underlying superspaces, though one can see that it corresponds (contravariantly) to a morphism of the respective coordinate superalgebras:

$$\{\text{morphisms } X \longrightarrow Y\} \qquad \longleftrightarrow \qquad \{\text{morphisms } \mathcal{O}(Y) \longrightarrow \mathcal{O}(X)\} .$$

We have what is called an *equivalence of categories* between the category of affine superalgebras and the category of affine supervarieties, in perfect analogy with the ordinary setting:

$$\begin{array}{ccc} \text{(affine superalgebras)} & \longleftrightarrow & \text{(affine supervarieties)} \\ A & \longrightarrow & (|X|, \mathcal{O}_X) \\ \mathcal{O}(X) & \longleftarrow & X\,, \end{array}$$

where $(|X|, \mathcal{O}_X)$ is obtained from A with the procedure described above.

We define an *algebraic supervariety* as a superspace which is locally isomorphic to an affine supervariety. ■

We now want to define the concept of *subvariety.*

Definition 1.6.3. Let X be an affine supervariety and $\mathcal{O}(X)$ the corresponding coordinate superalgebra. We say that S is a *subvariety* of X if S is the affine supervariety corresponding to the superalgebra $\mathcal{O}(X)/I$ for I an ideal of $\mathcal{O}(X)$.

If $X = (|X|, \mathcal{O}_X)$ is a supervariety not necessarily affine, we say that $S = (|S|, \mathcal{O}_S)$ is a *subvariety* of X if $|S|$ is a subspace of $|X|$ and $\mathcal{O}_S = \mathcal{O}_X/\mathcal{I}$, where $\mathcal{I}$ is an ideal sheaf with the following property: For any affine cover $\{U_i\}$ of X, $\mathcal{I}(U_i)$ is an ideal in $\mathcal{O}_X(U_i)$, $\mathcal{O}_S(U_i) = \mathcal{O}_X(U_i)/\mathcal{I}(U_i)$ and on such U_i the sheaf $\mathcal{O}_S|_{U_i}$ is obtained starting from the superalgebra $\mathcal{O}_S(U_i)$ as explained after Definition 1.6.1. ■

Let us now examine some relevant examples.

Example 1.6.4.

1. *The affine superspace.* We defined the *polynomial superalgebra* as

$$k[x_1, \dots, x_p, \theta_1, \dots, \theta_q] := k[x_1, \dots, x_p] \otimes \wedge(\theta_1, \dots, \theta_q)\,.$$

We want to interpret this superalgebra as the coordinate superring of a supervariety that we call the *affine superspace* of superdimension $p|q$, and we shall denote it with the symbol $k^{p|q}$. We want to distinguish it from the differential setting, so we will denote it as $\mathbb{A}^{p|q}$ to mean that we have chosen the Zariski topology. The underlying topological space is then $\mathbb{A}^p$, that is, k^p with the Zariski topology, and the sheaf is

$$\mathcal{O}_{\mathbb{A}^{p|q}}(U) := \mathcal{O}_{\mathbb{A}^p}(U) \otimes \wedge(\theta_1, \dots, \theta_q)\,.$$

2. The supersphere. The superalgebra

$$k[x_1, x_2, x_3, \eta_1, \eta_2, \eta_3]/(x_1^2 + x_2^2 + x_3^2 + \eta_1 x + \eta_2 x_2 + \eta_3 x_3 - 1)$$

is the superalgebra of the global sections of an affine supervariety, whose underlying topological space is the sphere in $\mathbb{A}^3$ of radius 1.

As usual, we are denoting with $(f_1, \dots, f_n)$ the ideal generated by the elements $(f_1, \dots, f_n)$. This supervariety is a subvariety of $\mathbb{A}^{3|3}$. ■

The simplest and perhaps most important example of a supervariety which is not affine is the *projective superspace.*

Example 1.6.5. *Projective Superspace.* We proceed in strict analogy with 1.4.7. We consider the ordinary projective space $\mathbb{P}^m$ as a quotient $k^{m+1} \setminus \{0\}/\sim$ with the Zariski topology. As in 1.4.7 we have a cover of $\mathbb{P}^m$ by open affine subsets $U_i \cong k^m$ the affine space. On each U_i we take the global ordinary coordinates $u_0^i, \dots, \hat{u}_i^i, \dots, u_m^i$, $u_k := x_k/x_i$ and we define the sheaf of superalgebras $\mathcal{O}_{\mathcal{U}_i}$:

$$\mathcal{O}_{\mathcal{U}_i}(V) = \mathcal{O}_{U_i}(V) \otimes \wedge(\xi_1^i, \dots, \xi_n^i), \qquad V \text{ open in } U_i$$

where now $\mathcal{O}_{U_i}$ is the sheaf of ordinary polynomial functions on U_i and $\xi_1^i, \dots, \xi_n^i$ are odd variables (i is a suffix, not an exponent, and we shall omit it whenever it is clear where the ξ_j^i lie). $\mathcal{U}_i = (U_i, \mathcal{O}_{\mathcal{U}_i})$ is a supervariety, isomorphic to the affine superspace $\mathbb{A}^{m|n}$. We want to define the morphisms $\phi_{ij} : \mathcal{U}_i \cap \mathcal{U}_j \to \mathcal{U}_i \cap \mathcal{U}_j$, where $\mathcal{U}_i \cap \mathcal{U}_j$ in the domain is an open subvariety of $\mathcal{U}_i$, while $\mathcal{U}_i \cap \mathcal{U}_j$ in the codomain is an open subvariety of $\mathcal{U}_j$, even if we still use the same notation. So in the domain $\mathcal{U}_i \cap \mathcal{U}_j = (U_i \cap U_j, \mathcal{O}_{\mathcal{U}_i}|_{U_i \cap U_j})$, while in the codomain: $\mathcal{U}_i \cap \mathcal{U}_j = (U_i \cap U_j, \mathcal{O}_{\mathcal{U}_j}|_{U_i \cap U_j})$.

In the equivalence of categories detailed above, ϕ_{ij} corresponds to a morphism of the superrings, $\phi_{ij}^* : \mathcal{O}_{\mathcal{U}_j}(U_i \cap U_j) \to \mathcal{O}_{\mathcal{U}_i}(U_i \cap U_j)$, where

$$\mathcal{O}_{\mathcal{U}_i}(U_i \cap U_j) = k[u_0^i, \dots, \hat{u}_i^i, \dots, u_m^i, \xi_1^i, \dots, \xi_n^i][{u_j^i}^{-1}]$$

and similarly for $\mathcal{O}_{\mathcal{U}_j}(U_j \cap U_j)$. We define:

$$(u_0^j, \dots, \hat{u}_j^j, \dots, u_m^j, \xi_1^j, \dots, \xi_n^j) \xrightarrow{\phi_{ij}^*} \left(\frac{u_0^i}{u_j^i}, \dots, \frac{1}{u_j^i}, \dots, \frac{u_m^i}{u_j^i}, \frac{\xi_1^i}{u_j^i}, \dots, \frac{\xi_n^i}{u_j^i} \right),$$

where on the right hand side the $1/u_j^i$ appears in the ith position and the jth position is omitted. One can easily check that the ϕ_{ij}'s satisfy the compatibility conditions:

$$\phi_{ij}\phi_{ki} = \phi_{jk}, \qquad \text{on} \quad \mathcal{U}_i \cap \mathcal{U}_j \cap \mathcal{U}_k,$$

hence they allow us to define uniquely a sheaf by 1.3.6, denoted with $\mathcal{O}_{\mathbb{P}^{m|n}}$ and then a supervariety structure on the topological space $\mathbb{P}^m$. The supervariety $(\mathbb{P}^m, \mathcal{O}_{\mathbb{P}^{m|n}})$ is called the *algebraic projective space of dimension* $m|n$. We shall omit 'algebraic' whenever it is clear the category that we are considering. ■

We are especially interested in *projective supervarieties*, namely, supervarieties which are subvarieties of the projective superspace.

In the ordinary setting, a projective variety X is defined as the set of zeros of an ideal I generated by $\mathbb{Z}$-homogeneous polynomials in the indeterminates $x_0, \dots, x_m$ in the projective space $\mathbb{P}^m$. If $\{U_i\}$ is the affine cover of $\mathbb{P}^m$ described above, $U_i \cap X$ is an affine variety and its coordinate ring is given by the *projective localization* of $k[x_0, \dots, x_m]/I$ at x_i, that is, it is the ring of the elements of degree zero in $(k[x_0, \dots, x_m]/I)[x_i^{-1}]$ (where we assign degree -1 to x_i^{-1}). In some sense the homogeneous ring $k[x_0, \dots, x_m]/I$ plays the role of the coordinate ring of an affine variety. It is however very important to remember that while in the affine setting we have a bijective correspondence between affine algebras and affine algebraic varieties, such correspondence fails for projective varieties (for more details see Chapter 1 in Ref. [85]).

We now want to define projective supervarieties. Consider the $\mathbb{Z}$-graded polynomial superalgebra $S = k[x_0, \dots, x_m, \xi_1, \dots, \xi_n]$, where the $\mathbb{Z}$-grading on the polynomials is given by the polynomial degree. As we shall presently see, subvarieties of $\mathbb{P}^{m|n}$ correspond to $\mathbb{Z}$-homogeneous super ideals of S, in analogy with what happens in the affine case, though there are some important differences that we will examine later. Before we go to such correspondence, for each r, $0 \leq r \leq m$, we consider the $\mathbb{Z}$-graded superalgebra

$$S[r] = k[x_0, \dots, x_m, \xi_1, \dots, \xi_n][x_r^{-1}], \qquad \deg(x_r^{-1}) = -1\,.$$

The subalgebra $S[r]^0 \subset S[r]$ of $\mathbb{Z}$-degree 0 elements is

$$S[r]^0 \approx k[u_0, \dots, \hat{u}_r, \dots, u_m, \eta_1, \dots, \eta_n], \qquad u_j = \frac{x_j}{x_r},\ \eta_\alpha = \frac{\xi_\alpha}{x_r}, \qquad (1.3)$$

(the 'ˆ' means that this generator is omitted) and it is called the *projective localization* of S at x_r. This is an affine superalgebra and it corresponds to the affine open supervariety $\mathcal{U}_r = (U_r, \mathcal{O}_{U_r}) \cong \mathbb{A}^{m|n}$ described in Example 1.6.5. The sheaves $\mathcal{O}_{\mathcal{U}_r}$ glue to give the sheaf $\mathcal{O}_{\mathbb{P}^{m|n}}$ according to the superalgebra morphism, on the coordinate superalgebras:

$$\begin{array}{ccc} \mathcal{O}_{\mathcal{U}_r}(U_r \cap U_s) & \longrightarrow & \mathcal{O}_{\mathcal{U}_s}(U_r \cap U_s) \\ u_j & \longrightarrow & u_j/u_s \\ \xi_\alpha & \longrightarrow & \xi_\alpha/u_s\,, \end{array}$$

where

$$\mathcal{O}_{\mathcal{U}_r}(U_r \cap U_s) = k[u_0, \dots, \hat{u}_r, \dots u_m, \eta_1, \dots \eta_n][u_s^{-1}]\,,$$

(refer again to Example 1.6.5).

Let $I \subset S = k[x_1, \dots, x_m, \xi_1, \dots, \xi_n]$ be a $\mathbb{Z}$-homogeneous super ideal (i.e. also $\mathbb{Z}_2$-graded); then S/I is also a graded superalgebra and we can repeat the same construction as above. In this case we obtain a supervariety structure on the topological space $|X|$ corresponding to the ordinary projective variety associated with the reduced homogeneous ring $(S/I)_r$. In fact, consider the superalgebra of $\mathbb{Z}$-degree zero elements in $(S/I)[x_i^{-1}]$, $x_i \notin I$, i.e. the *projective localization*:

$$\left(\frac{k[x_0, \dots x_m, \xi_1 \dots \xi_n]}{I} [x_i^{-1}] \right)_0 \cong \frac{k[u_0, \dots, \hat{u}_i, \dots u_m, \eta_1, \dots, \eta_n]}{I_{\mathrm{loc}}}\,,$$

where I_{loc} are the even elements of $\mathbb{Z}$-degree zero in $I[x_i^{-1}]$. This affine superalgebra defines an affine supervariety $\mathcal{V}_i = (V_i, \mathcal{O}_{V_i})$, which is a subvariety of $\mathcal{U}_i$. One can check that the sheaves $\mathcal{O}_{\mathcal{V}_i}$ glue to give a sheaf on the topological space $|X|$, which is the ordinary projective variety corresponding to the graded algebra $(S/I)_r$.

Vice-versa, if $X = (|X|, \mathcal{O}_X)$ is a subvariety of the projective superspace $\mathbb{P}^{m|n}$, then for each $V_i = U_i \cap |X|$, by the very definition of subvariety, we have an ideal I_i of the polynomial superalgebra $k[u_0, \dots, \hat{u}_i, \dots, u_m, \eta_1, \dots, \eta_n]$. This ideal defines an ideal sheaf on each U_i. Such sheaves glue together to give an ideal sheaf on the topological space $\mathbb{P}^m$ and they define a unique homogeneous super ideal I in $k[x_0, \dots, x_m, \eta_1, \dots, \eta_n]$, very much in the same way as in the classical setting. Let us see an illustration of this correspondence with an example.

Example 1.6.6. Consider the projective superspace $\mathbb{P}^{2|3}$. It admits the affine cover $\{U_0, U_1, U_2\}$. In the homogeneous superalgebra $k[x_0, x_1, x_2, \xi_1, \xi_2, \xi_3]$, consider the ideal generated by the following relations:

$$x_1^2 x_2 = 4x_0^3 - a_1 x_0^2 x_2 - a_2 x_2^3, \qquad 2(x_0 x_2 - e_1 x_2^2)\xi_2 = x_1 x_2 \xi_1,$$

$$x_1 x_2 \xi_2 = 2(x_0 - e_2 x_2)(x_0 - e_3 x_2)\xi_1, \qquad \xi_3 x_2 = x_0 \xi_1\,,$$

where e_i and a_i are constants. This ideal gives immediately the ideals I_0, I_1 and I_2 in the projective localizations corresponding to the U_i. For example I_2 is generated by

$$y^2 = 4x^3 - a_1 x^2 - a_2, \qquad 2(x - e_1)\eta_2 = y\eta_1,$$

$$y\eta_2 = 2(x - e_2)(x - e_3)\eta_1, \qquad \eta_3 = x\eta_1 \, ,$$

where $x = x_0/x_2$, $y = x_1/x_2$, $\eta_i = \xi_i/x_2$. It is immediate to see how to recover I from I_1, I_2, I_3.

This ideal gives the projective embedding of a super Riemann surface in $\mathbb{P}^{2|3}$. For more details, see Ref. [127]. ∎

It is very important to remark that, contrary to the affine case, there is no coordinate superring intrinsically associated to a projective supervariety, but there is a coordinate superring associated with the projective supervariety and its projective embedding. In other words, we can have the same projective supervariety admitting non isomorphic coordinate superrings with respect to two different projective embeddings.

This is a phenomenon already observed at the classical level. We can see this with a simple example. Let us consider $\mathbb{P}^1(\mathbb{C})$, the complex projective space, naturally associated with the homogeneous ring $\mathbb{C}[x_0, x_1]$. We consider the embedding

$$\begin{array}{ccc} \mathbb{P}^1(\mathbb{C}) & \longrightarrow & \mathbb{P}^2(\mathbb{C}) \\ [a_0, a_1] & \longrightarrow & [a_0^2, a_0 a_1, a_1^2] \, . \end{array}$$

$\mathbb{P}^1(\mathbb{C})$ is isomorphic to its image Y in $\mathbb{P}^2(\mathbb{C})$, which is a conic. The homogeneous coordinate ring of Y in its embedding in $\mathbb{P}^2(\mathbb{C})$ is not isomorphic to $\mathbb{C}[x_0, x_1]$. We invite the reader to consult Chapter 1 in Ref. [85] for more details.

1.7 The functor of points

We now want to introduce the functor of points approach to supergeometry. In the ordinary setting, the functor of points is a powerful, yet technical device mainly employed in the theory of schemes; in supergeometry it becomes an essential tool both in the differential and in the algebraic context because it allows us to recover the geometric intuition, otherwise not available in the superspace formulation of supermanifolds and supervarieties.

Classically we can examine the points of a variety over different fields and rings. For example we can look at the rational points of the ordinary complex sphere defined by the equation $x^2+y^2+z^2=1$ in $\mathbb{C}^3$. They are in one to one correspondence with the morphisms: $\mathbb{C}[x,y,z]/(x^2+y^2+z^2-1) \longrightarrow \mathbb{Q}$. In fact, each such morphism is specified by the knowledge of the images of the generators. The idea behind the functor of points is to extend this picture and consider *all* morphisms from the coordinate ring of the affine variety X to *all* affine algebras at once. Because of the equivalence of categories between affine varieties and reduced affine algebras described in the previous section, this is the same than considering all the morphisms from $T \longrightarrow X$ for all varieties T. Such morphisms are called the T-points of the variety X.

In the supergeometric context, we shall take a further step into abstraction and define the T-points for a superspace S. This will enable us to talk about T-points of a supermanifold and a supervariety at once.

Recall from Section 1.4 that a superspace S consists of a pair $(|S|, \mathcal{O}_S)$, where $|S|$ is a topological space and $\mathcal{O}_S$ is a sheaf of commutative superalgebras with $\mathcal{O}_{S,x}$ being a local superalgebra for all $x \in |S|$.

Definition 1.7.1. Let S and T be superspaces. A T-*point* of S is a morphism $T \longrightarrow S$. We denote the set of all T-points by $S(T)$. Equivalently,

$$S(T) = \mathrm{Hom}(T, S)\,.$$

Let (sspaces) be the category of superspaces and, as usual, we denote as $(\text{sspaces})^o$ the opposite category. We define the *functor of points* of the superspace S as the functor $S : (\text{sspaces})^o \to (\text{sets})$ given as follows. On objects it is given by:

$$S(T) = \mathrm{Hom}(T, S), \qquad T \in (\text{sspaces})\,,$$

while on morphisms

$$\begin{array}{ccc} S(T) & \xrightarrow{S(\phi)} & S(T') \\ f & \longrightarrow & S(\phi)f = f \circ \phi\,, \end{array}$$

(notice that if $\phi : T' \to T$, we have $f \circ \phi : T' \longrightarrow T \longrightarrow S$, hence for all $f \in S(T)$ we obtain an element $f \circ \phi \in S(T')$). ■

By a common abuse of notation, the superspace S and the functor of points of S are denoted with the same letter. Whenever is necessary to make a distinction, we shall write h_S for the functor of points of the superspace S.

We have defined the functor of points of a superspace. Clearly, we can also define the functor of points of a supermanifold or a supervariety, just by changing the category we start from.

Let (smflds) and (svar) denote respectively the categories of supermanifolds and supervarieties introduced above.

Definition 1.7.2. We define the *functor of points of the supermanifold* M as the functor $M : (\mathrm{smflds})^o \to (\mathrm{sets})$ such that

$$T \to M(T) = \mathrm{Hom}_{(\mathrm{smflds})}(T, M), \qquad M(\phi)f = f \circ \phi\,.$$

Similarly we define the *functor of points of the supervariety* X the functor $X : (\mathrm{svar})^o \to (\mathrm{sets})$ such that

$$T \to X(T) = \mathrm{Hom}_{(\mathrm{svar})}(T, X), \qquad X(\phi)f = f \circ \phi\,.$$

■

Given two supermanifolds (or supervarieties) X and Y, Yoneda's lemma establishes a one-to-one correspondence between the morphisms $X \longrightarrow Y$ and the natural transformations between the corresponding functors of points.

Lemma 1.7.3. *(Yoneda's Lemma) Let M and N be two superspaces (resp. supermanifolds or supervarieties). There is a bijection from the set of morphisms $\varphi : M \longrightarrow N$ to the set of maps $\varphi_T : M(T) \longrightarrow N(T)$, functorial in T. In particular, M and N are isomorphic if and only if their functors of points are isomorphic.*

Proof. Given a map $\varphi : M \longrightarrow N$, for any morphism $t : T \longrightarrow M$ the composition $\varphi \circ t$ is a morphism $T \longrightarrow N$. Conversely, we attach to the system (φ_T) the image of the identity map from $\varphi_M : M(M) \longrightarrow N(M)$. ■

The following proposition is very useful when we want to explicitly describe the functor of points of a supermanifold or a supervariety. To ease the notation we write $\mathcal{O}(T)$ for the global sections of T. For the proof, we refer to Chapters 4 and 10 of Ref. [25].

Proposition 1.7.4. *Let $M = (|M|, \mathcal{O}_M)$ and $T = (|T|, \mathcal{O}_T)$ be differentiable supermanifolds or affine supervarieties. Then*

$$\mathrm{Hom}(T, M) = \mathrm{Hom}(\mathcal{O}(M), \mathcal{O}(T))\,.$$

■

Notice that this result will *not* hold for complex analytic supermanifolds, since their global functions may not contain sufficient information about the supermanifold. This is a completely classical phenomenon.

Let us give some examples of T-points.

Example 1.7.5.

(1) Let T be just an ordinary topological point viewed as real differentiable supermanifold, i.e. $T = \mathbb{R}^{0|0} = (\mathbb{R}^0, \mathbb{R})$. By definition a T-point of a manifold M is a morphism $\phi : \mathbb{R}^{0|0} \longrightarrow M$. ϕ consists of a continuous map $|\phi| : \mathbb{R}^0 \longrightarrow |M|$, which corresponds to the choice of a point x in the topological space $|M|$, and a sheaf morphism $\phi^* : \mathcal{O}_M \longrightarrow \phi_*(\mathbb{R})$, which assigns to a section its value in x. Then a T-point of M is an ordinary topological point of $|M|$. Similarly, if T is a topological point, $T = \mathbb{C}^{0|0}$, and M a complex analytic supermanifold $M(T)$ is in one to one correspondence with $|M|$.

(2) Let M be the supermanifold $k^{p|q}$ (k real or complex according to M differentiable or analytic supermanifold) with global coordinates $t^1, \dots, t^p, \theta^1, \dots, \theta^q$ and let T be a supermanifold. A T-point of $k^{p|q}$ is a morphism $T \longrightarrow k^{p|q}$, which by the Chart Theorem 1.4.6 corresponds to a choice of p even and q odd global sections on T. Thus

$$k^{p|q}(T) := \{(u_1, \dots, u_p, \nu_1, \dots, \nu_q) \,|\, u_i \in \mathcal{O}(T)_0,\ \nu_j \in \mathcal{O}(T)_1\}$$

$$= \mathcal{O}(T)_0^p \oplus \mathcal{O}(T)_1^q\,.$$

(3) Let X be the supervariety $\mathbb{A}^{m|n}$ as in Example 1.6.4 and let T be an affine supervariety. By definition, a T-point of X is a morphism $\phi : T \longrightarrow \mathbb{A}^{m|n}$, which corresponds to a super ring morphism $\psi : \mathcal{O}(\mathbb{A}^{m|n}) \longrightarrow \mathcal{O}(T)$, that is $\psi : k[x_1, \dots, x_m, \xi_1, \dots, \xi_n] \longrightarrow \mathcal{O}(T)$. Hence, as before, ψ amounts to a choice of m even global sections in $\mathcal{O}(T)$ and n odd ones:

$$\mathbb{A}^{m|n}(T) = \mathcal{O}(T)_0^m \oplus \mathcal{O}(T)_1^n$$

$$= \{(a_1, \dots, a_m, \alpha_1, \dots, \alpha_n) \quad | \quad a_i \in \mathcal{O}(T)_0, \quad \alpha_j \in \mathcal{O}(T)_1\}\,.$$

■

Remark 1.7.6. Yoneda's lemma allows us to replace a superspace (resp. a supermanifold or a supervariety) S with its set of T-points, $S(T)$. We can now think of S as a *representable functor*[2] from the category of superspaces (resp. supermanifolds or supervarieties) to the category of sets. In fact, when constructing a superspace, it is often convenient to construct first its functor of points and then to prove that the functor is representable in the appropriate category. We shall not pursue this point of view in these notes; more along this line of thought can be found in the Appendix B. ■

We end this section by giving the construction of the functor of points of the complex analytic supermanifold $\mathbb{P}^{m|n}$. This will turn out to be extremely important in the sequel.

For a supermanifold T we define $P(T)$ as the set of locally free sheaves $\mathcal{F}_T \subset \mathcal{O}_T^{m+1|n} := (k^{m+1|n} \otimes \mathcal{O}_T)_0$ of rank $1|0$.

Proposition 1.7.7. *There is a one-to-one correspondence between the two sets:*

$$P(T) \longleftrightarrow \mathbb{P}^{m|n}(T), \qquad T \in (\text{smflds})$$

which is functorial in T.

Proof. We briefly sketch the proof, leaving to the reader the routine checks. Let us start with an element in $P(T)$, that is, a locally free sheaf $\mathcal{F}_T \subset \mathcal{O}_T^{m+1|n}$ of rank $1|0$. We want to associate to $\mathcal{F}_T$ a T-point of $\mathbb{P}^{m|n}$, that is, a morphism $T \longrightarrow \mathbb{P}^{m|n}$. First, we cover T with V_i so that $\mathcal{F}_T|_{V_i}$ is free. Hence

$$\mathcal{F}_T(V_i) = \text{span}\left\{(t_0, \dots, t_m, \theta_1, \dots, \theta_n)\right\},$$

where we assume that the section $t_i \in \mathcal{O}_T(V_i)$ is invertible without loss of generality, since the rank of $\mathcal{F}_T$ is $1|0$ (this assumption may require to change the cover). We have

$$\mathcal{F}_T(V_i) = \text{span}\left\{(t_0/t_i, \dots, 1, \dots, t_m/t_i, \theta_1/t_i, \dots, \theta_n/t_i)\right\}.$$

We have then immediately a morphism of supermanifolds $f_i : \mathcal{V}_i \longrightarrow \mathcal{U}_i \subset \mathbb{P}^{m|n}$:

$$f_i(u_1, \dots, u_m, \xi_1, \dots, \xi_n) = (t_0/t_i, \dots, t_m/t_i, \theta_1/t_i, \dots, \theta_n/t_i),$$

where $\mathcal{V}_i = (V_i, \mathcal{O}_T|_{V_i})$ and $\mathcal{U}_i = (U_i, \mathcal{O}_{\mathbb{P}^{m|n}}|_{U_i})$. It is immediate to check that the f_i's agree on $\mathcal{V}_i \cap \mathcal{V}_j$ so they glue to give a morphism $f : T \longrightarrow \mathbb{P}^{m|n}$.

[2]A functor $F : (\text{sspaces})^o \longrightarrow (\text{sets})$ is representable if $F(T) = \text{Hom}(T, S)$ for a suitable $S \in (\text{sspaces})$ (see Appendix B for more details).

Vice-versa, consider $f : T \longrightarrow \mathbb{P}^{m|n}$ and define $V_i = |f|^{-1}(U_i)$. We can define immediately the free sheaves $\mathcal{F}_{V_i}$ on each of the V_i as

$$\mathcal{F}_{V_i}(V) := \text{span}\{ \ (t_0|_V, \dots, 1, \dots, t_m|_V, \theta_1|_V, \dots, \theta_n|_V) \quad | \\ t_k = f^*|_{V_i}(u_k), \quad \theta_j = f^*|_{V_i}(\xi_j)\},$$

where u_i and ξ_j are as in Example 1.6.5. As one can readily check, the $\mathcal{F}_{V_i}$ glue to give a locally free subsheaf of $\mathcal{O}_T^{m+1|n}$. ■

The assignment $T \longrightarrow P(T)$ defines a functor P : (smflds) $\longrightarrow$ (sets), once we give its obvious definition on the arrows. The content of Proposition 1.7.7 is then that P is the functor of points of the projective superspace. This result has an obvious generalization to the algebraic setting; we leave to the reader the routine checks concerning all of these statements. We shall come back with more details to this, in the supergeometric setting, in the next chapters.

1.8 Supergroups

Supergroups are group objects in the category of supermanifolds (respectively, supervarieties). Together with their homogeneous superspaces, they occupy a prominent position in supergeometry. Ultimately, the study of supersymmetry in physical theories is the study of certain supergroups and their homogeneous spaces, and it was supersymmetry that prompted the development of supergeometry.

Let our ground field k be $\mathbb{R}$ or $\mathbb{C}$.

Definition 1.8.1. A *Lie supergroup* is a supermanifold G whose functor of points (also denoted by G)

$$G : (\text{smflds})^o \longrightarrow (\text{sets})$$

is group valued; this means that for any supermanifold T, $G(T)$ is a group, and for $f : S \longrightarrow T$, the map $G(f) : G(T) \longrightarrow G(S)$ is a group homomorphism.

Saying that G is a Lie supergroup or that the functor G is group valued is equivalent to have the following commutative diagrams of natural transformations:

(1) *Multiplication* $\mu : G \times G \longrightarrow G$, such that $\mu \circ (\mu \times \mathrm{id}) = (\mu \times \mathrm{id}) \circ \mu$, i.e.

$$\begin{array}{ccc} G \times G \times G & \xrightarrow{\mu \times \mathrm{id}} & G \times G \\ {\scriptstyle \mathrm{id} \times \mu} \big\downarrow & & \big\downarrow {\scriptstyle \mu} \\ G \times G & \xrightarrow{\mu} & G \end{array}$$

(2) *Unit* $e : e_k \longrightarrow G$, where e_k is the functor e_k : $(\mathrm{smflds})^o \longrightarrow (\mathrm{sets})$ associating to each object T the unique morphism $k \longrightarrow \mathcal{O}(T)$ sending 1_k to $1_{\mathcal{O}(T)}$. e must be such that $\mu \circ (\mathrm{id} \otimes e) = \mu \circ (e \times \mathrm{id})$, i.e.

$$\begin{array}{ccccc} G \times e_k & \xrightarrow{\mathrm{id} \times e} & G \times G & \xrightarrow{e \times \mathrm{id}} & e_k \times G \\ \big\downarrow & & {\scriptstyle \mu} \big\downarrow & & \big\downarrow \\ G & = & G & = & G \end{array}$$

(3) *Inverse* $i : G \longrightarrow G$, such that $\mu \circ (\mathrm{id} \times i) = e \circ \mathrm{id}$, i.e.

$$\begin{array}{ccc} G & \xrightarrow{(\mathrm{id}, i)} & G \times G \\ \big\downarrow & & \big\downarrow {\scriptstyle \mu} \\ e_k & \xrightarrow{e} & G \end{array}$$

■

Definition 1.8.1 goes unchanged in the algebraic category.

Definition 1.8.2. An *algebraic supergroup* is an algebraic supervariety whose functor of points (as before also denoted by G)

$$G : (\mathrm{svar})^o \longrightarrow (\mathrm{sets})$$

is group valued. Again saying that G is an algebraic supergroup corresponds to the diagrams in Definition 1.8.1. ■

If G is an affine algebraic supergroup, being the category of affine supervarieties contravariantly equivalent to the category of affine superalgebras, G is equivalently viewed as the functor:

$$G : (\mathrm{salg}) \longrightarrow (\mathrm{sets}), \qquad G(A) = \mathrm{Hom}_{(\mathrm{salg})}(\mathcal{O}(G), A)\,,$$

where $\mathcal{O}(G)$ is the superalgebra representing G. The superalgebra $\mathcal{O}(G)$ will satisfy the dual diagrams as in 1.8.1, hence carrying a natural *Hopf superalgebra structure* (see Section 1.11 for its definition).

Let us now examine some examples of Lie and algebraic supergroups.

Example 1.8.3.

1. Let $k^{m|n} = (k^m, \mathcal{O}_{k^{m|n}})$ as in 1.4.2 and let T be a supermanifold. The set $k^{m|n}(T)$ consists of $m|n$-tuples $(t_1, \dots, t_m, \theta_1, \dots, \theta_n)$, where the t_i's and θ_j's are, respectively, even and odd global sections of $\mathcal{O}_T$ (see 1.7.5). Since $k^{m|n}(T)$ is an additive group for all T, $k^{m|n}$ is a Lie supergroup.

2. Consider the supermatrices as in Example 1.4.5:

$$\mathrm{M}(m|n) = (\mathrm{M}_m \times \mathrm{M}_n, \mathcal{O}_{k^{m^2+n^2|2mn}}) .$$

Based on (1) we see that for any supermanifold T,

$$\mathrm{M}(m|n)(T) = \{(t_{ij}, \theta_{kl}) \mid t_{ij} \in \mathcal{O}(T)_0,\ \theta_{kl} \in \mathcal{O}(T)_1\},$$

where $1 \leq i, j \leq m$ or $m+1 \leq i, j \leq m+n$ and $1 \leq k \leq m$, $m+1 \leq l \leq m+n$ or $m+1 \leq k \leq m+n$, $1 \leq l \leq m$. It is customary to rearrange the variables into a matrix form:

$$\mathrm{M}(m|n)(T) = \left\{ \begin{pmatrix} t_{ij} & \theta_{il} \\ \theta_{kj} & t_{kl} \end{pmatrix} \mid t_{ij} \in \mathcal{O}(T)_0,\ \theta_{kl} \in \mathcal{O}(T)_1 \right\}.$$

Again, since $\mathrm{M}(m|n)(T)$ is an additive group for all T, we have that $\mathrm{M}(m|n)$ is a Lie supergroup.

3. Similarly to (1) we have that the affine superspace $\mathbb{A}^{m|n}$ is an additive algebraic supergroup and similarly to (2) we can define the algebraic supermatrices, that we denote, with an abuse of notation, with the same symbol $\mathrm{M}(m|n)$ as above. ∎

We want to pause here a moment to define the supertranspose of a supermatrix. We will do it in a more general context than the square, even matrices of $\mathrm{M}(m|n)(T)$.

Let R be any superalgebra and consider the free modules $R^{p|q}$, $R^{r|s}$. In the standard basis of both modules, a linear map $L : R^{p|q} \to R^{r|s}$ is given by a matrix

$$X = \begin{pmatrix} A_{r\times p} & B_{r\times q} \\ C_{s\times p} & D_{s\times q} \end{pmatrix},$$

with entries in R. If L is an even map (respecting parity) then the entries in A and D are even and the entries in B and C are odd. If L is an odd map (reversing parity) then the entries in A and D are odd and the entries in B and C are even. So, in the previous example, we have that:

$$\mathrm{M}(m|n)(T) = \{L : \mathcal{O}^{m|n}(T) \longrightarrow \mathcal{O}^{m|n}(T) \mid L \quad \text{even}\}.$$

As in ordinary algebra, one can define the dual map between the dual spaces $L^* : A^{r|s^*} \to A^{p|q^*}$ as

$$L^*(\alpha)(v) = \alpha(L(v)), \qquad \alpha \in A^{r|s^*}, \quad v \in A^{p|q}.$$

Now, suppose that L is an homogeneous map, (even or odd, $p(L) = 0, 1$). Then the matrix of L^* in the dual basis is the *supertranspose matrix* (or just *transpose* for short)

$$X^t = \begin{pmatrix} A^t_{r\times p} & (-1)^{p(L)} C^t_{r\times q} \\ -(-1)^{p(L)} B^t_{s\times p} & D^t_{s\times q} \end{pmatrix}. \tag{1.4}$$

The transpose satisfies

$$(X_1 X_2)^t = (-1)^{p(X_1)p(X_2)} X_2^t X_1^t.$$

For this rule to hold, one should take the convention that transposing the product of submatrices with odd entries introduces a sign, for example,

$$(BC)^t = -C^t B^t, \qquad \text{for } B, C \text{ odd}. \tag{1.5}$$

We now revisit the Example 1.4.5 about the *general linear supergroup*.

Definition 1.8.4. We define the functor $\mathrm{GL}(m|n) : (\text{smflds})^o \longrightarrow (\text{sets})$ as:

$$\mathrm{GL}(m|n)(T) := \left\{ \begin{pmatrix} p & q \\ r & s \end{pmatrix} \mid p,\ s \text{ are invertible} \right\} \subset \mathrm{M}(m|n)(T),$$

where p and s are $m \times m$ and $n \times n$ matrices respectively with entries in $\mathcal{O}(T)_0$, and q and r are $m \times n$, $n \times m$ matrices respectively with entries in $\mathcal{O}(T)_1$.

If $f : T \longrightarrow S$ is a supermanifold morphism, we define $\mathrm{GL}(m|n)(f) : \mathrm{GL}(m|n)(S) \longrightarrow \mathrm{GL}(m|n)(T)$ as the map which assigns to the entry $u_{ij} \in \mathcal{O}(S)$ of a matrix in $\mathrm{GL}(m|n)(S)$ the (i,j) entry $f(u_{ij})$ in $\mathcal{O}(T)$.

The functor $\mathrm{GL}(m|n)$ is called the *general linear supergroup functor*. ■

The condition that p and q are ordinary invertible matrices is equivalent to the invertibility of the matrix $\begin{pmatrix} p & q \\ r & s \end{pmatrix}$. This is a simple exercise that we leave to the reader, based on the fact that the entries in the matrices q and r consist of nilpotent elements.

$\mathrm{GL}(m|n)$ is the functor of points of a supermanifold. Its underlying topological space is an open set U in the ordinary space $k^{m^2+n^2}$ consisting of the matrices with invertible diagonal blocks (see Example 1.4.5).

Let us verify that the functor of points of the supermanifold

$$M = (U, \mathcal{O}_{k^{m^2+n^2|2mn}}\big|_U)$$

is indeed $\mathrm{GL}(m|n)$. Assume for the moment that we are in the differentiable setting, so that $k = \mathbb{R}$. By Proposition 1.7.4, we have that

$$M(T) = \mathrm{Hom}(T, M) = \mathrm{Hom}(\mathcal{O}(M), \mathcal{O}(T)).$$

If x_{ij}, ξ_{kl} are the global coordinates on M, with

$$\begin{array}{lll} 1 \leq i, j \leq m, & \text{or} \quad m+1 \leq i, j \leq m+n, & \text{and} \\ 1 \leq k \leq m,\ m+1 \leq l \leq m+n, & \text{or} \quad 1 \leq l \leq m,\ m+1 \leq k \leq m+n, & \end{array}$$

we have:

$$\mathcal{O}(M) = \mathcal{O}(\mathbb{R}^{m^2+n^2|2mn})\big|_U \cong \mathcal{O}(\mathbb{R}^{m^2+n^2|2mn})[a,b]/(a\det(p)-1, b\det(s)-1).$$

We shall refer informally to the latter ring also as

$$\mathcal{O}(\mathbb{R}^{m^2+n^2|2mn})[\det(p)^{-1}, \det(s)^{-1}].$$

In our notation, $\mathcal{O}(\mathbb{R}^{m^2+n^2|2mn}) = C^\infty(x_{ij}) \otimes \wedge(\xi_{kl})$ are the global sections of the supermanifold of real supermatrices $\mathrm{M}(m|n)$ (see Example 1.8.3) and

$$\det(p) = \det(x_{ij})_{1 \leq i,j \leq m}, \qquad \det(s) = \det(x_{ij})_{m+1 \leq i,j \leq m+n}.$$

In order to assign a morphism $T \to M$ or, equivalently, a morphism $\mathcal{O}(M) \to \mathcal{O}(T)$, we need to specify the images of x_{ij} and ξ_{kl} in such a way that the corresponding determinants of the matrices p and s are invertible,

$$\begin{array}{ccc} \mathcal{O}(M) & \xrightarrow{\phi} & \mathcal{O}(T) \\ x_{ij} & \longrightarrow & t_{ij}, \\ \xi_{kl} & \longrightarrow & \theta_{kl}. \end{array}$$

It is convenient to rearrange the images to form a matrix:

$$\begin{array}{ccc} \mathcal{O}(M) & \xrightarrow{\phi} & \mathcal{O}(T) \\ & & \\ \begin{pmatrix} x_{ij} & \xi_{il} \\ \xi_{kj} & x_{kl} \end{pmatrix} & \longrightarrow & \begin{pmatrix} t_{ij} & \theta_{il} \\ \theta_{kj} & t_{kl} \end{pmatrix}, \end{array}$$

where $1 \leq i, j \leq m$, $m+1 \leq k, l \leq m+n$ and $\det(t_{ij})$, $\det(t_{kl})$ are invertible. We then see that specifying a morphism ϕ is equivalent to give an element in $\mathrm{GL}(m|n)(T)$. We then write $\mathrm{GL}(m|n)$ also for the Lie supergroup whose functor of points is $T \to \mathrm{GL}(m|n)(T)$.

The same is true when we consider $\mathrm{GL}(m|n)$ in the complex analytic category, though the proof has to be done using the Chart Theorem 1.4.6. This is because the easy characterization of morphisms in Proposition 1.7.4 that we have used, does not hold anymore in such generality. It does hold, however, in our special case.

We now want to introduce the *general linear supergroup* in the algebraic context.

Definition 1.8.5. In analogy with the supermanifold setting, we define the functor $\mathrm{GL}(m|n)^{alg} : (\mathrm{svar})^o \longrightarrow (\mathrm{sets})$ as:

$$\mathrm{GL}(m|n)^{alg}(T) := \left\{ \begin{pmatrix} p & q \\ r & s \end{pmatrix} \mid p,\ s \,\text{are invertible} \right\},$$

where p and s are $m \times m$, $n \times n$ matrices respectively with entries in $\mathcal{O}(T)_0$, and q and r are $m \times n$, $n \times m$ matrices respectively with entries in $\mathcal{O}(T)_1$, for T an affine supervariety.

If $f : T \longrightarrow S$ is a supervariety morphism, we define $\mathrm{GL}(m|n)^{alg}(f)$: $\mathrm{GL}(m|n)^{alg}(S) \longrightarrow \mathrm{GL}(m|n)^{alg}(T)$ as the map which assigns to the entry $u_{ij} \in \mathcal{O}(S)$ of a matrix in $\mathrm{GL}(m|n)^{alg}(S)$ the (i, j) entry $f(u_{ij})$ in $\mathcal{O}(T)$ of a matrix in $\mathrm{GL}(m|n)^{alg}(T)$. ■

Again, the condition that p and s are ordinary invertible matrices is equivalent to the invertibility of the matrix $\begin{pmatrix} p & q \\ r & s \end{pmatrix}$.

The functor $\mathrm{GL}(m|n)^{alg}$ corresponds to a supervariety with coordinate superalgebra

$$\mathcal{O}(\mathrm{GL}(m|n)^{alg}) := k[x_{ij}, \xi_{kl}][\det(p)^{-1}, \det(s)^{-1}],$$

where

$$\begin{aligned} &1 \le i, j \le m, && \text{or} \quad m+1 \le i, j \le m+n, && \text{and} \\ &1 \le k \le m,\ m+1 \le l \le m+n, && \text{or} \quad 1 \le l \le m,\ m+1 \le k \le m+n, \end{aligned}$$

and

$$\det(p) = \det(x_{ij})_{1 \le i,j \le m}, \qquad \det(s) = \det(x_{ij})_{m+1 \le i,j \le m+n},$$

(for the notation refer also to Example 1.8.4 set in the differential setting).

In fact, giving a morphism

$$\begin{aligned} \mathcal{O}(\mathrm{GL}(m|n)^{alg}) &\xrightarrow{\ \phi\ } \mathcal{O}(T) \\ x_{ij} &\longrightarrow t_{ij} \\ \xi_{ij} &\longrightarrow \theta_{kl} \end{aligned}$$

(T an affine supervariety) is equivalent to assign t_{ij}, $\theta_{kl} \in \mathcal{O}(T)$ in such way that the parity is respected and the corresponding determinants of p and s are invertible. Rearranging the generators in convenient matrix form we have

$$\begin{array}{ccc} \mathcal{O}(\mathrm{GL}(m|n)^{alg}) & \longrightarrow & \mathcal{O}(T) \\ & & \\ \begin{pmatrix} x_{ij} & \xi_{il} \\ \xi_{kj} & x_{kl} \end{pmatrix} & \longrightarrow & \begin{pmatrix} t_{ij} & \theta_{il} \\ \theta_{kj} & t_{kl} \end{pmatrix} . \end{array}$$

Hence, a morphism ϕ corresponds uniquely to a matrix in $\mathrm{GL}(m|n)^{alg}(T)$ when T is affine. The functor of points of any supervariety is uniquely determined once we know its restriction to the category of affine supervarieties, so we do not lose generality by considering only those (see Chapter 10 of Ref. [25] for more details).

We will call the supervariety $\mathrm{GL}(m|n)^{alg}$ the *general linear algebraic supergroup* and we may drop the label "*alg*" whenever it is clear that we are discussing the algebraic category.

We are now ready to define the super analogue of the determinant: the *Berezinian*, which makes sense in both, the category of supermanifolds and the category of supervarieties.

Definition 1.8.6. We define the *Berezinian* of a matrix in $\mathrm{GL}(m|n)(T)$ as the following element in $\mathcal{O}(T)_0$:

$$\mathrm{Ber}\begin{pmatrix} p & q \\ r & s \end{pmatrix} = \det(s^{-1})\det(p - qs^{-1}r) . \tag{1.6}$$

■

Hence the Berezinian is a supermanifold (respectively, a supervariety) morphism $\mathrm{Ber} : \mathrm{GL}(m|n) \longrightarrow (k^{1|0})^{\times}$, where $(k^{1|0})^{\times}(T)$ are the invertible elements in $k^{1|0}(T)$.

Notice that in contrast to the ordinary setting, we have not defined the Berezinian for a generic supermatrix, but only for an invertible one. Remarkably the Berezinian will not reduce to the determinant when we reduce modulo the odd variables, and this happens because its infinitesimal counterpart is the *supertrace* versus the trace. In fact, let us compute heuristically the differential of the Berezinian at the identity, $(\mathrm{dBer})_I$: $T_I(\mathrm{GL}(m|n) \cong M(m|n) \longrightarrow k^{1|0}$. We can substitute

$$\begin{pmatrix} p & q \\ r & s \end{pmatrix} \approx \begin{pmatrix} 1 + ta & tb \\ tc & 1 + td \end{pmatrix} ,$$

so

$$(\mathrm{dBer})_I \begin{pmatrix} a & b \\ c & d \end{pmatrix} = \frac{\mathrm{d}}{\mathrm{d}t} \det(1 - td)\det(1 + ta - tb(1 - td)tc)\Big|_0$$
$$= \mathrm{tr}(a) - \mathrm{tr}(d)\,.$$

Hence we have that $(\mathrm{dBer})_I = \mathrm{str}$, where the supertrace of a matrix is

$$\mathrm{str}\begin{pmatrix} a & b \\ c & d \end{pmatrix} := \mathrm{tr}(a) - \mathrm{tr}(d)\,.$$

One of the features the Berezinian shares with its ordinary counterpart, the determinant, is the multiplicativity. The interested reader can find the proof of this fact in Section 1.5 of Ref. [25].

Proposition 1.8.7. *Let* A, $B \in \mathrm{GL}(m|n)(T)$. *Then*

$$\mathrm{Ber}(AB) = \mathrm{Ber}(A)\mathrm{Ber}(B)\,.$$

Then $\mathrm{Ber} : \mathrm{GL}(m|n)(T) \longrightarrow (k^{1|0})^{\times}$ *is a supergroup morphism.* ■

We now introduce another supergroup, which will be extremely important in the sequel.

Definition 1.8.8. We define the T-points of the *special linear supergroup* $\mathrm{SL}(m|n)(T)$ as the subgroup of $\mathrm{GL}(m|n)(T)$ consisting of all matrices with Berezinian equal to 1. ■

The functor $T \longrightarrow \mathrm{SL}(m|n)$, defined in an obvious way on the arrows, is the functor of points of a supergroup that we call $\mathrm{SL}(m|n)$. This fact is obtained in the supermanifold category as an application of the Submersion Theorem (see Theorem 1.5.10 and Chapter 5 of Ref. [25] for more details). In the supervariety category, we have immediately that the superalgebra of its global sections is $\mathcal{O}(\mathrm{SL}(m|n)) = \mathcal{O}(\mathrm{GL}(m|n))/(\mathrm{Ber} - 1)$, where Ber is the berezinian function of the generators of $\mathcal{O}(\mathrm{GL}(m|n))$.

We now turn our attention to the supermanifold category and we define the Lie superalgebra associated with a Lie supergroup (see Ref. [25] for more details).

Definition 1.8.9. Let G be a Lie supergroup and V a vector field on G. We say that V is *left invariant* if

$$(\mathrm{id} \otimes V) \circ \mu^* = \mu^* \circ V\,,$$

where μ denotes the multiplication in G and μ^* as usual it is the corresponding sheaf morphism. The *Lie superalgebra* of G, $\mathrm{Lie}(G)$ is the superspace of left invariant vector fields together with the super bracket inherited from the composition of vector fields

$$[X, Y] := X \circ Y - (-1)^{|X||Y|} Y \circ X .$$

As in the ordinary case we can identify Lie(G) with the tangent space of G at the identity (which is a topological point). For more details see Chapter 4 of Ref. [25] and pg. 276 of Ref. [146]. ■

We now turn to examine the Lie superalgebras associated with the Lie supergroups GL($m|n$) and SL($m|n$).

The Lie superalgebra Lie(GL($m|n$))(T) consists on the $m|n \times m|n$ matrices with entries in $\mathcal{O}(T)$, M($m|n$)($\mathcal{O}(T)$). This fact is immediate, since $|\mathrm{GL}(m|n)|$ is open in the super vector space M($m|n$), hence the tangent space Lie(GL($m|n$)) becomes identified with M($m|n$) (which we also denote with End($k^{m|n}$)). Lie(GL($m|n$)) is a Lie superalgebra with bracket (see Section 1.1):

$$[X, Y] = XY - (-1)^{|X||Y|} YX .$$

The Lie superalgebra Lie(SL($m|n$))(T) is the subalgebra of Lie(GL($m|n$)) consisting of matrices with zero supertrace, that is matrices such that

$$\mathrm{str} \begin{pmatrix} p_{m\times m} & q_{m\times n} \\ r_{n\times m} & s_{n\times n} \end{pmatrix} = \mathrm{tr}(p) - \mathrm{tr}(s) = 0 .$$

This is coming from a corollary of the Submersion Theorem (see Section 5.3 in Ref. [25]).

In the next section we are going to the study the theory of homogeneous superspaces for Lie supergroups.

1.9 Homogeneous superspaces

In this section we want to understand the notion of *quotient* of Lie supergroups and the related notion of *homogeneous superspace*. Our main reference is Chapter 9 of Ref. [25] (for the ordinary theory we refer the reader to Ref. [146]).

We start with the definition of *action* of a Lie supergroup on a supermanifold.

Definition 1.9.1. Let G be a Lie supergroup with multiplication, inverse and unit given by μ, i and e respectively, and M a supermanifold. We say that G *acts* on M if we have a morphism:

$$a \colon G \times M \longrightarrow M, \qquad a_T(g,x) = g \cdot x, \quad \forall g \in G(T), \forall x \in M(T),$$

such that, for all supermanifolds T:

(1) $1 \cdot x = x, \forall x \in M(T)$, with 1 the unit in $G(T)$,
(2) $(g_1 g_2) \cdot x = g_1 \cdot (g_2 \cdot x)$, $\forall x \in M(T)$, and $\forall g_1, g_2 \in G(T)$.

These correspond to the commutativity of the following diagrams

$$\begin{array}{ccc} G \times G \times M & \xrightarrow{\mu \times id_M} & G \times M \\ \Big\downarrow{\scriptstyle id_G \times a} & & \Big\downarrow{\scriptstyle a} \\ G \times M & \xrightarrow{\quad a \quad} & M \end{array} \qquad \begin{array}{ccc} M \simeq \mathbb{R}^{0|0} \times M & \xrightarrow{\quad e \times \mathrm{id} \quad} & G \times M \\ & {\scriptstyle \mathrm{id}} \searrow \quad \swarrow {\scriptstyle a} & \\ & M & \end{array}$$

where $e_T : \mathbb{R}^{0|0}(T) \longrightarrow G(T)$ is $e_T(p) = 1_T$ (see also to Example 1.7.5).

If the morphism $G \longrightarrow M$, $g \mapsto g \cdot m$ is a surjective submersion, we say that G acts *transitively* and that M is an *homogeneous superspace*. ■

Let G be a Lie supergroup of dimension $m|n$ and let H be a closed Lie subsupergroup of dimension $r|s$. The quotient $|G|/|H|$ has a unique manifold structure such that it is an homogeneous space for the natural action of the ordinary Lie group $|G|$. We now want to build on the topological space $|G|/|H|$ a supermanifold structure which carries an action of the Lie supergroup G.

Let $\mathfrak{g} = \mathrm{Lie}(G)$ and $\mathfrak{h} = \mathrm{Lie}(H)$. For each $Z \in \mathfrak{g}$, let D_Z be the left invariant vector field on G defined by Z. For $x_0 \in |G|$ let ℓ_{x_0} and r_{x_0} be the left and right translations of G given by x_0. We denote by $i_{x_0} = \ell_{x_0} \circ r_{x_0}^{-1}$ the inner automorphism defined by x_0. It fixes the identity and induces the transformation Ad_{x_0} on $\mathfrak{g}_0$.

For any open subset $U \subset |G|$ and any subsuperalgebra $\mathfrak{k}$ of $\mathfrak{g}$ we define $\mathcal{O}_{\mathfrak{k}}(U)$ by

$$\mathcal{O}_{\mathfrak{k}}(U) = \{ f \in \mathcal{O}_G(U) \mid D_Z f = 0 \text{ on } U \text{ for all } Z \in \mathfrak{k} \}.$$

Then $\mathcal{O}_{\mathfrak{k}}$ is a subsheaf of $\mathcal{O}_G$ (and, in particular, so is $\mathcal{O}_{\mathfrak{h}}$). On the other hand, for any open subset $W \subset |G|$, invariant under right translations by elements of $|H|$, we put

$$\mathcal{O}_H(W) = \{ f \in \mathcal{O}_G(W) \mid f \text{ is invariant under } r_{x_0} \text{ for all } x_0 \in |H| \}.$$

Notice that if $|H|$ is connected we have

$$\mathcal{O}_H(W) = \mathcal{O}_{\mathfrak{h}}(W).$$

Definition 1.9.2. Let $|X| := |G|/|H|$ and $\pi : |G| \longrightarrow |G|/|H|$ the natural projection. For any open set $W_0 \subset |X| = |G|/|H|$ with $W = |\pi|^{-1}(W_0)$ we set

$$\mathcal{O}_X(W_0) := \mathcal{O}_H(W) \cap \mathcal{O}_{\mathfrak{h}}(W).$$

The subsheaf $\mathcal{O}_X$ is a sheaf of superalgebras on $|X|$ and we define the superspace $G/H := X = (|X|, \mathcal{O}_X)$. ■

We first want to show that X is a supermanifold. It is clear that the left action of the group $|G|$ on $|X|$ leaves $\mathcal{O}_X$ invariant and so it is enough to prove that there is an open neighborhood W_0 of $|\pi|(1) \equiv \bar{1}$ with the property that $(W_0, \mathcal{O}_X|_{W_0})$ is a super domain, i.e., isomorphic to an open submanifold of $k^{p|q}$.

We will do this using the local Frobenius Theorem (see Theorem 1.5.12). On $|G|$ we have a distribution spanned by the vector fields in $\mathfrak{h}$. We denote it by $\mathcal{D}_{\mathfrak{h}}$. On each $|H|$-coset $x_0|H|$ we have a supermanifold structure, which is a closed subsupermanifold of G. It is an integral manifold of $\mathcal{D}_{\mathfrak{h}}$, i.e. the tangent space at any point is the subspace $\mathfrak{h}$ at that point. By the local Frobenius Theorem 1.5.12, there is an open neighborhood U of 1 and coordinates x_i, $1 \leq i \leq m$ and θ_α, $1 \leq \alpha \leq n$ on U such that at each point of U, $\mathcal{D}_{\mathfrak{h}}$ is spanned by $\partial/\partial x_i, \partial/\partial\theta_\alpha (1 \leq i \leq r, 1 \leq \alpha \leq s)$. Moreover, from the ordinary theory on the Lie group $|G|$ we may assume that the slices

$$L(\mathbf{c}) := \{(x_1, \dots, x_m) \mid x_j = c_j, \ \ r+1 \leq j \leq m\}$$

are open subsets of distinct $|H|$-cosets for distinct $\mathbf{c} = (c_{r+1}, \dots, c_m)$. These slices are therefore supermanifolds with coordinates

$$x_i, \theta_\alpha, \qquad 1 \leq i \leq r, \ \ 1 \leq \alpha \leq s\,.$$

We have a subsupermanifold W' of U defined by $x_i = 0$ with $1 \leq i \leq r$ and $\theta_\alpha = 0$ with $1 \leq \alpha \leq s$. By the ordinary theory, we may assume that the map $|\pi| : |G| \longrightarrow |X|$ is a diffeomorphism of $|W'|$ with its image $|W|$ in $|X|$.

Lemma 1.9.3. *The map*

$$\begin{array}{ccc} W' \times H & \xrightarrow{\ \gamma\ } & G \\ (w, h) & \longrightarrow & wh \end{array}$$

is a super diffeomorphism of $W' \times H$ onto the open subsupermanifold of G with reduced manifold the open subset of $|G|$ isomorphic to $|W'| \times |H|$.

Proof. The map γ is actually the map $\mu \circ (i_{W'} \times i_H)$ where i_M refers to the canonical inclusion $M \hookrightarrow G$ of a subsupermanifold of G into G, and $\mu : G \times G \longrightarrow G$ is the multiplication morphism of the supergroup G.

From the classical theory, the reduced map $|\gamma|$ is a diffeomorphism of $|W'| \times |H|$ onto the open set $U = |W'||H|$. This uses the fact that the cosets $w|H|$ are distinct for distinct $w \in |W'|$. It is thus enough to show that $\mathrm{d}\gamma$ is surjective at all points of $|W'| \times |H|$. For any $h \in |H|$, the right translation by h (on the second factor in $W' \times H$ and simply r_h on G) is a super diffeomorphism commuting with γ and so it is enough to prove this at $(w, 1)$. If $X \in \mathfrak{g}$ is tangent to W' at w and $Y \in \mathfrak{h}$, then

$$\mathrm{d}\gamma(X, Y) = \mathrm{d}\gamma(X, 0) + \mathrm{d}\gamma(0, Y) = \mathrm{d}\mu(X, 0) + \mathrm{d}\mu(0, Y) = X + Y\,.$$

Hence the range of $\mathrm{d}\gamma$ is all of $\mathfrak{g}$ since, from the coordinate chart at 1, we see that the tangent spaces to W' and wH at w are transversal and span the tangent space to G at w which is $\mathfrak{g}$. ■

Lemma 1.9.4. *Let us consider the open set $|W| = |\pi|(|W'|)$ in $|X|$ and the subsuperspace $W = (|W|, \mathcal{O}_X|_{|W|})$ of X. We have*

$$\gamma^* \mathcal{O}_X\Big|_{|W|} = \mathcal{O}_{W'} \otimes 1\,,$$

where $\gamma^* : \mathcal{O}_G \longrightarrow \gamma_* \mathcal{O}_{W' \times H}$.

Proof. To ease the notation we drop the open set in writing a sheaf superalgebra, that is, we will write $\mathcal{O}_X$ instead of $\mathcal{O}_X(U)$.

We want to show that for any g in $\mathcal{O}_X\big|_U$, $\gamma^* g$ is of the form $f \otimes 1$ and that the map $g \longmapsto f$ is bijective with $\mathcal{O}_{W'}$. Now γ^* intertwines $D_Z (Z \in \mathfrak{h})$ with $1 \otimes D_Z$ and so $(1 \otimes D_Z)\gamma^* g = 0$. Since the D_Z span all the super vector fields on $|H|$ it follows using charts that for any $p \in |H|$ we have $\gamma^* g = f_p \otimes 1$ locally around p for some $f_p \in \mathcal{O}_{W'}$. Clearly f_p is locally constant in p. Hence f_p is independent of p if $|H|$ is connected. If we do assume that $|H|$ is connected, the right invariance under $|H|$ shows that f_p is independent of p. In the other direction, it is obvious that if we start with $f \otimes 1$, it is the image of an element of $\mathcal{O}_X\big|_U$. ■

Theorem 1.9.5. *The superspace $(|X|, \mathcal{O}_X)$ is a supermanifold.*

Proof. At this stage by the previous lemmas we know that $(|X|, \mathcal{O}_X)$ is a supermanifold at $\bar{1}$. The left invariance of the sheaf under $|G|$ shows this to be true at all points of $|X|$. ■

Clearly $|G|$ acts on $|X|$ but there is more: there is a natural action of the supergroup G itself on X. We shall now describe how this action comes about.

Proposition 1.9.6. *There is a map $\beta : G \times X \longrightarrow X$ such that the following diagram*

$$\begin{array}{ccc} G \times G & \xrightarrow{\mu} & G \\ {\scriptstyle 1\times\pi}\big\downarrow & & \big\downarrow{\scriptstyle \pi} \\ G \times X & \xrightarrow{\beta} & X \end{array}$$

commutes.

Proof. Let $\alpha := \pi \circ \mu : G \times G \longrightarrow X$. The action of $|G|$ on $|X|$ shows that such a map β exists at the reduced level. So it is a question of constructing the pull-back map

$$\beta^* : \mathcal{O}_X \longrightarrow \mathcal{O}_{G\times X}$$

such that

$$(1 \times \pi)^* \circ \beta^* = \alpha^* .$$

Now π^* is an *isomorphism* of $\mathcal{O}_X$ onto the sheaf $\mathcal{O}_G$ restricted to a sheaf on X $(W \longmapsto \mathcal{O}_G(|\pi|^{-1}(W)))$, so to prove the *existence and uniqueness* of β^* it is a question of proving that α^* and $(1 \times \pi)^*$ have the same image in $\mathcal{O}_{G\times G}$. It is easy to see that $(1 \times \pi)^*$ has as its image the subsheaf of sections f killed by $1 \otimes D_X (X \in \mathfrak{h})$ and invariant under $1 \times r_h$, $(h \in |H|)$. It is not difficult to see that this is also the image of α^*. ■

Notice that in the course of the proof of the previous proposition we showed the existence of the surjective submersion

$$\pi : G \longrightarrow X = G/H .$$

Hence X is an homogeneous superspace for the supergroup G.

The uniqueness for the supermanifold X may then be formulated as follows.

Proposition 1.9.7. *Let X' be a supermanifold with $|X'| = |X|$ and let π' be a morphism $G \longrightarrow X'$. Suppose that*

(1) π' is a surjective submersion.
(2) The fibers of π' are the supermanifolds which are the cosets of H.

Then there is a natural isomorphism

$$X \simeq X'.$$

Proof. From the local description of submersions as projections it is clear that for any open $W_0 \subset |X|$, the elements of $\pi'^*(\mathcal{O}_{X'}(W_0))$ are invariant under $r(|H|)$ and killed by $D_X (X \in \mathfrak{h})$. Hence we have a natural map $X \longrightarrow X'$ commuting with π and π'. This is a submersion, and by dimension considerations it is clear that this map is an isomorphism. ■

We end this section with a characterization of the homogeneous superspaces.

Definition 1.9.8. Let G be a Lie supergroup acting transitively on a supermanifold M and let $p \in |M|$ be a topological point. We define the *stabilizer functor* as:

$$\mathrm{Stab}_p : (\mathrm{smflds})^o \longrightarrow (\mathrm{sets}), \qquad \mathrm{Stab}_p(T) = \{g \in G(T) \,|\, g \cdot p = p\}.$$

We leave to the reader as an exercise the definition of Stab_p on the arrows. ■

The following theorem is given without proof, the reader can consult it in Chapter 9 of Ref. [25] and in Ref. [6].

Theorem 1.9.9. *Let G be a Lie supergroup acting transitively on a supermanifold M and let $p \in |M|$ be a topological point.*

- *The stabilizer functor* Stab_p *is the functor of points of a closed Lie subsupergroup of G.*
- *The supermanifold M is diffeomorphic to G/Stab_p via the morphism* $(G/\mathrm{Stab}_p)(T) \longrightarrow M(T)$, $g\mathrm{Stab}_p \mapsto g \cdot p$.

■

1.10 Real structures and real forms

We are now ready to discuss real forms of complex analytic supermanifolds. A major character in this play is the *complex conjugate of a supermanifold.* We first review the classical setting and put it in a form that is suitable for the generalization.

Let $M = (|M|, \mathcal{H}_M)$ be a complex manifold, with $\mathcal{H}_M$ the sheaf of holomorphic functions on $|M|$. The *complex conjugate* of M is the complex manifold $\mathbb{M} = (|M|, \mathcal{H}_{\mathbb{M}})$, where $\mathcal{H}_{\mathbb{M}}$ is the sheaf of *antiholomorphic functions* on M. They are immediately defined once we have $\mathcal{H}_M$ and the complex structure on M. We have the map

$$\begin{array}{ccc} \mathcal{H}_M & \xrightarrow{\sigma} & \mathcal{H}_{\mathbb{M}} \\ f & \longrightarrow & \sigma(f), \end{array} \qquad \sigma(f)(m) := \overline{f(m)}, \quad m \in |M|\,. \tag{1.7}$$

This map has all of the properties of a sheaf morphism, except for the fact that it is *antilinear*. It is also a bijective map between the complex manifold M and its complex conjugate $\mathbb{M}$.

A *real structure* on M is an isomorphism of ringed spaces $\tau : \mathbb{M} \to M$, such that $|\tau|^2 = \mathrm{id}$. For a suitable choice of local coordinates $z = (z^1, \ldots, z^n)$, it is

$$\begin{array}{ccc} |M| & \xrightarrow{|\tau|} & |M| \\ z & \longrightarrow & \bar{z} \end{array} \tag{1.8}$$

and on the sheaves

$$\begin{array}{ccc} \mathcal{H}_M & \xrightarrow{\tau^*} & \mathcal{H}_{\mathbb{M}} \\ f & \longrightarrow & \tau^*(f), \end{array} \qquad \tau^*(f)(\bar{z}) = f(z)\,. \tag{1.9}$$

The map $|\tau|$ is an involutive map and we can consider the set of fixed points

$$|M|^{|\tau|} = \{m \in |M| \mid |\tau|(m) = m\}\,,$$

which in local coordinates is just the set of points such that $\bar{z} = z$.

The composition $\rho^* := \sigma^{-1} \circ \tau^* : \mathcal{H}_M \to \mathcal{H}_M$ is also involutive. Since σ is antilinear and τ^* is linear, the composition ρ^* is an antilinear, involutive map. We then have a map

$$(|\rho|, \rho^*) : (M, \mathcal{O}_M) \to (M, \mathcal{O}_M) \tag{1.10}$$

that has all the properties of a ringed space isomorphism, except that it is antilinear. It is also an involutive map.

The set of fixed points of ρ^* consists of the functions

$$\rho^*(f) = f, \quad \text{that is,} \quad \overline{f(\bar{z})} = f(z) \quad \text{in local coordinates.}$$

We denote the sheaf of these functions as $\mathcal{O}_{M^\rho}$. On $|M|^{|\rho|}$, the functions of $\mathcal{O}_{M^\rho}$ acquire real values.

For example, if $M = \mathbb{C}$, then the holomorphic functions satisfying $\overline{f(\bar{z})} = f(z)$ are functions whose Taylor expansion has real coefficients and then, on $z = \bar{z}$ acquire real values.

The real ringed space $M^\rho = (|M|^{|\rho|}, \mathcal{O}_{M^\rho}\big|_{|M|^{|\rho|}})$ is a real manifold and it is called a *real form* of $M = (|M|, \mathcal{O}_M)$.

Remark 1.10.1. Let z be the complex, global coordinate of $M = \mathbb{C}$ with the usual complex structure. Forgetting the complex structure, $M_\mathbb{R} = \mathbb{R}^2$ is the underlying real manifold of M and $z = x + \mathrm{i}y$, where x and y are global, real coordinates in $\mathbb{R}^2$.

The variable $\bar{z} = x - \mathrm{i}y$ is the complex, global coordinate of $\mathbb{M}$, the complex conjugate manifold of $M = \mathbb{C}$. It is $|\mathbb{C}|$ itself with the complex conjugated complex structure. As real manifolds, $M_\mathbb{R}$ and $\mathbb{M}_\mathbb{R}$ are identical. Their complexification is just $(M_\mathbb{R})^\mathbb{C} = (\mathbb{M}_\mathbb{R})^\mathbb{C} = \mathbb{C} \times \mathbb{C}$, with global coordinates $(z, \bar{z})$.

Based on this local model we can always find the underlying, real manifold of a complex manifold M. If $z = (z_1, \dots, z_n)$ are local coordinates of M, then $z = (\bar{z}_1, \dots, \bar{z}_n)$ are local coordinates of $\mathbb{M}$ and $(z, \bar{z})$ are local coordinates of the complexification of the underlying, real manifold.

As for the antilinear 'isomorphism' (1.10), it is a true isomorphism of the underlying real manifold $M_\mathbb{R}$. ■

In the super context it is not possible to speak directly of antiholomorphic functions and for this reason we need to generalize this procedure.

Definition 1.10.2. Let $M = (|M|, \mathcal{O}_M)$ be a complex supermanifold. We define a *complex conjugate* of M as a complex supermanifold $\mathbb{M} = (|M|, \mathcal{O}_\mathbb{M})$ together with a map

$$\begin{array}{ccc} \mathcal{O}_M & \stackrel{\sigma}{\longrightarrow} & \mathcal{O}_\mathbb{M} \\ f & \longrightarrow & \sigma(f) \end{array} \tag{1.11}$$

which has all the properties of a sheaf isomorphism, except that it is *antilinear*. On the reduced manifold, it is required to behave as in (1.7). ■

Let us see an easy example:

Example 1.10.3. Let us take $M = \mathbb{C}^{1|1}$. Then $\mathcal{O}_M = \mathcal{O}_\mathbb{C}[\theta]$, with $\mathcal{O}_\mathbb{C}$ the sheaf of holomorphic functions on $\mathbb{C}$. A function $f \in \mathcal{O}_\mathbb{C}[\theta]$ is of the form

$$f(z) = f_0(z) + f_1(z)\theta, \qquad f_0, f_1 \in \mathcal{O}_\mathbb{C}\,.$$

The complex conjugate manifold is $\mathbb{M}$ with $\mathcal{O}_{\mathbb{M}} = \mathcal{O}_{\bar{\mathbb{C}}}[\bar{\theta}]$ and $\mathcal{O}_{\bar{\mathbb{C}}}$ the sheaf of antiholomorphic functions on $\mathbb{C}$. So $g \in \mathcal{O}_{\bar{\mathbb{C}}}[\bar{\theta}]$ will be of the form

$$g(\bar{z}) = g_0(\bar{z}) + g_1(\bar{z})\bar{\theta} \qquad g_0, g_1 \in \mathcal{O}_{\bar{\mathbb{C}}} .$$

We can define now

$$\begin{array}{ccc} \mathcal{O}_{\mathbb{C}}[\theta] & \xrightarrow{\sigma} & \mathcal{O}_{\bar{\mathbb{C}}}[\bar{\theta}] \\ f & \longrightarrow & \sigma(f), \end{array} \qquad \text{such that } \sigma(f)(\bar{z}) = \overline{f_0(z)} + \overline{f_1(z)}\,\bar{\theta}\,,$$

where, $\overline{f_i(z)}$ is just ordinary complex conjugation (1.7). The supermanifold $\mathbb{M} = (\mathbb{C}, \mathcal{O}_{\bar{\mathbb{C}}}[\bar{\theta}])$ is a complex conjugate supermanifold of $M = \mathbb{C}^{1|1}$.

Notice that while $\bar{z}$ has the meaning of the complex conjugate of z, the variable $\bar{\theta}$ is simply another odd variable that we introduce to define the complex conjugate. Practically one can think of the complex conjugate supermanifold as a way of giving a meaning to $\sigma(f)$ as the complex conjugate of a super holomorphic function. ■

Remark 1.10.4. In Definition 1.10.2 we have taken for σ the convention of Ref. [37]

$$\sigma(fg) = \sigma(f)\sigma(g)\,. \tag{1.12}$$

The one used in physics is instead

$$\sigma'(fg) = \sigma'(g)\sigma'(f)\,,$$

and they both differ by a sign when f and g are odd elements. This gives rise to minus signs and 'i' factors that we would like to avoid. But the main reason to choose (1.12) is that it has categorical significance (see pg. 89 of Ref. [37] for a detailed explanation).

One convention can be reverted into the other by the change

$$\sigma(f) = \begin{cases} \sigma'(f) \text{ if } f \text{ is even}\,, \\ \\ \mathrm{i}\sigma'(f) \text{ if } f \text{ is odd}\,. \end{cases}$$

■

We are ready to define a real structure on a complex supermanifold. To avoid heavy notation, we use the same terminology and letters as we did for the classical setting.

Definition 1.10.5. Let $M = (|M|, \mathcal{O}_M)$ be a complex supermanifold and let $\mathbb{M}$ be a complex conjugate of M. We define a *real structure* on M as an isomorphism of super ringed spaces $\tau : \mathbb{M} \longrightarrow M$, which behaves as (1.8), (1.9) on the reduced manifolds and such that the map $\rho^* = \sigma^{-1} \circ \tau^*$ is involutive. ■

More explicitly, on the sheaves, the map

$$\begin{array}{ccc} \mathcal{O}_M & \xrightarrow{\tau^*} & \mathcal{O}_{\mathbb{M}} \\ f & \longrightarrow & \tau^*(f) \end{array}$$

is linear. If we consider the composition

$$\mathcal{O}_M \xrightarrow{\tau^*} \mathcal{O}_{\mathbb{M}} \xrightarrow{\sigma^{-1}} \mathcal{O}_M, \qquad \rho^* = \sigma^{-1} \circ \tau^* .$$

We are requiring: ${\rho^*}^2 = \mathrm{id}$.

The map

$$\rho := (|\rho|,\ \rho^*) : (M, \mathcal{O}_M) \to (M, \mathcal{O}_M)$$

with $|\rho| = |\tau|$ has all the properties of a ringed space automorphism except that it is antilinear. It is also involutive.

We are ready for the definition of real form.

Definition 1.10.6. The *real form* defined by ρ of a supermanifold M is the supermanifold $(|M|^{|\rho|}, \mathcal{O}_{M_\rho})$ where $|M|^{|\rho|}$ are the fixed points of $|\rho| : |M| \to |M|$ and $\mathcal{O}_{M_\rho}$ consists of the restrictions[3] to $|M|^\rho$ of all the functions $f \in \mathcal{O}_M$ such that $\rho^*(f) = f$. ■

This definition becomes clearer by returning to the Example 1.10.3.

Example 1.10.7. Let us consider $M = \mathbb{C}^{1|1}$, and $\mathbb{M}$ and σ as in Example 1.10.3. We define τ^* as

$$\begin{array}{ccc} \mathcal{O}_{\mathbb{C}}[\theta] & \xrightarrow{\tau^*} & \mathcal{O}_{\bar{\mathbb{C}}}[\bar{\theta}] \\ f & \longrightarrow & \tau^*(f), \end{array} \qquad \text{such that } \ \tau^*(f)(\bar{z}) = f_0(\bar{z}) + f_1(\bar{z})\bar{\theta}\,.$$

The composition $\rho^* = \sigma^{-1} \circ \tau^*$ gives on $f \in \mathcal{O}_{\mathbb{C}}[\theta]$

$$\rho^*(f)(z) = \sigma^{-1}\big(\tau^*(f)\big)(z) = \sigma^{-1}\Big(f_0(\bar{z}) + f_1(\bar{z})\bar{\theta}\Big) = \overline{f_0(\bar{z})} + \overline{f_1(\bar{z})}\theta.$$

It is clear that ρ^* is involutive and that it is also an antilinear map. The set of fixed points $\rho^*(f) = f$ are clearly of the form

$$f(z) = f_0(z) + f_1(z)\theta\,,$$

where f_i are holomorphic functions whose Taylor expansion has *real* coefficients. On the submanifold $z = \bar{z}$, which is $\mathbb{R}$, f_i takes real values.

We have then found that $M^\rho = \mathbb{R}^{1|1}$ is a real form of $\mathbb{C}^{1|1}$. ■

[3]We are taking the restrictions of a sheaf on a non open set, however we shall not insist on the technicalities of this construction, the reader can consult Ref. [75].

We want to see how to describe the real form of a complex supermanifold in terms of the functor of points.

Let $M = (|M|, \mathcal{O}_M)$ be a complex supermanifold. One can always give the complex conjugate $\mathbb{M}$ by taking the sheaf $\mathcal{O}_{\mathbb{M}}$ to be the structural sheaf $\mathcal{O}_M$ of M with the complex conjugate $\mathbb{C}$-algebra structure. This means that the multiplication by scalars is

$$\begin{array}{ccc} \mathbb{C} \times \mathcal{O}_{\mathbb{M}}(U) & \longrightarrow & \mathcal{O}_{\mathbb{M}}(U) \\ (\lambda, f) & \longrightarrow & \bar{\lambda} f\,, \end{array}$$

where U is open in $|M|$.

If M is a complex analytic supermanifold, we can introduce $M_{\mathbb{R}}$, the *real underlying supermanifold.* In terms of the functor of points, for an affine[4] complex analytic supermanifold M we define

$$M_{\mathbb{R}}(T) = \mathrm{Hom}_{(\mathrm{salg})_{\mathbb{C}}}(\mathcal{O}(M), \mathcal{O}(T) \otimes \mathbb{C})\,,$$

where T is a real supermanifold. One can show that this is indeed the functor of points of a real supermanifold and this definition can be adapted to the general case, i.e. when M is not affine (hence $\mathcal{O}(M)$ may not contain all of the informations to retrieve M). We shall however not need this generality, since for us it is actually enough to examine the case in which we have global coordinates for M. Let us see two important examples.

Example 1.10.8.

(1) Let us consider the analytic supermanifold $\mathbb{C}^{m|n}$ and let us fix global coordinates $z_1, \ldots, z_m$, $\zeta_1, \ldots, \zeta_n$. A T-point ψ of $\mathbb{C}^{m|n}_{\mathbb{R}}$ consists of a morphism:

$$\psi : \mathcal{O}(\mathbb{C}^{m|n}) \longrightarrow \mathcal{O}(T) \otimes \mathbb{C}\,.$$

ψ amounts to the choice of $2m$ even and $2n$ odd global sections of $\mathcal{O}(T)$:

$$\psi(z_j) = t_j + \mathrm{i}t_{m+j}\,, \qquad \psi(\zeta_k) = \theta_k + \mathrm{i}\theta_{n+k}\,.$$

One could write this more suggestively as

$$\mathbb{C}^{m|n}_{\mathbb{R}}(T) = \{\, (t_1 + \mathrm{i}t_{m+1}, \ldots, t_m + \mathrm{i}t_{2m}, \theta_1 + \mathrm{i}\theta_{n+1}, \ldots, \theta_n + \mathrm{i}\theta_{2n}) \mid$$
$$t_i \in \mathcal{O}(T)_0,\ \theta_j \in \mathcal{O}(T)_1 \,\}\,.$$

[4] The word "affine" means that our manifold is obtained by considering global algebraic relations in some $\mathbb{C}^{m|n}$.

(2) Let us consider the general linear supergroup $\mathrm{GL}(m|n)$ as complex analytic supermanifold. A T-point ϕ of its underlying real supermanifold $\mathrm{GL}(m|n)_{\mathbb{R}}$ is again a morphism

$$\mathcal{O}(\mathrm{GL}(m|n)) \xrightarrow{\ \phi\ } \mathcal{O}(T) \otimes \mathbb{C}\,.$$

Once we fix global coordinates z_{ij} and ζ_{kl} corresponding to the diagonal block and off diagonal block matrices respectively, we have that we can effectively represent ϕ in a matrix form

$$\phi \cong \begin{pmatrix} t & \theta \\ \eta & s \end{pmatrix}, \qquad t, s \text{ being invertible matrices,}$$

where as before we have

$$\phi(z_{ij}) = t_{ij} + \mathrm{i}t'_{ij}, \qquad 1 \leq i, j \leq m,$$

$$\phi(z_{ij}) = s_{ij} + \mathrm{i}s'_{ij}, \qquad m+1 \leq i, j \leq m+n,$$

$$\phi(\zeta_{kl}) = \theta_{kl} + \mathrm{i}\theta'_{kl},\ 1 \leq k \leq m,\ m+1 \leq l \leq m+n,$$

$$\phi(\zeta_{kl}) = \eta_{kl} + \mathrm{i}\eta'_{kl},\ 1 \leq l \leq m,\ m+1 \leq k \leq m+n\,.$$

■

In general, in order to give a real structure, we need to give the antilinear map $\rho : M \to M$. This is a super ringed space morphism of the underlying real manifold $M_{\mathbb{R}}$ (although not every morphism will do the job: it has to be antilinear with respect to the complex structure).

By Yoneda's Lemma, giving $\rho : M_{\mathbb{R}} \to M_{\mathbb{R}}$ is equivalent to give an invertible natural transformation between the functors of points, that is, a functorial map

$$M_{\mathbb{R}}(R) \xrightarrow{\ \rho_R\ } M_{\mathbb{R}}(R)\,,$$

for R a *real* supermanifold.

Then, the functor of points of the real form can be computed as the set of fixed points of ρ.

Let us see an example to explain better the situation.

Example 1.10.9. Consider the complex analytic supergroup[5] $G = \mathrm{GL}(m|n, \mathbb{C})_{\mathbb{R}}$, taken as a real manifold. In order to define a real form through the functor of points formalism, we need an involution of $G(R)$ for

[5]In this section we need to stress the ground field, since both $\mathbb{C}$ and $\mathbb{R}$ appear.

any real supermanifold R, which is functorial in R. An element of $G(R)$ can be written as

$$\begin{pmatrix} t & \theta \\ \eta & s \end{pmatrix} = \begin{pmatrix} t_0 + \mathrm{i}t_1 & \theta_0 + \mathrm{i}\theta_1 \\ \eta_0 + \mathrm{i}\eta_1 & s_0 + \mathrm{i}s_1 \end{pmatrix}, \tag{1.13}$$

where t_i, θ_i, η_i and s_i have entries in $\mathcal{O}(R)$ and therefore belong to a real algebra. We trust the reader to understand the formulation (1.13).

On $G(R)$ we look at the involution:

$$G(R) \xrightarrow{\rho_R} G(R)$$

$$\begin{pmatrix} t & \theta \\ \eta & s \end{pmatrix} \longrightarrow \begin{pmatrix} \overline{t} & \overline{\theta} \\ \overline{\eta} & \overline{s} \end{pmatrix},$$

where $\overline{t} = t_0 - \mathrm{i}t_1$ and similarly for $\overline{\theta}$, $\overline{\eta}$ and $\overline{s}$. Notice that ρ_R preserves the group multiplication

$$\rho_R(gg') = \rho_R(g)\rho_R(g'),$$

thus the set of fixed points of this involution is, not only a real supermanifold, but a real supergroup. It is $G = \mathrm{GL}(m|n, \mathbb{R})$ which, as expected, is a real form of $\mathrm{GL}(m|n, \mathbb{C})$. ■

Remark 1.10.10. A real form of a supermanifold is then defined by the antilinear, involutive isomorphism of ringed spaces $\rho : M \to M$. However, having the notion of complex conjugate $\overline{\mathbb{M}}$, and in particular, having attached a concrete meaning to the 'complex conjugate' of the odd variables is most useful in many explicit examples. It is the same situation that one encounters when describing the real group SU(2) in terms of complex, unitary matrices.

■

1.11 Hopf algebras and Hopf superalgebras

In this section we briefly discuss *Hopf algebras* and *Hopf superalgebras* and some relevant examples. This is a different point of view to describe groups, supergroups and their actions on homogeneous spaces.

For us, it will become most relevant in Chapter 5, when we construct deformations of certain spaces and superspaces together with the deformations of the groups and supergroups acting on them. The reason why we take such approach is that Hopf algebras are possibly the most straightforward way to understand deformations of groups known as *quantum groups*. The extension to the super setting presents no difficulty.

We start with some definitions. Let k be our ground field.

Definition 1.11.1. We say that module A is a *Hopf algebra* if it has the following properties:

(1) A is an *algebra* (not necessarily commutative), that is, there are linear maps, the *multiplication* $\mu : A \otimes A \longrightarrow A$ and the *unit* $i : k \longrightarrow A$ such that the following diagrams commute

$$\begin{array}{ccc} A \otimes k & \xrightarrow{\mathrm{id} \otimes i} & A \otimes A \\ \cong \downarrow & & \downarrow \mu \\ A & \xrightarrow{\mathrm{id}} & A \end{array} \qquad \begin{array}{ccc} k \otimes A & \xrightarrow{i \otimes \mathrm{id}} & A \otimes A \\ \cong \downarrow & & \downarrow \mu \\ A & \xrightarrow{\mathrm{id}} & A \end{array}$$

$$\begin{array}{ccc} A \otimes A \otimes A & \xrightarrow{\mu \otimes \mathrm{id}} & A \otimes A \\ \mathrm{id} \otimes \mu \downarrow & & \downarrow \mu \\ A \otimes A & \xrightarrow{\mu} & A \end{array}$$

A *morphism* $\phi : A \longrightarrow B$ of two algebras, with multiplication μ_A and μ_B and unit i_A and i_B respectively, is a linear map such that

$$\mu_B \circ (\phi \otimes \phi) = \phi \circ \mu_A, \qquad \phi \circ i_A = i_B .$$

(2) A is a *coalgebra*, that is, we can define two linear maps called *comultiplication* $\Delta : A \longrightarrow A \otimes A$ and *counit* $\epsilon : A \longrightarrow k$ with the following properties:

$$\begin{array}{ccc} A \otimes A & \xrightarrow{\mathrm{id} \otimes \epsilon} & A \otimes k \\ \Delta \uparrow & & \uparrow \cong \\ A & \xrightarrow{\mathrm{id}} & A \end{array} \qquad \begin{array}{ccc} A \otimes A & \xrightarrow{\epsilon \otimes \mathrm{id}} & k \otimes A \\ \Delta \uparrow & & \uparrow \cong \\ A & \xrightarrow{\mathrm{id}} & A \end{array}$$

$$\begin{array}{ccc} A \otimes A & \xrightarrow{\Delta \otimes \mathrm{id}} & A \otimes A \otimes A \\ \Delta \uparrow & & \uparrow \mathrm{id} \otimes \Delta \\ A & \xrightarrow{\Delta} & A \otimes A \end{array}$$

A *morphism* $\phi : A \longrightarrow B$ of two coalgebras, with comultiplication Δ_A, Δ_B and counit ϵ_A, ϵ_B respectively, is a linear map such that

$$(\phi \otimes \phi) \cdot \Delta_A = \Delta_B \cdot \phi, \qquad \epsilon_B \cdot \phi = \epsilon_A .$$

(3) The multiplication μ and the unit i are coalgebra morphisms.
(4) The comultiplication Δ and the counit ϵ are algebra morphisms.
(5) A is equipped with a bijective linear map $S : A \longrightarrow A$ called the *antipode* such that the following diagrams commute:

$$\begin{array}{ccc} A\otimes A & \xrightarrow{S\otimes \mathrm{id}} & A\otimes A \\ {\scriptstyle\Delta}\uparrow & & \downarrow{\scriptstyle\mu} \\ A & \xrightarrow{i\cdot\epsilon} & A \end{array} \qquad \begin{array}{ccc} A\otimes A & \xrightarrow{\mathrm{id}\otimes S} & A\otimes A \\ {\scriptstyle\Delta}\uparrow & & \downarrow{\scriptstyle\mu} \\ A & \xrightarrow{i\cdot\epsilon} & A \end{array}$$

(Conditions (3) and (4) are equivalent).

A *Hopf algebra morphism* is a linear map $\phi : A \to B$ which is a morphism of both the algebra and coalgebra structures of A and B and in addition it commutes with the antipodes

$$S_B \circ \phi = \phi \circ S_A\,,$$

where S_A and S_B denote, respectively, the antipodes in A and B.

A module A satisfying the first four properties is called a *bialgebra.* ■

Bialgebras do not have necessarily an antipode, but if an antipode exists, it is unique.

Let A be a coalgebra with comultiplication $\Delta : A \to A\otimes A$ and consider the linear map

$$\begin{array}{ccc} A\otimes A & \xrightarrow{\sigma} & A\otimes A \\ a\otimes b & \longrightarrow & b\otimes a\,. \end{array}$$

We say that A is *cocommutative* if

$$\Delta = \sigma \circ \Delta\,.$$

In the same way, if A is an algebra with multiplication $\mu : A\otimes A \to A$, we can express the commutativity condition as

$$\mu = \mu \circ \sigma\,.$$

We will say that a Hopf algebra is commutative or cocommutative if the underlying algebra and coalgebra structures are so.

Let A be a coalgebra with comultiplication Δ and counit ϵ. We say that a subspace $I \subset A$ is a *coideal* if

$$\Delta(I) \subset I\otimes A + A\otimes I, \quad \epsilon(I) = 0\,.$$

If A is a Hopf algebra, we say that $I \subset A$ is a *Hopf ideal* if I is an ideal of the algebra structure, a coideal of the coalgebra structure and

$$S(I) \subset I .$$

One can check immediately that in that case the algebra A/I inherits naturally a Hopf algebra structure from A.

The square of the antipode of a Hopf algebra, S^2, is an isomorphism of Hopf algebras. However, it is not true in general that $S^2 = \mathrm{id}$. For the cases in which the Hopf algebra is commutative or cocommutative, then we have that $S^2 = \mathrm{id}$.

Let A be an algebra with multiplication $\mu : A \otimes A \to A$ and unit $i : k \to A$. Let us consider its dual space A^*. Then, the dual maps

$$\begin{array}{cccccc} A^* & \xrightarrow{\mu^*} & (A \otimes A)^* & A^* & \xrightarrow{i^*} & k \\ a & \longrightarrow & \mu^*(\alpha) = \alpha \circ \mu , & \alpha & \longrightarrow & i^*(\alpha) = \alpha \circ i \end{array}$$

define on A^* a coalgebra structure provided we can identify $(A \otimes A)^* \cong A^* \otimes A^*$. If A is finite dimensional, this is always the case. The concepts of algebra and coalgebra are then seen to be dual concepts. This prompts the following definition:

Definition 1.11.2. We say that the two Hopf algebras H and H' are in *duality* with each other if we have a non degenerate pairing $\langle \, , \, \rangle : H \times H' \longrightarrow k$ satisfying the properties:

$$\langle uv, x \rangle = \langle u \otimes v, \Delta'(x) \rangle, \qquad \langle u, xy \rangle = \langle \Delta(u), x \otimes y \rangle,$$

$$\langle 1, x \rangle = \epsilon'(x), \qquad \langle u, 1 \rangle = \epsilon(u) ,$$

$$\langle S(u), x \rangle = \langle u, S'(x) \rangle \qquad \text{for} \quad u, v \in H, \quad x, y \in H' .$$

Δ, ϵ, S and Δ', ϵ', S' denote the comultiplication, counit and antipode in H and H' respectively. ■

There are many interesting examples of Hopf algebras; we refer the reader to Refs. [28, 93, 113, 141] for a comprehensive treatment of the subject. We are going to study here two important Hopf algebras that can be associated to a group in ordinary geometry: $\mathcal{O}(G)$, the commutative Hopf algebra of algebraic functions on G, and $\mathcal{U}(\mathrm{Lie}(G))$, the universal enveloping algebra of the Lie algebra $\mathrm{Lie}(G)$.

Example 1.11.3.

(1) Let us consider the algebra $\mathcal{O}(G)$ of an affine algebraic group G. Then we have that $\mathcal{O}(G \times G) \cong \mathcal{O}(G) \otimes \mathcal{O}(G)$ and we can define a coproduct

$$\begin{array}{ccc} \mathcal{O}(G) & \xrightarrow{\Delta} & \mathcal{O}(G) \otimes \mathcal{O}(G) \cong \mathcal{O}(G \times G) \\ f & \longrightarrow & \Delta(f)\,, \end{array}$$

such that $\Delta(f)(g, g') = f(gg')$ for $g, g' \in G$. In *Sweedler's notation* one expresses the coproduct as

$$\Delta(f) = \sum f^{(1)} \otimes f^{(2)}\,,$$

so

$$\Delta(f)(g, g') = \sum f^{(1)}(g) f^{(2)}(g') = f(gg')\,,$$

where we have used the identification $\mathcal{O}(G \times G) \cong \mathcal{O}(G) \otimes \mathcal{O}(G)$ explicitly. The counit and the antipode are defined as

$$\begin{array}{ccc} \mathcal{O}(G) & \xrightarrow{\epsilon} & k \\ f & \longrightarrow & f(\mathrm{id}_G)\,, \end{array} \qquad \begin{array}{ccc} \mathcal{O}(G) & \xrightarrow{S} & \mathcal{O}(G) \\ f & \longrightarrow & S(f)\,, \end{array}$$

such that $S(f)(g) = f(g^{-1})$. One can check that these maps satisfy all the axioms of Definition 1.11.1, so we have given to $\mathcal{O}(G)$ the structure of Hopf algebra. The multiplication is commutative, so the Hopf algebra is commutative, but it is not cocommutative unless the group is abelian. The coproduct that we have defined encodes, indeed, all the information about the product in the group, and one can prove that the category of affine agebraic groups and the category of affine, commutative Hopf algebras are equivalent. We will prove this statement for the super case in Proposition 1.11.5.

(2) Let $\mathcal{U}(\mathrm{Lie}(G))$ be the enveloping algebra of the Lie algebra of the affine, algebraic group G. We denote $\mathfrak{g} = \mathrm{Lie}(G)$. Then $\mathcal{U}(\mathfrak{g})$ is the quotient of the tensor algebra

$$\mathcal{U}(\mathfrak{g}) = T(\mathfrak{g})/(X \otimes Y - Y \otimes X - [X, Y]), \qquad \forall X, Y \in \mathfrak{g}\,,$$

with $[\,,\,]$ the bracket in $\mathfrak{g}$. It is a non commutative algebra. We can define the coproduct, counit and antipode on the generators $X \in \mathfrak{g}$ and extend it by multiplication:

$$\Delta(X) = X \otimes 1 + 1 \otimes X, \quad \epsilon(X) = 0, \quad S(X) = -X\,.$$

One can check the axioms of Definition 1.11.1 and prove that this defines a Hopf algebra structure on $\mathcal{U}(\mathfrak{g})$. Δ is cocommutative, so $\mathcal{U}(\mathfrak{g})$ is a cocommutative Hopf algebra. ■

The two Hopf algebras that we have just described are dual to each other in the sense of Definition 1.11.2. The pairing between $\mathcal{O}(G)$ and $\mathcal{U}(\mathfrak{g})$ with $\mathfrak{g} = \mathrm{Lie}(G)$ is realized as follows:

$$\begin{aligned} \mathcal{U}(\mathfrak{g}) \times \mathcal{O}(G) &\longrightarrow k \\ (u, f) &\longrightarrow \langle u, f \rangle := u(f)\,, \end{aligned} \tag{1.14}$$

where $u(f)$ is the left invariant differential operator u applied to the algebraic function f. When $u \in \mathfrak{g}$, $u(f)$ reads

$$u(f) = \frac{\mathrm{d}}{\mathrm{d}t} f(\exp tu)\Big|_{t=0}\,.$$

The reader can look at the details of the proof in Chapter 7 of Ref. [28].

We now go about the definition of *Hopf superalgebra*, which is most relevant for our purposes. We have to keep in mind the rule (1.1), which defines the multiplication in the tensor product of superalgebras $A \otimes B$, namely

$$(a \otimes b)(c \otimes d) = (-1)^{|b||c|}(ac \otimes bd), \quad a, c \in A,\ b, d \in B\,.$$

Taking this into account, the definition of Hopf superalgebra is essentially the same as Definition 1.11.1, with all the linear maps being now $\mathbb{Z}_2$-graded maps.

Definition 1.11.4. We say that the superalgebra A (not necessarily commutative) is a *Hopf superalgebra* if A has the following properties.

(1) A is a *superalgebra*, with multiplication $\mu : A \otimes A \to A$ and unit $i : k \to A$, i.e. the multiplication and unit respect the parity and satisfy the diagrams detailed in Definition 1.11.1.
(2) A is a super *coalgebra*, that is, we can define two linear maps called *comultiplication* $\Delta : A \to A \otimes A$ and *counit* $\epsilon : A \to k$ respecting parity and with the properties detailed in Definition 1.11.1.
(3) The multiplication μ and the unit i are super coalgebra morphisms.
(4) The comultiplication Δ and the counit ϵ are superalgebra morphisms.
(5) A is equipped with a bijective linear map $S : A \to A$ called the *antipode* such that the diagrams in Definition 1.11.4 commute.

■

We want to end this section with an important result giving the equivalence between the categories of affine supergroups and Hopf superalgebras.

Proposition 1.11.5. *Let $G = (|G|, \mathcal{O}(G))$ be an affine supervariety. Then G is a supergroup if and only if $\mathcal{O}(G)$ is a Hopf superalgebra. Moreover, we can identify the category of affine supergroups with the category of commutative affine Hopf superalgebras.*

Proof. We first observe that if G is a supervariety and $\mathcal{O}(G)$ has a Hopf superalgebra structure with comultiplication Δ, counit ϵ and antipode S, then $G(T)$ has a natural group structure, for any affine supervariety T [6].

Let $A = \mathcal{O}(T)$ and μ_A the multiplication in the superalgebra A. Let x and y two elements in $G(T)$, so they are morphisms $x, y : \mathcal{O}(G) \to A$. We can define the product of x and y as the composition

$$\mathcal{O}(G) \xrightarrow{\quad\Delta\quad} \mathcal{O}(G) \otimes \mathcal{O}(G) \xrightarrow{\quad x\otimes y\quad} A \otimes A \xrightarrow{\quad\mu_A\quad} A$$

$$\text{so} \quad x \cdot y = \mu_A \circ x \otimes y \circ \Delta \,.$$

One can immediately check that the multiplication is a morphism, that is

$$(x \cdot y)(ab) = (x \cdot y)(a)(x \cdot y)(b), \qquad \forall a, b \in A \,,$$

(though hidden, the sign rule plays a crucial role here). This multiplication in $G(T)$ gives rise to the comultiplication Δ in $\mathcal{O}(G)$, as one can readily see.

Let η_A be the unit in $A = \mathcal{O}(T)$. The unit e_T and the inverse i_T in $G(T)$ are defined as follows:

$$\mathcal{O}(G) \xrightarrow{\quad\epsilon\quad} k \xrightarrow{\quad\eta_A\quad} A,, \qquad \mathcal{O}(G) \xrightarrow{\quad S\quad} \mathcal{O}(G) \xrightarrow{\quad x\quad} A,$$

$$e_T = \eta_A \circ \epsilon, \qquad\qquad i_T(x) = x \circ S \,.$$

We leave to the reader the routine checks.

Vice-versa, if G is a supergroup, we can define the comultiplication $\Delta : \mathcal{O}(G) \longrightarrow \mathcal{O}(G) \otimes \mathcal{O}(G)$ as the dual of the multiplication $\mu \in \mathrm{Hom}(G \times G, G)$ using the identification:

$$\mathrm{Hom}(G \times G, G) \cong \mathrm{Hom}(\mathcal{O}(G), \mathcal{O}(G \times G)) \,,$$

(one can check that $\mathcal{O}(G \times G) \cong \mathcal{O}(G) \otimes \mathcal{O}(G)$ as in the non super case). Similarly one defines the counit and the antipode ϵ and S as the duals of unit e and inverse i.

[6] The functor of points of G is determined once we know it on the affine supervarieties (see Chapter 10 in Ref. [25]).

A careful look shows that, formally, the diagrams defining a supergroup (see 1.8.1) are essentially the same as those defining a Hopf superalgebra with arrows reversed. We leave to the reader the routine verification that G satifies all the diagrams in Definition 1.8.1 if and only if $\mathcal{O}(G)$ satisfies the diagrams in Definition 1.11.4.

Finally, an immediate consequence of the previous discussion is the equivalence between the categories of affine supergroups and commutative Hopf superalgebras. ■

We now turn to examine how to interpret supergroup actions and homogeneous superspaces via the Hopf superalgebra formalism. We start with linear representations.

Definition 1.11.6. Let V be a super vector space and G an affine supergroup. We say that G *acts* on V if we have a morphism $G \longrightarrow \mathrm{GL}(V)$. We say that V is a *left G-comodule* if there exists a linear map

$$V \xrightarrow{\Delta_V} \mathcal{O}(G) \otimes V\,,$$

called a *comodule map*, with the properties

(1) $(\mathrm{id}_G \otimes \Delta_V)\Delta_V = (\Delta \otimes \mathrm{id}_V)\Delta_V$,
(2) $(\epsilon \otimes \mathrm{id}_V)\Delta_V = \mathrm{id}_V$, where $\mathrm{id}_G : \mathcal{O}(G) \longrightarrow \mathcal{O}(G)$ is the identity map.

■

One can also define a right G-comodule in the obvious way.

Observation 1.11.7. The two notions of G acting on V and V being a (left) G-comodule are essentially equivalent. In fact, given a representation $\rho : G \longrightarrow \mathrm{GL}(V)$, it defines immediately a comodule map

$$\Delta_V(v) = \rho_G(\mathrm{id}_G)v, \qquad v \in V,\ \mathrm{id}_G \in G(G) = \mathrm{Hom}_{(\mathrm{salg})}(\mathcal{O}(G), \mathcal{O}(G))\,,$$

where we are using the natural identification

$$\mathrm{GL}(V)(G) \subset \mathrm{End}(V)(G) \cong \mathrm{Hom}_{(\mathrm{smod})}(V, \mathcal{O}(G) \otimes V).$$

(Note: to show such identification one can notice that $\mathrm{End}(V) \cong \mathrm{M}(m|n)$ once we fix a homogeneous basis for V and hence $\mathrm{End}(V)(G)$ becomes identified with the $m|n$ matrices with coefficients in $\mathcal{O}(G)$. Refer also to Examples 1.4.5, 1.8.5 and 1.8.3).

Vice-versa, if we have a comodule map Δ_V, we can define a representation in the following way:

$$\begin{array}{ccc} G(T) & \xrightarrow{\rho_T} & \mathrm{GL}(V)(T) \subset \mathrm{Hom}_{(\mathrm{smod})}(V, \mathcal{O}(T) \otimes V) \\ g & \longrightarrow & v \to (g \otimes \mathrm{id})(\Delta_V(v)), \end{array}$$

where $g \in G(T) = \mathrm{Hom}_{(\mathrm{salg})}(\mathcal{O}(G), \mathcal{O}(T))$. ■

For example, we can consider the natural action of $\mathrm{GL}(m|n)$ on $V = k^{m|n}$. Let $\{e_k\}_{\{k=1,\dots,m+n\}}$ be the standard homogeneous basis for $k^{m|n}$ and let g_{ij} be the generators for $\mathcal{O}(\mathrm{GL}(m|n))$ (refer to Example 1.8.5, where such generators were denoted as x_{ij}, ξ_{kl}). We have:

$$\Delta_V(e_j) = \sum_i g_{ij} \otimes e_i \in \mathcal{O}(\mathrm{GL}(m|n)) \otimes V.$$

The representation ρ_T takes the form

$$\begin{array}{ccc} V & \xrightarrow{\rho_T(g)} & \mathcal{O}(T) \otimes V \\ e_j & \longrightarrow & \sum_i t_{ij} \otimes e_i, \end{array}$$

where $g = (t_{ij}) : \mathcal{O}(\mathrm{GL}(m|n)) \longrightarrow \mathcal{O}(T)$ is a T-point of $\mathrm{GL}(m|n)$ defined as $g(g_{ij}) = t_{ij}$.

We now turn to actions of G on supervarieties.

Definition 1.11.8. We say that an algebraic supergroup G *acts* on a supervariety X if we have a morphism $\rho : G \times X \longrightarrow X$ corresponding to the functorial family of morphisms:

$$\begin{array}{ccc} G(T) \times X(T) & \xrightarrow{\rho_T} & X(T) \\ (g, x) & \longrightarrow & g \cdot x, \end{array} \qquad \forall g \in G(T),\ x \in X(T),$$

$(T \in (\mathrm{svar}))$, satisfying the following properties

(1) $1 \cdot x = x$, for all $x \in X(T)$,
(2) $(g_1 g_2) \cdot x = g_1 \cdot (g_2 \cdot x)$, for all $g_1, g_2 \in G(T)$ and $x \in X(T)$.

■

Observation 1.11.9. The action of G on an affine supervariety X is equivalent to the existence of a morphism:

$$\mathcal{O}(X) \xrightarrow{\Delta_X} \mathcal{O}(G) \otimes \mathcal{O}(X),$$

such that the following diagrams commute:

$$\begin{array}{ccc} \mathcal{O}(X) & \xrightarrow{\Delta_X} & \mathcal{O}(G)\otimes\mathcal{O}(X) \\ {\scriptstyle \Delta_X}\big\downarrow & & \big\downarrow{\scriptstyle \mathrm{id}\otimes\Delta_X} \\ \mathcal{O}(G)\otimes\mathcal{O}(X) & \xrightarrow{\Delta\otimes\mathrm{id}} & \mathcal{O}(G)\otimes\mathcal{O}(G)\otimes\mathcal{O}(X) \end{array}$$

$$\begin{array}{ccc} \mathcal{O}(X) & \xrightarrow{\Delta_X} & \mathcal{O}(G)\otimes\mathcal{O}(X) \\ {\scriptstyle \mathrm{id}}\big\downarrow & & \big\downarrow{\scriptstyle \epsilon\otimes\mathrm{id}} \\ \mathcal{O}(X) & = & k\otimes\mathcal{O}(X) \end{array}$$

This fact is immediate once one translates the conditions from the supergroups and supervarieties to their superalgebras of global sections.

The morphism Δ_X is called a *coaction* of $\mathcal{O}(G)$ on $\mathcal{O}(X)$. Coactions of Hopf (super)algebras on (super)algebras can be defined even if they are not commutative. As we will see in Chapter 5, the coaction point of view is the one that can be used to generalize to quantum groups (neither commutative nor cocommutative Hopf algebras) and quantum spaces (non commutative algebras) the concept of action of a group on a space. ■

1.12 $*$-Structures and real forms

In this section we introduce the notion of Hopf $*$-algebras and Hopf $*$-superalgebras (refer to [28, 93] for more details). We also discuss real forms of Lie superalgebras.

Definition 1.12.1. Let A be a complex superalgebra. A $*$-*structure* on A is an antilinear map $*: A \to A$ (it is a linear map of the underlying, real super vector space $A_{\mathbb{R}}$) such that

- $*$ is involutive, i.e. $(a^*)^* = a$,
- $*$ is antimultiplicative $(ab)^* = b^* a^*$.

If A is a Hopf algebra we futherly require that:

- It is a morphism of real coalgebras, $\Delta(a^*) = \Delta(a)^*$.
- If S is the antipode of A, then $S(x)^* = S^{-1}(x^*)$ for all $x \in A$.

We say that a morphism $\phi : A \longrightarrow B$ of Hopf algebras with $*$-structures $*_A$ and $*_B$, preserves the $*$-structures if $\phi(a^{*_A}) = \phi(a)^{*_B}$.

We say that a Hopf algebra is a *Hopf $*$-algebra*, if it has a $*$-structure.

If A is just a complex associative algebra (no coalgebra structure on it), then an antilinear map $*$ satisfying 1 and 2 is also called a $*$-structure on A.

■

This definition has an obvious extension to the super context.

The condition on the antipode does follow indeed from the other conditions, due the uniqueness of S.

Let us examine some simple, but very significant examples.

Example 1.12.2. Let us consider the Hopf algebra $H = \mathcal{O}(\mathrm{SL}_2(\mathbb{C})) = \mathbb{C}[a,b,c,d]/(ad - bc - 1)$. We have two different $*$-structures on H:

$$H \xrightarrow{*_1} H \qquad\qquad H \xrightarrow{*_2} H$$

$$\begin{pmatrix} a & b \\ c & d \end{pmatrix} \longrightarrow \begin{pmatrix} a & b \\ c & d \end{pmatrix}, \qquad \begin{pmatrix} a & b \\ c & d \end{pmatrix} \longrightarrow \begin{pmatrix} d & -c \\ -b & a \end{pmatrix}.$$

As customary, we have organized the generators a, b, c, d in a matrix form. The maps $*_1$ and $*_2$ define two different $*$-structures on the Hopf algebra $H = \mathcal{O}(\mathrm{SL}_2(\mathbb{C}))$. They correspond to the following two real forms of $\mathrm{SL}_2(\mathbb{C})$:

$$\mathrm{SL}_2(\mathbb{R}) = \left\{ \begin{pmatrix} a & b \\ c & d \end{pmatrix} \in \mathrm{SL}_2(\mathbb{C}) \,\middle|\, a = \overline{a},\ b = \overline{b},\ c = \overline{c},\ d = \overline{d} \right\},$$

$$\mathrm{SU}(2) = \left\{ \begin{pmatrix} a & b \\ c & d \end{pmatrix} \in \mathrm{SL}_2(\mathbb{C}) \,\middle|\, d = \overline{a},\ c = -\overline{b}, \right\},$$

where $\overline{z}$ stands for the complex conjugate of z.

We can consider now the $*$-morphisms $H \to \mathbb{C}$. It is not difficult to see that the points of the group $\mathrm{SL}_2(\mathbb{R})$ are recovered as the $*_1$-morphisms $H \longrightarrow \mathbb{C}$: they are all of the form $\phi \to \phi(x)$ for $x \in \mathrm{SL}_2(\mathbb{R})$.

Similarly, the points of the group $\mathrm{SU}(2)$ are the $*_2$-morphisms $H \longrightarrow \mathbb{C}$ and they are all of the type $f \mapsto f(x)$, $x \in \mathrm{SU}(2)$.

So the pair $(\mathcal{O}(\mathrm{SL}_2(\mathbb{C})), *_1)$ defines the real algebraic group $\mathrm{SL}_2(\mathbb{R})$, while the pair $(\mathcal{O}(\mathrm{SL}_2(\mathbb{C})), *_2)$ the real algebraic group $\mathrm{SU}(2)$. Both of these groups are real forms of $\mathrm{SL}_2(\mathbb{C})$. One may retrieve the real algebraic functions on the real forms by looking at the fixed elements of the involutions, as one can readily check. ■

What we have just shown for $\mathrm{SL}_2(\mathbb{C})$ is a general fact for any complex algebraic group. We state the result in the following observation (for more details, see Chapter 4 of Ref. [28]).

Observation 1.12.3. Let G be a complex algebraic group and let $\mathcal{O}(G)$ be the associated Hopf algebra. There is a one to one correspondence between the real forms of G and the Hopf $*$-structures on $\mathcal{O}(G)$. The $\mathbb{R}$-points of a given real form of G are recovered as the $*$-morphisms $\mathcal{O}(G) \to \mathbb{C}$.

Notice that $\mathcal{O}(G)$ is a commutative algebra, so the condition 2 in Definition 1.12.1 becomes $(ab)^* = a^*b^*$. Since the map $*$ is involutive, it is a *conjugation* of A compatible with the algebra and coalgebra structures and the antipode. One can recover the interpretation of the real form of A as the set of fixed points of a conjugation. ■

The following observation relates the notion of $*$-structure and conjugation in a case which is especially relevant for us.

Observation 1.12.4. Let $\mathfrak{g} = \mathrm{Lie}(G)$, G an algebraic group as above. Through the pairing (1.14), a $*$-structure $*$ on the Hopf algebra $\mathcal{O}(G)$ defines a map in $\mathcal{U}(\mathfrak{g})$ with the same properties than in Definition 1.12.1, but with the roles of the multiplication and the comultiplication interchanged. We denote it as $\sigma : \mathcal{U}(\mathfrak{g}) \to \mathcal{U}(\mathfrak{g})$. In particular, this means that σ is an *algebra morphism* of the Lie algebra $\mathcal{U}(\mathfrak{g})$, taken as a *real* Lie algebra, that is,

$$\sigma(XY) = \sigma(X)\sigma(Y), \qquad X, Y \in \mathcal{U}(\mathfrak{g}).$$

The map σ can be restricted to $\mathfrak{g} \subset \mathcal{U}(\mathfrak{g})$ and then the properties $\sigma^2 = 1$, antilinearity and

$$\sigma([X, Y]) = [\sigma(X), \sigma(Y)], \qquad \forall X, Y \in \mathfrak{g},$$

mean that σ is a *conjugation* of the Lie algebra $\mathfrak{g}$. The set of fixed points of σ

$$\mathfrak{g}^\sigma = \{X \in \mathfrak{g} \mid \sigma(X) = X\},$$

is a *real form*, $\mathfrak{g}^\sigma$ of $\mathfrak{g}$ in the ordinary sense, that is, $\mathfrak{g}^\sigma$ is a real Lie algebra whose complexification is $\mathfrak{g}$. ■

Let us see an example, clarifying the meaning of our previous observation.

Example 1.12.5. If $\mathfrak{g} = \mathfrak{sl}_2(\mathbb{C})$, we have two conjugations corresponding to the two real forms of $\mathrm{SL}_2(\mathbb{C})$ described in Example 1.12.2:

$$\begin{array}{ccc} \mathfrak{sl}_2(\mathbb{C}) & \xrightarrow{\sigma_1} & \mathfrak{sl}_2(\mathbb{C}) \\ \\ \begin{pmatrix} a & b \\ c & -a \end{pmatrix} & \longrightarrow & \begin{pmatrix} \bar{a} & \bar{b} \\ \bar{c} & -\bar{a} \end{pmatrix} \end{array}, \qquad \begin{array}{ccc} \mathfrak{sl}_2(\mathbb{C}) & \xrightarrow{\sigma_2} & \mathfrak{sl}_2(\mathbb{C}) \\ \\ \begin{pmatrix} a & b \\ c & -a \end{pmatrix} & \longrightarrow & \begin{pmatrix} -\bar{a} & -\bar{c} \\ -\bar{b} & \bar{a} \end{pmatrix} \end{array}.$$

We have then the two non isomorphic real forms of $\mathfrak{sl}_2(\mathbb{C})$: $\mathfrak{sl}_2(\mathbb{C})^{\sigma_1} = \mathfrak{sl}_2(\mathbb{R})$ and $\mathfrak{sl}_2(\mathbb{C})^{\sigma_2} = \mathfrak{su}(2)$. ■

We end this section with a discussion on how to interpret in the supergeometric setting the concept of the real form of a (Lie) superalgebra. We just need to repeat Observation 1.12.4 in the super setting; for clarity, we write here the formal definition.

Definition 1.12.6. Let $\mathfrak{g}$ be a complex Lie superalgebra. We say that $\sigma : \mathfrak{g} \to \mathfrak{g}$ is a *conjugation* if

(1) σ is an involution, i.e. $\sigma^2 = \mathrm{id}$;
(2) σ is antilinear, that is,

$$\sigma(aX + bY) = \bar{a}\sigma(X) + \bar{b}\sigma(Y), \qquad \forall X, Y \in \mathfrak{g}, \ \ a, b \in \mathbb{C}\,;$$

(3) σ preserves the superbracket, that is

$$\sigma([X, Y]) = [\sigma(X), \sigma(Y)], \qquad \forall X, Y \in \mathfrak{g}\,.$$

As before, σ is an automorphism of $\mathfrak{g}_{\mathbb{R}}$, the real Lie superalgebra underlying $\mathfrak{g}$.

The set of fixed points of a conjugation of $\mathfrak{g}$, $\mathfrak{g}^{\sigma}$, is a real form of $\mathfrak{g}$, that is, the complexification of $\mathfrak{g}^{\sigma}$ is $\mathfrak{g}$.

■

Let us examine an example, which is the super counterpart of Example 1.12.5 in arbitrary dimensions.

Example 1.12.7. Let $\mathfrak{g} = \mathfrak{sl}(m|n)(\mathbb{C})$ be the Lie superalgebra of $m|n \times m|n$ matrices with supertrace zero and let

$$F = \begin{pmatrix} \mathrm{id}_r & 0 \\ 0 & -\mathrm{id}_{m-r} \end{pmatrix}, \qquad G = \begin{pmatrix} \mathrm{id}_s & 0 \\ 0 & -\mathrm{id}_{s-r} \end{pmatrix}.$$

We define the following conjugations

$$\begin{array}{ccc} \mathfrak{sl}(m|n)(\mathbb{C}) & \xrightarrow{\tau} & \mathfrak{sl}(m|n)(\mathbb{C}) \\ \begin{pmatrix} a & b \\ c & d \end{pmatrix} & \longrightarrow & \begin{pmatrix} \overline{a} & \overline{b} \\ \overline{c} & \overline{d} \end{pmatrix}, \end{array}$$

$$\begin{array}{ccc} \mathfrak{sl}(m|n)(\mathbb{C}) & \xrightarrow{\tau_{r,s}} & \mathfrak{sl}(m|n)(\mathbb{C}) \\ \begin{pmatrix} a & b \\ c & d \end{pmatrix} & \longrightarrow & \begin{pmatrix} -Fa^\dagger F & -iFc^\dagger G \\ -\mathrm{i}Gb^\dagger F & -Gd^\dagger G \end{pmatrix}. \end{array}$$

where now a, b, c, d are complex matrices of the appropriate dimensions, $\mathrm{tr}(a) = \mathrm{tr}(d)$ and as usual $x^\dagger = \overline{x}^t$. We have then the non isomorphic real forms of $\mathfrak{sl}(m|n)(\mathbb{C})$

$$\mathfrak{sl}(m|n)(\mathbb{R}) := \mathfrak{sl}(m|n)(\mathbb{C})^\tau, \qquad \mathfrak{su}(r, m-r|s, n-s) := \mathfrak{sl}(m|n)(\mathbb{C})^{\tau_{r,s}}.$$

The matrices of $\mathfrak{sl}(m|n)(\mathbb{R})$ have all their entries real, while

$$\mathfrak{su}(r, m-r|s, n-s) = \left\{ \begin{pmatrix} a & b \\ c & d \end{pmatrix} \mid a = -Fa^\dagger F,\ d = -Gd^\dagger G,\ b = -iFc^\dagger G \right\}.$$

We shall encounter $\mathfrak{su}(r, m-r|s, n-s)$ in Chapter 4, its role being crucial in the study of the real conformal superspace.

We can also generalize Definition 1.12.1 to the super setting. One just has to choose a certain convention, as in the next definition.

Definition 1.12.8. A *star structure* on a complex associative superalgebra $\mathcal{A}$ (not necessarily commutative) is an antilinear involutive map satisfying

$$(f \cdot g)^* = (-1)^{p(f)p(g)} g^* f^*, \qquad f, g \in \mathcal{A}, \tag{1.15}$$

where as usual $p(f)$ is the parity of the element f.

The definition of *Hopf $*$-superalgebra* it is then straightforward, along the lines of Definition 1.12.1. ■

1.13 References

For our brief introduction and overview of supergeometry we send the reader to the works by Berezin [14], Kostant [97], Manin [108], Varadarajan [147] and the notes of Bernstein [37]. In such papers, especially at the beginning, the exposition contains more details. In particular in [108] there is a treatment of supervarieties, but for a more detailed account of the algebraic setting we send the reader to the Chapters 10, 11 of the monograph [25].

Chapter 2

The Ordinary Minkowski and Conformal Spaces

In this chapter we want to introduce the ordinary Minkowski and the conformal spacetimes as homogeneous spaces for the actions of their symmetry groups, namely the Poincaré and conformal group respectively.

We start with a brief introduction to the special theory of relativity from a historical point of view: it is done in such a way to stress the deep impact that experimental facts, as the relativity of motion or the constancy of the velocity of light, have in our fundamental notions about spacetime. The special theory of relativity in fact, imposes restrictions on the structure of spacetime: the Poincaré group arises naturally as the group transforming inertial coordinate systems.

Conformal symmetry is then introduced as a generalization of the symmetry of special relativity, and although it is not universally valid as the Poincaré symmetry is, it certainly plays a fundamental role from many different points of view. Taking into consideration a more general type of symmetry forces us to enlarge the spacetime, so conformal spacetime emerges.

Once the physical motivation and the physical point of view have been explained in detail, we give a rigorous mathematical treatment, first of the complex Minkowski and conformal spaces and then of their real forms. In describing the geometry of the conformal space, we shall take the functorial point of view: this is not strictly necessary to discuss the ordinary (non super) picture, but it becomes essential when we go to the corresponding supergeometric one.

2.1 Relativity and the Poincaré group

Since the days of Galilei, we are aware that there is no absolute motion and that every measure of the position and velocity of a body depends on the observer. We say that motion is relative. In fact, all observations of physical systems have to be referred to an observer, so they are always relative. The structure of spacetime in classical (non quantum) physics is determined by the relativity of motion, widely confirmed by the experiments; even if their consequences challenge our preconceived ideas on the nature of space and time, they are unavoidable.

For us, the observers are devices that measure distances and intervals of time. This means that, at least locally, spacetime is described by four coordinates[1], three for space and one for time.

We want first to describe galilean relativity.

There is a special kind of observer, called inertial observer, for which the laws of motion take a particularly simple form. Let the position vector $\vec{x}(t)$ of a point-like particle[2] be the vector linking the observer with the particle. It may depend on time, as the relative position of the particle and the observer varies. For an inertial observer, $\vec{x}(t)$ follows *Newton's law*, which is a law of motion of the type

$$\vec{F}(\vec{x}, \vec{v}, t, c_1, \dots, c_r) = m\vec{a}, \tag{2.1}$$

where

$$\vec{v} = \frac{\mathrm{d}\vec{x}}{\mathrm{d}t}, \qquad \vec{a} = \frac{\mathrm{d}^2\vec{x}}{\mathrm{d}t^2}.$$

The parameter m is the *inertial mass* of the particle and it is the same for all experiments with the same particle. The vector $\vec{F}$ is the *force*. It describes mathematically the interactions of the particle with other bodies. Each contribution to the force should be unequivocally attached to an interaction. Also, the force depends only on the position and velocity of the body and, possibly, on some parameters $c_1, \dots, c_r$.

It is clear from (2.1) that any two inertial observers move one with respect to another with constant velocity.

[1] Or more: string theory and supergravity make use of extra spacetime coordinates that are supposed to be curled over themselves in a very small volume, not to conflict with observations. These extra dimensions could be detected with energies of the order of the Planck mass, $m_P = \sqrt{\hbar c/G} = (1.220932 \pm 0.00073) \times 10^{19}\mathrm{GeV}/c^2$, much beyond the reach of any conceivable particle accelerator.

[2] A point-like particle is an idealization of the situation that we find in experiments.

In physics, all measurements are affected by an error that limits its precision, so the most we can say is that a certain observer is an inertial observer up to some level of precision. The more an observer is far from other bodies that could interact with it, the more inertial it will become.

An observer on the surface of the Earth may be inertial in good approximation for certain experiments, while in other cases the interaction of the Earth with the Sun may disturb the results of the experiment.

Perhaps paradoxically, it is impossible to construct experimentally a perfect inertial observer, but the ideal concept of it is most relevant in physics.

The *principle of relativity* says that the laws of physics like (2.1) are the same for all inertial observers. Let A and B be two inertial observes and let $\vec{V}$ be the velocity of A with respect to B. Then, the position of A with respect to B is

$$\vec{x}_{AB} = \vec{V} t_B + \vec{T},$$

where t_B is the time measured by the observer B and $\vec{T}$ is a constant vector that marks the position of A with respect to B in the instant $t_B = 0$. In galilean relativity, it is assumed that the measurements of the two observers on the position of a particle and time are related by

$$\begin{aligned} \vec{x}_B &= \vec{x}_A + \vec{x}_{AB} \qquad\qquad t_B = t_A + \tau\,. \\ &= \vec{x}_A + \vec{V} t_B + \vec{T}\,, \end{aligned} \tag{2.2}$$

The crucial hypothesis here is that the time measured by the two observers differs only by a constant τ, which simply marks a difference in the instant chosen by the two observers to start their chronometers; otherwise, the time is the same for all inertial observers. If Newton's law (2.1) holds for the observer A it also holds for observer B.

The quantities τ and $\vec{T}$ are called *temporal* and *spacial translations* respectively, since they represent global shifts on the choice of the origin of the time and the position vector.

The velocities measured by the two observers are related simply by

$$\vec{v}_B = \vec{v}_A + \vec{V}\,. \tag{2.3}$$

Each observer may have a system of three mutually perpendicular axes to which it refers vectors. The observer is located at the origin of the coordinate axes and we say that we have a *reference frame* or *reference system*. Just as for the observers, there is a special kind of reference frames called *inertial frames* for which Newton's law decomposes in a particularly simple form. Let $\{\vec{e}_1, \vec{e}_2, \vec{e}_3\}$ denote perpendicular basis vectors along the three axes of an inertial frame. Then

$$\vec{F} = F^1\vec{e}_1 + F^2\vec{e}_2 + F^3\vec{e}_3\,, \qquad \vec{x} = x^1\vec{e}_1 + x^2\vec{e}_2 + x^3\vec{e}_3\,,$$

and

$$F^1 = m\frac{\mathrm{d}^2x^1}{\mathrm{d}t^2}\,, \quad F^2 = m\frac{\mathrm{d}^2x^2}{\mathrm{d}t^2}\,, \quad F^3 = m\frac{\mathrm{d}^2x^3}{\mathrm{d}t^2}\,.$$

Notice that two inertial frames with the same origin may only differ by a constant rotation.

Let us now consider two inertial observers, A and B. To the observer A we can attach an inertial frame with the basis vectors $\{\vec{e}_1, \vec{e}_2, \vec{e}_3\}$, and to the observer B we attach an inertial frame with the basis vectors $\{\vec{e'}_1, \vec{e'}_2, \vec{e'}_3\}$ (see Fig. 1). Then, the position vectors of a point particle with respect to A and B have coordinates

$$\vec{x}_A = x^1_A\vec{e}_1 + x^2_A\vec{e}_2 + x^3_A\vec{e}_3\,,$$

$$\vec{x}_B = {x'}^1_B\vec{e'}_1 + {x'}^2_B\vec{e'}_2 + {x'}^3_B\vec{e'}_3\,.$$

We may also consider the position vector $\vec{x}_A$ decomposed with respect to the basis $\{\vec{e'}_1, \vec{e'}_2, \vec{e'}_3\}$, and the position vector $\vec{x}_B$ decomposed with respect to the basis $\{\vec{e}_1, \vec{e}_2, \vec{e}_3\}$ (see Fig. 1)

$$\vec{x}_A = {x'}^1_A\vec{e'}_1 + {x'}^2_A\vec{e'}_2 + {x'}^3_A\vec{e'}_3\,,$$

$$\vec{x}_B = x^1_B\vec{e}_1 + x^2_B\vec{e}_2 + x^3_B\vec{e}_3\,.$$

Then we have

$$\begin{pmatrix} {x'}^1_B \\ {x'}^2_B \\ {x'}^3_B \end{pmatrix} = R\begin{pmatrix} x^1_B \\ x^2_B \\ x^3_B \end{pmatrix}, \qquad \begin{pmatrix} {x'}^1_A \\ {x'}^2_A \\ {x'}^3_A \end{pmatrix} = R\begin{pmatrix} x^1_A \\ x^2_A \\ x^3_A \end{pmatrix},$$

where R is the matrix of SO(3) relating both coordinate axes as

$$\vec{e}_i = {R^T}_i^{\ j}\vec{e'}_j\,, \qquad i, j = 1, 2, 3\,.$$

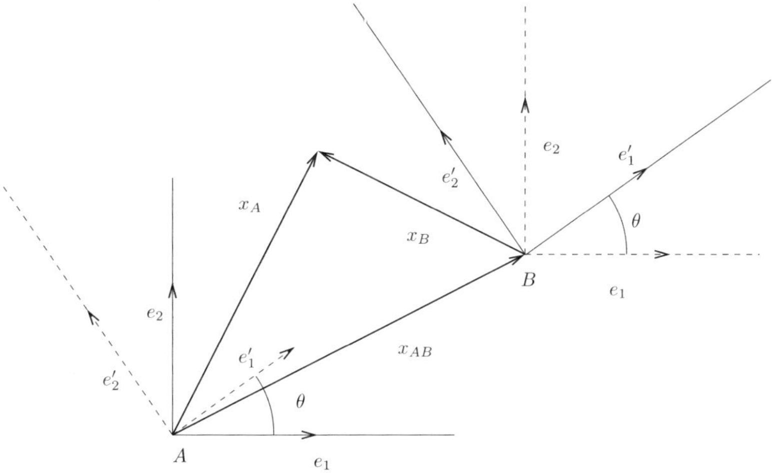

Fig. 1: Two observers with rotated coordinate axes.

The *galilean transformations* (2.2) can be written in components as

$$\begin{pmatrix} t_B \\ {x'}^1_B \\ {x'}^2_B \\ {x'}^3_B \end{pmatrix} = \begin{pmatrix} 1 & 0 & 0 & 0 \\ {V'}^1 & & & \\ {V'}^2 & & R & \\ {V'}^3 & & & \end{pmatrix} \begin{pmatrix} t \\ x^1_A \\ x^2_A \\ x^3_A \end{pmatrix} + \begin{pmatrix} \tau \\ {T'}^1 \\ {T'}^2 \\ {T'}^3 \end{pmatrix}. \tag{2.4}$$

The picture that arises here is that spacetime is the affine space $\mathbb{R}^4$. Transformations between inertial observers are affine transformations of the form (2.4). There is a choice of $\mathbb{R} \subset \mathbb{R}^4$ representing time that is preserved by these transformations.

The *Galilei group* is the group of transformations of the form (2.4). It has the structure of a semidirect product

$$G = \widetilde{G} \ltimes \mathbb{R}^4,$$

where $\widetilde{G}$ is the set of matrices of the form

$$\widetilde{G} = \left\{ \begin{pmatrix} 1 & 0 & 0 & 0 \\ {V'}^1 & & & \\ {V'}^2 & & R & \\ {V'}^3 & & & \end{pmatrix} \;\middle|\; R \in \mathrm{SO}(3),\ \vec{V} \in \mathbb{R}^3 \right\},$$

and $\mathbb{R}^4$ is the abelian subgroup of spacetime translations. G is a Lie group of dimension 10.

The group of euclidean motions of $\mathbb{R}^3$, $E_3 = SO(3) \ltimes \mathbb{R}^3$, where $\mathbb{R}^3$ corresponds to the spatial translations, sits as a subgroup in G. These transformations do not depend on the velocity of the observer and do not transform time. It seems obvious that E_3 has to be part of any relativity theory.

Notice that $\widetilde{G}$ itself is also isomorphic to the euclidean group E_3, the group of rotations and spacial translations.

There is however something wrong with this simple picture. In the 1860's Maxwell showed how the equations for the electromagnetic field predicted waves traveling at the speed of light. This really meant that light was a wave of electromagnetic nature. Other kinds of waves like sound waves, for example, had a speed that was referred to the medium in which they propagate, or, said in other words, the medium they agitate. The air is the medium of sound waves, but what is the medium for light? No clue could be extracted from Maxwell's equations, so no theoretical model was available to test. Experimentally, light travels through outer space and through a vacuum chamber, so what could be the intangible medium of light waves? It should be filling the whole space, although physicists could not manage to detect it in any way.

At the end of the nineteenth century, that medium had a name (and nothing else): the *aether*. Michelson and Morley designed a very precise interferometry experiment to measure differences in the velocity of light traveling in perpendicular directions. If the Earth had a certain velocity with respect to the aether[3], they should be able to detect these velocity differences.

They were prepared to check different theories on aether and aether winds, that is, models on how the Earth would drag the aether in its motion. Their experiment had a disappointing result: no differences at all where detected. It is perhaps the most famous 'failed' experiment in physics.

In 1905 Einstein was the one that dared to assume that light was actually propagating in the vacuum. Then, since no natural reference frame could be associated with the measured velocity of light[4], this velocity must be the same for every observer. This directly contradicted the galilean rule for the composition of velocities (2.3), but it seemed to Einstein that the conclusion was unavoidable. According to the relativity principle, Maxwell's equations

[3]And after Copernicus, it was not reasonable to assume that the aether was moving together with the Earth!

[4]The velocity of light was measured previously by Michelson himself with high accuracy.

should be the same for all inertial observers, like any physical law, so the Galilei group should be modified to take into account that the velocity of light was the same for all inertial observers.

It is reasonable to assume that the group of transformations between inertial frames also acts as affine transformations on $\mathbb{R}^4$, so this new group would be a semidirect product

$$\mathcal{P} = \mathcal{L} \ltimes \mathbb{R}^4 \tag{2.5}$$

in analogy with the Galilei group.

According to (2.5), translations have the same form as in galilean relativity. On the other hand, if we are able to deduce that the rotations are a subgroup of $\mathcal{L}$, the part of the Galilei group that is independent of the relative velocity of the observers will survive in the new group.

Let us now consider a particle that moves at the velocity of light, c, between two points or *events* in spacetime, event p and event q. In the two reference frames one has

$$|\vec{x}_A{}^p - \vec{x}_A{}^q| = c(t_A{}^p - t_A{}^q), \qquad |\vec{x}_B{}^p - \vec{x}_B{}^q| = c(t_B{}^p - t_B{}^q). \tag{2.6}$$

Defining the quantity $x^0 = ct$, with dimension of length, we have the *quadrivectors*

$$x^\mu_A = (x^0_A, x^1_A, x^2_A, x^3_A), \qquad x'^\mu_B = (x'^0_B, x'^1_B, x'^2_B, x'^3_B).$$

This notation may seem confusing at first, but it is standard in physics for quadrivectors. Notice than now we can suppress the labels A and B, since the 'prime' index indicates unequivocally the observer. We also define

$$\Delta x^\mu = x^{\mu p} - x^{\mu q}, \qquad \mu = 0, 1, 2, 3\,.$$

The vector Δx^μ is invariant under spacetime translations, so the transformation properties of Δx^μ will determine $\mathcal{L}$. The equations (2.6) become

$$(\Delta x^0)^2 - (\Delta x^1)^2 - (\Delta x^2)^2 - (\Delta x^3)^2 = 0,$$
$$(\Delta x'^0)^2 - (\Delta x'^1)^2 - (\Delta x'^2)^2 - (\Delta x'^3)^2 = 0\,.$$

The vectors Δx^μ and $\Delta x'^\mu$ belong then to the zero set of the indefinite, non degenerate quadratic form

$$Q(y) = y_0^2 - y_1^2 - y_2^2 - y_3^2\,, \tag{2.7}$$

so the transformations that we are seeking preserve the zero set of Q. This set is a cone in $\mathbb{R}^4$ (see Fig. 2).

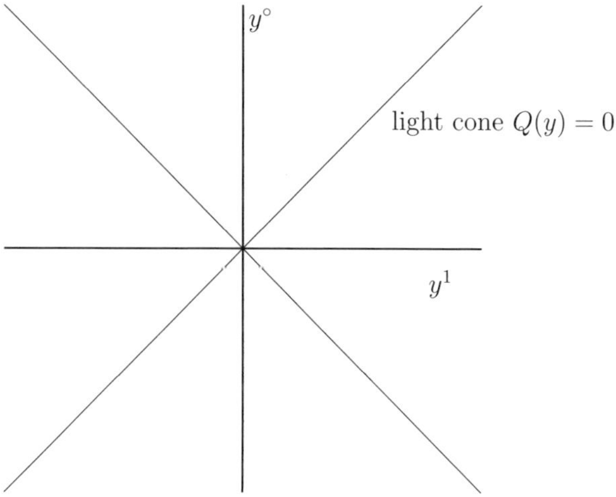

Fig. 2: Light cone of $Q(y) = y_0^2 - y_1^2$.

Since the gradient of Q

$$\vec{\nabla} Q = 2(y_0, -y_1, -y_2, -y_3) \neq 0 \qquad \text{except at } 0 \in \mathbb{R}^4\,,$$

then the cone is a regular submanifold of $\mathbb{R}^4$ except at the vertex. It follows that any other differentiable function that is 0 on the points of the cone should be proportional to Q (for this result, see, for example, Theorem 2.8, pg 82 of Ref. [118]). In particular, if $y' = \Lambda(y)$ denotes the transformation between inertial systems

$$Q'(y) \equiv Q(y') = Q\big(\Lambda(y)\big) = \lambda(y) Q(y)\,, \tag{2.8}$$

for some function $\lambda(y)$, even on the points where $Q(y) \neq 0$. Since the transformations are linear, we can assume that Q' is again a quadratic polynomial and then λ is constant and well defined also for $y = 0$.

The affine space $\mathbb{R}^4$ together with he quadratic form Q, or the corresponding symmetric bilinear form

$$a \cdot b = (a^0, a^1, a^2, a^3) \begin{pmatrix} 1 & 0 & 0 & 0 \\ 0 & -1 & 0 & 0 \\ 0 & 0 & -1 & 0 \\ 0 & 0 & 0 & -1 \end{pmatrix} \begin{pmatrix} b^0 \\ b^1 \\ b^2 \\ b^3 \end{pmatrix}, \tag{2.9}$$

is called the *Minkowski space* and it is denoted as $\mathbf{M}_0$ or $\mathbb{R}^{1,3}$. Here $n_- = 1$ is the number of negative eigenvalues of Q and $n_+ = 3$ is the number of positive eigenvalues. The number $s = n_+ - n_-$ is the *signature* of Q. It is customary in physics to write the bilinear product in several forms, namely

$$a \cdot b = a^T \eta b = a^\mu \eta_{\mu\nu} b^\nu, \qquad \eta = \mathrm{diag}(+1, -1, -1, -1)\,,$$

where summation over repeated indices is understood (Einstein's convention).

Let us consider for a moment only those transformations that preserve the quadratic form (we will comment on the factor λ in Section 2.2). They satisfy

$$\Lambda^T \eta \Lambda = \eta\,, \tag{2.10}$$

so they form the orthogonal group $\mathrm{O}(1,3)$. (See for example [87], Chapter X, for a full description of such Lie groups and their Lie algebras.) Notice that it is a Lie group of dimension six, just as $\widetilde{G}$ was, and that it contains the rotation group $\mathrm{O}(3)$ as a subgroup.

In general, the orthogonal groups $\mathrm{O}(m,n)$, with $m, n \geq 1$ have four connected components. Obviously there are two sets that are disconnected, namely, the ones with $\det \Lambda = \pm 1$, $\Lambda \in \mathrm{O}(m,n)$. Moreover, writing Λ in block form

$$\Lambda = \begin{pmatrix} A_{m\times m} & B_{m\times n} \\ C_{n\times m} & D_{n\times n} \end{pmatrix},$$

one can prove that $\det A \neq 0$ and $\det D \neq 0$. Then the subset with $\det \Lambda > 0$ decomposes in two connected components, corresponding to

$$\det A > 0,\ \det D > 0 \qquad \text{and} \qquad \det A < 0,\ \det D < 0\,.$$

The first one contains the identity and it is a subgroup (in the Lorentz case it is called the *ortochronus Lorentz group*). The set with $\det \Lambda < 0$ also decomposes in two connected components,

$$\det A < 0,\ \det D > 0 \qquad \text{and} \qquad \det A > 0,\ \det D < 0\,.$$

In the compact case, $\mathrm{O}(n)$ has only two connected components, characterized by the sign of the determinant.

It is instructive to see an example of such transformation that involves the velocity $\vec{V}$.

Example 2.1.1. *Boost in the direction x^1.* Let us consider two observers A and B as before. We will assume that

$$x^\mu = 0 \quad \Leftrightarrow \quad x'^\mu = 0,$$

so we discard translations. We also assume that the two systems of axis are parallel, so we discard rotations. The observer A is traveling with velocity $\vec{V} = (V, 0, 0)$ with respect to the observer B.

If we define

$$\beta = \frac{V}{c}, \qquad \gamma = \frac{1}{\sqrt{1-\beta^2}},$$

then the matrix

$$\Lambda = \begin{pmatrix} \gamma & \gamma\beta & 0 & 0 \\ \gamma\beta & \gamma & 0 & 0 \\ 0 & 0 & 1 & 0 \\ 0 & 0 & 0 & 1 \end{pmatrix} \tag{2.11}$$

satisfies (2.10) and the additional condition

$$x'^i = 0, \; i = 1, 2, 3 \quad \Rightarrow \quad x^1 = Vt, \; x^2 = x^3 = 0,$$

which expresses the condition that the observer B, sitting at the origin of the reference frame B, moves with constant velocity V in the direction of x^1 with respect to the reference frame A.

The matrix Λ is a Lorentz transformation called a *boost* in the direction of x^1 with velocity V. ■

Lorentz transformations were written by Lorentz and others when looking for transformations that left Maxwell's equations invariant, always in the spirit of deducing the properties of the aether. Later on, Poincaré realized that these transformations were a group and called it the Lorentz group. It was Einstein, though, the one that derived the Lorentz transformations from the principle of relativity and abandoned the idea of the aether.

Composing two 'boosts' of velocities β and β' in the same direction means multiplying the boost matrices as (2.11). In this way we obtain the law for the composition of velocities

$$\beta'' = \frac{\beta + \beta'}{1 + \beta\beta'}, \qquad \gamma'' = \frac{1}{\sqrt{1 - \beta''^2}}. \tag{2.12}$$

If the velocities involved are small, that is, if

$$\beta, \beta' << 1,$$

one can expand (2.12) to the first order on the dimensionless quantities β and β' and one recovers (2.3). In fact, performing the same expansion in

(2.11) one also recovers (2.4). When performing the expansion one should take into account that in (2.11) $x^0 = ct$. This is the *correspondence principle* of physics for relativity: Einstein's relativity reproduces Galilei's relativity in the regime of velocities that are small compared with the velocity of light. Newton's law, which started our discussion, also gets appropriately modified in Einstein's theory. In the limit when velocities are small, it is reproduced in its usual form.

We have then identified the factor $\mathcal{L}$ in (2.5) as $\mathrm{O}(1,3)$, the Lorentz group. The semidirect group of the Lorentz group with translations is known as the *Poincaré group*

$$\mathcal{P} = \mathcal{L} \ltimes \mathbb{R}^4 \,. \tag{2.13}$$

2.2 Conformal symmetry in physics

Let us now come back to the factor λ in (2.8). A linear transformation $R : \mathbf{M} \to \mathbf{M}$ changing Q by this factor (2.8) satisfies

$$R^T \eta R = \lambda \eta \,. \tag{2.14}$$

It is known that a transformation as (2.14) on a symmetric bilinear form cannot change the numbers of positive and negative eigenvalues, so this means that λ has to be positive. There is an exception when these numbers are the same and

$$\eta' = \begin{pmatrix} \mathrm{id}_{n\times n} & 0 \\ 0 & -\mathrm{id}_{n\times n} \end{pmatrix} .$$

In this case, a transformation exchanging the first n coordinates with the last n coordinates would change $Q \to -Q$ without changing the numbers of positive and negative eigenvalues. Nevertheless, this case will not be relevant for us and we can assume $\lambda > 0$.

There are linear transformations in $\mathbb{R}^4$ that change the quadratic form Q by a factor $\lambda > 0$, for example

$$y^\mu \longrightarrow \lambda^{1/2} y^\mu \,. \tag{2.15}$$

These are called *dilations*. Maxwell's theory of electromagnetism in the vacuum is invariant not only under the Poincaré group, but also under dilations. The symmetry group, including dilations, would then be

$$(\mathcal{L} \times \mathbb{R}^+) \ltimes \mathbb{R}^4 \,.$$

This is not true for every physical theory, so the principle of relativity does not say that all laws of physics are invariant under dilations, but it is so for electromagnetism.

More than that: the scale invariance or invariance under dilations of the Maxwell theory is enhanced to the whole *conformal group*. To introduce the concept of *conformal transformation*, we need to go beyond the framework of quadratic forms in linear spaces and consider the more general concept of metric on a manifold. Our aim is to introduce the conformal group without losing sight of the physical meaning. It is somehow cumbersome, because we have to resort to local descriptions that may be occasionally unsatisfactory. Nevertheless, the complete mathematical picture will be clear at the end. Instead, we can build better our understanding if we start from a less sophisticated point of view.

Let us introduce first the concept of pseudoriemannian manifold together with the local notation that will be used in the sequel.

Let M be an oriented manifold and let $U \subset M$ be an open set with coordinates $\{x^\mu,\ \mu = 1, \dots, n\}$. Let $\{\partial_\mu\}$ be the coordinate basis of the tangent space at a point $x \in M$ and let $\{\mathrm{d}x^\mu\}$ be its dual basis in the cotangent space T^*M, so

$$\mathrm{d}x^\mu(\partial_\nu) = \delta^\mu_\nu\,.$$

A *pseudoriemannian metric* on M is a tensor field

$$\begin{aligned} M &\xrightarrow{\mathrm{d}s^2} \mathrm{Sym}(T^*M \otimes T^*M) \\ x &\longrightarrow \mathrm{d}s^2_x = g_{\mu\nu}(x)\mathrm{d}x^\mu \mathrm{d}x^\nu\,, \end{aligned}$$

where

$$\mathrm{d}x^\mu \mathrm{d}x^\nu = \frac{1}{2}(\mathrm{d}x^\mu \otimes \mathrm{d}x^\nu + \mathrm{d}x^\mu \otimes \mathrm{d}x^\nu)\,, \qquad g_{\mu\nu} = g_{\nu\mu}\,,$$

and $g_{\mu\nu}(x)$ is non degenerate for each x, so $\det\big(g_{\mu\nu}(x)\big) \neq 0$. The notation $\mathrm{d}s^2$ for the metric is used in physics because it reminds of the *length element* $\mathrm{d}s$ and we have decided to keep it: it has no hidden meaning.

The number of positive eigenvalues (n_+) and of negative eigenvalues (n_-) of $g_{\mu\nu}(x)$ is independent of the basis and, because of continuity, it is constant on M. So it is a characteristic of $\mathrm{d}s^2$. The number $s = n_+ - n_-$ is the *signature* of the pseudoriemannian metric, and $n = n_+ + n_-$ is the dimension of M. If $s = n_+ = n$ we say that the signature is *riemannian* and if $s = 1 - (n-1) = 2 - n$ we say that the signature is *minkowskian.*

A pseudoriemannian metric is then a continuous (and differentiable) choice of symmetric, non degenerate, bilinear form on the tangent space of each point. On any paracompact manifold M there exists a riemannian metric. This is because the tangent bundle can always be reduced to a bundle with structure group the orthogonal group $\mathrm{SO}(n)$. A pseudoriemannian metric of arbitrary signature $s = n_+ - n_-$ exists always locally, but in order to have it globally defined, the tangent bundle should be reducible to a bundle with structure group $\mathrm{SO}(n_+) \times \mathrm{SO}(n_-)$, and this does not happen always. For example, the sphere S^2 does not admit a metric with signature $s = 0$.

We can always give a flat pseudoriemannian metric on an open set U by putting on the tangent space at each point the bilinear form (2.9)

$$\begin{aligned} U &\longrightarrow \mathrm{Sym}(T^*U \otimes T^*U) \\ x^\mu &\longrightarrow \mathrm{d}s_x^2 = \eta_{\mu\nu}\mathrm{d}x^\mu \mathrm{d}x^\nu\,, \end{aligned} \qquad \eta = \mathrm{diag}(+1,-1,-1,-1,)\,, \tag{2.16}$$

so, on two vectors of the tangent space at x, $X_x, Y_x \in T_xU$

$$\mathrm{d}s^2(X_x, Y_x) = X_x^\mu Y_x^\nu \eta_{\mu\nu}\,, \qquad X_x = X_x^\mu \partial_\mu\,, \quad Y_x = Y_x^\mu \partial_\mu\,.$$

Here and in what follows we will use Einstein's convention, so whenever in a product there are repeated indices we should understand them as summed.

Let us now consider a diffeomorphism between two open sets (we may think of U and U' as the Minkowski space or portions of it).

$$\begin{aligned} U &\xrightarrow{\ \phi\ } \quad U' \\ x &\longrightarrow x' = \phi(x)\,. \end{aligned} \tag{2.17}$$

On U' we have the induced metric

$$\mathrm{d}s'^2_{x'}(X'_{x'}, Y'_{x'}) = \mathrm{d}s_x^2(\mathrm{d}\phi^{-1}X'_x, \mathrm{d}\phi^{-1}Y'_{x'})\,, \qquad X'_{x'}, Y'_{x'} \in T_{y'}U'\,. \tag{2.18}$$

If the local coordinates in U' are denoted as $\{x'^\mu\}$ we have

$$\mathrm{d}s'^2_{x'} = g_{\rho\sigma}(x')\mathrm{d}x'^\rho \mathrm{d}x'^\sigma = \eta_{\mu\nu}\frac{\partial x^\mu}{\partial x'^\rho}\frac{\partial x^\nu}{\partial x'^\sigma}\mathrm{d}x'^\rho \mathrm{d}x'^\sigma\,.$$

We say that a transformation ϕ is a *conformal transformation* if the induced metric on U' is proportional to the Minkowski metric, that is,

$$\mathrm{d}s'^2_{x'} = \lambda(x')\eta_{\mu\nu}\mathrm{d}x'^\mu \mathrm{d}x'^\nu\,, \qquad \lambda(x') > 0\,,$$

which reduces to

$$\eta_{\mu\nu}\frac{\partial x^\mu}{\partial x'^\rho}\frac{\partial x^\nu}{\partial x'^\sigma} = \lambda(x')\eta_{\rho\sigma}\,. \tag{2.19}$$

Conformal transformations do not necessarily preserve lengths, but they preserve angles. Obviously, Poincaré transformations and dilations are particular cases of conformal transformations, but there are other conformal transformations. An example of a conformal transformation is the *inversion*

$$x^\mu \to \frac{x^\mu}{x^2}, \qquad \text{for which} \qquad \lambda(x) = \frac{1}{(x^2)^2}. \tag{2.20}$$

This is not the only one. As we shall see in Section 2.4, there are conformal transformations of the form

$$x^\mu \longrightarrow x'^\mu = \frac{x^\mu + b^\mu x^2}{1 + 2b \cdot x + b^2 x^2}, \tag{2.21}$$

where the components of the vector b^μ are real numbers, the parameters of the transformation and $b^2 = b \cdot b$. These are called the *special conformal transformations*, and the factor $\lambda(x)$ is in this case

$$\lambda(x) = (1 - 2b \cdot x + b^2 x^2)^2.$$

The geometric meaning of these transformations can be seen by decomposing them as an inversion, a translation and another inversion, that is,

$$x^\mu \quad \to \quad \frac{x^\mu}{x^2} \quad \to \quad \frac{x^\mu}{x^2} + b^\mu \quad \to \quad \frac{x^\mu/x^2 + b^\mu}{1 + 2b \cdot x + b^2 x^2}\, x^2 = x'^\mu .$$

At this point we can interpret a transformation like for example (2.21) either as a change of coordinates of the same point in Minkowski space (*passive picture*), as we have been doing up to now, or as a map that sends one point in Minkowski space with coordinates x^μ to another point in Minkowski space with coordinates x'^μ (*active picture*). We will take the active point of view in the developments of the next sections.

A couple of comments about the special conformal transformations: First of all, they are not linear transformations of spacetime. Second, for each point x^μ in Minkowski space there are values of the parameters b^μ for which the transformation is singular. This is also true for the inversion (2.20), which is not defined for $x = 0$. From this we learn that the special conformal transformations act non linearly on some sort of extended Minkowski space. This extended space is what we call *conformal spacetime* and we will denote it as **C**. It must be the union of the Minkowski space with some *points at infinity* that can be reached from a point x^μ by one of the singular special conformal transformations. We will study the mathematical structure of conformal spacetime in the following sections, but we can anticipate that conformal spacetime is a *compactification* of Minkowski space and that Minkowski space is the *big cell* of conformal spacetime. This is a special open set, which is dense in $\mathbf{C}$ and with a precise algebraic meaning that we shall detail in later sections.

The conformal group is distinguished in physics not only because it is a symmetry of Maxwell's theory, but for more general reasons. A careful exposition of all these reasons requires a thorough knowledge of modern Quantum Field Theory, the current framework were high energy physics is described, and general relativity, which is Einstein's theory of gravity. These are much beyond the scope of this book. Nevertheless, we think that it could be interesting to give a very quick overview of the current ideas that bring the conformal group into focus in physics.

The Coleman-Mandula theorem [33] of quantum field theory states that, under some technical assumptions, the conformal symmetry is the maximal symmetry that spacetime can have. All the other symmetries of a physical theory must be internal symmetries (i.e. not involving spacetime) and they have to be a factor in a direct product with the spacetime symmetry. So even when conformal symmetry is not an exact symmetry in the physical world, the introduction of conformal spacetime may be a useful tool: one goes from Minkowski to conformal spacetime by compactifying and viceversa, by restricting to the big cell. For example, free particles are described by certain unitary representations of the Poincaré group, which are classified by two numbers: the *mass*, a real non negative number, and the *spin*, a non negative half-integer. The representations of the Poincaré group that lift to representations of the conformal group are only the zero mass particles. In our world we see coexisting particles of zero mass, as the photons of the electromagnetic field, with particles of mass bigger than zero. It seems then that the basic framework would be conformal and when restricting to the Minkowski space one would allow terms in the lagrangian that break explicitly the conformal invariance, like mass terms. This was the point of view taken by Penrose in his *twistor space* description of Minkowski space [124]. One can also envisage regimes where the terms breaking the invariance become negligible. For example, in a regime of energies that are much higher than the mass of the particles, conformal symmetry may become an approximate symmetry.

One speaks in physics about scale invariance as a global invariance because the factor λ does not depend on the point of spacetime, as opposed to conformal invariance that is a local invariance because the factor λ depends on the point as in (2.19). As for electromagnetism, all relevant quantum field theories which have nice properties as unitarity, invariance under the Poincaré group, etc, that are scale invariant are also conformally invariant. One says then that the global invariance is enhanced to a local invariance. This is a non trivial phenomenon and the reason why it occurs has not been understood completely yet, at least not from a quantum, non perturbative point of view (we refer the reader to some of the original papers [23,32,126] and a recent review [115]). This kind of symmetry enhancements always have a deep meaning. For example in Einstein's theory of gravitation (General Relativity) the solutions that are static and have symmetry under rotations around an axis are also spherically symmetric, so they become the Schwartzschild solution. This is the content of the no-hair theorem for black holes. Moreover, black hole physics has a thermodynamic-like interpretation, and the microstates giving rise to a statistical macrostate (the black hole) are supposed to be quantum gravity microstates. The no-hair theorem would then have profound implications for a (still unknown) theory of quantum gravity.

The AdS/CFT conjecture (or gauge/gravity duality) [107] connects conformal field theories with gravity theories through the so called holographic principle [140, 142, 143]. Presumably, the enhancement of the scale symmetry to the conformal group at the quantum level has also its reflection in the consistency of a theory of quantum gravity. It is conjectured that renormalization group properties in conformal field theories are responsible of this enhancement. Whatever is the mechanism, the holographic principle would then give very valuable clues towards a formulation of the long sought consistent theory of quantum gravity.

The study of the Minkowski space in terms of the conformal group has so far been motivated from the origins of Einstein's theory of relativity to the most recent trends in high energy physics. We go now to the detailed mathematical description of it.

2.3 Conformal transformations and their Lie algebra

In this section we want to derive the conformal transformations as symmetries of the Minkowski metric starting from the requirement (2.19). We will first derive the corresponding Lie algebra, that is, the *infinitesimal transformations*, and then we will integrate them to finite transformations. This derivation is not strictly needed for the rest of the book, and one can skip it and go directly to the next section, where, starting from the Lie algebra, we show that the conformal group of Minkowski space is $O(2,4)$.

As we have discussed in the previous section, conformal transformations form a group G that acts, not on Minkowski space, but on the conformal space $\mathbf{C}$

$$G \times \mathbf{C} \longrightarrow \mathbf{C}\,.$$

All we know about $\mathbf{C}$ is that it is a manifold of dimension four containing the Minkowski space plus some points at infinity. We shall determine it together with G itself. The procedure, as it is common in physics, will be to start from a local picture. The global picture will emerge a posteriori.

It is easier to determine first the Lie algebra of G, which is realized as a subalgebra of the algebra of vector fields on $\mathbf{C}$, denoted by $\mathcal{X}(\mathbf{C})$. In order to do that, we need to introduce some concepts of differential geometry and its relation with the notation used in physics.

Let $X \in \mathcal{X}(\mathbf{C})$ be a vector field on $\mathbf{C}$ and $x \in \mathbf{C}$. Let $c_x : I \subset \mathbb{R} \to \mathbf{C}$ be an integral curve of X through x, that is, $c_x(\epsilon)$ is a solution of the first order differential equation

$$\frac{\mathrm{d}}{\mathrm{d}\epsilon} c_x(\epsilon) = X\big(c_x(\epsilon)\big) \qquad \text{with initial condition} \qquad c_x(0) = x\,. \tag{2.22}$$

We can assume that I is an interval containing the 0. Any two such integral curves coincide in their domains of definition.

Let $U \subset \mathbf{C}$ an open set with $x \in U$. A *flow box* of X is a map

$$\begin{array}{ccc} I \times U & \xrightarrow{F} & U \\ (\epsilon, x) & \longrightarrow & F_\epsilon(x) = c_x(\epsilon)\,. \end{array}$$

It is not difficult to see that $F_\epsilon \circ F_\sigma = F_{\epsilon+\sigma}$ whenever the composition is possible (that is, if $F_\sigma(x) \in U$). The map

$$F_\epsilon : U \to \mathbf{C}$$

is a diffeomorphism of U into its image.

A vector field X is said to be *complete* if each integral curve can be extended to the whole $\mathbb{R}$, so

$$\begin{array}{ccc} \mathbb{R} \times \mathbf{C} & \xrightarrow{F} & \mathbf{C} \\ (\epsilon, x) & \longrightarrow & F_\epsilon(x)\,. \end{array}$$

In this case, the set of transformations of $\mathbf{C}$ $\{F_\epsilon,\ \epsilon \in \mathbb{R}\}$ is a one-parameter group of transformations of $\mathbf{C}$.

We search now for vector fields whose flows are conformal transformations. This means that the local diffeomorphisms $F_\epsilon : U \to U'$ satisfy the equation (2.19), that we rewrite here

$$\eta_{\mu\nu} \frac{\partial x^\mu}{\partial x'^\rho} \frac{\partial x^\nu}{\partial x'^\sigma} = \lambda(x')\eta_{\rho\sigma} \,, \tag{2.23}$$

where x^μ and x'^μ are local coordinates in U and U' respectively and $x' = F_\epsilon(x)$.

So we have to look for the restrictions that (2.23) imposes on the vectors $X \in \mathcal{X}(U)$. The easiest way is to write the flow, to the first order in ϵ, as

$$x'^\mu \approx x^\mu + \epsilon X^\mu(x), \qquad X(x) = X^\mu(x)\partial_\mu. \tag{2.24}$$

This expression is what physicists call infinitesimal transformations. Then we substitute (2.24) in (2.23) and keep only the first order terms in ϵ. We obtain

$$\eta_{\mu\sigma}\partial_\rho X^\mu + \eta_{\rho\nu}\partial_\sigma X^\nu = \partial_\mu \lambda\, X^\mu\, \eta_{\rho\sigma} \,. \tag{2.25}$$

Notice that (2.25) can be rewritten in terms of the Lie derivative with respect to X

$$\mathcal{L}_X \eta = \chi\, \eta \,, \qquad \text{with } \chi = \partial_\mu \lambda\, X^\mu \,, \tag{2.26}$$

which is in fact saying that X is a *homothetic Killing vector* for the metric η. Infinitesimal conformal transformations are then homothetic Killing vectors of the flat metric η of the Minkowski space. Notice that homothetic Killing vectors of a metric form a Lie algebra. We will compute explicitly the structure constants of the conformal Lie algebra later on.

The idea now is to determine the form of the functions $X^\mu(x)$ and $\lambda(x)$.

Remark 2.3.1. *Notation.* It is customary in physics to define the matrix $\eta^{\mu\nu}$ as the inverse of $\eta_{\mu\nu}$, so

$$\eta^{\mu\nu}\eta_{\nu\sigma} = \delta^\mu_\sigma \,, \qquad \eta^{\mu\nu} = \mathrm{diag}(+1, -1, -1, -1) \,.$$

One can also define, for a vector $X^\mu \partial_\mu$, the covector

$$X_\mu \mathrm{d}x^\mu \in T^* U \quad \text{where} \quad X_\mu = \eta_{\mu\nu} X^\nu \,,$$

so the position of the index indicates if the object is a vector or a covector. Of course, this notation is only valid when working in a chosen coordinate system and the indices are written explicitly, but it is very common in physics, because it eases the calculations. For a general metric $g_{\mu\nu}$, one lowers and raises indices with the metric $g_{\mu\nu}$ and its inverse $g^{\mu\nu}$.

■

Using Remark 2.3.1 we can rewrite (2.25) as

$$\partial_\rho X_\sigma + \partial_\sigma X_\rho = \chi\, \eta_{\rho\sigma}\,. \tag{2.27}$$

We are going to perform some manipulations to solve these equations for X^μ and λ. First, we can multiply (2.27) by $\eta^{\rho\sigma}$ and then sum over repeated indices, which gives

$$2\chi = \partial_\rho X^\rho\,. \tag{2.28}$$

This allows us to compute χ given X_σ.

We still do not know the form of X_σ. To compute it, we will first obtain an equation for χ that is straightforward to solve. We start by applying the derivative ∂_μ to (2.27), so

$$\partial_\mu\partial_\rho X_\sigma + \partial_\mu\partial_\sigma X_\rho = \partial_\mu\chi\, \eta_{\rho\sigma}\,.$$

From here we can solve for $\partial_\mu\partial_\rho X_\sigma$: It is enough to permute the indices (ρ, σ, μ) and write a linear combination of the corresponding (equivalent) equations

$$(\rho, \sigma, \mu) + (\mu, \sigma, \rho) - (\rho, \mu, \sigma)\,.$$

The result is the also equivalent equation

$$2\partial_\mu\partial_\rho X_\sigma = \eta_{\rho\sigma}\partial_\mu\chi + \eta_{\mu\sigma}\partial_\rho\chi - \eta_{\rho\mu}\partial_\sigma\chi\,. \tag{2.29}$$

We multiply now this equation by $\eta^{\mu\rho}$ and sum over repeated indices to obtain

$$\partial_\mu\partial^\mu X_\sigma = -\partial_\sigma\chi\,. \tag{2.30}$$

Applying now ∂_ν to (2.30), $\partial_\mu\partial^\mu$ to (2.27) and combining the results, we get

$$-2\partial_\rho\partial_\sigma\chi = \eta_{\rho\sigma}\partial_\mu\partial^\mu\chi\,. \tag{2.31}$$

Finally, multiplying (2.31) by $\eta^{\rho\sigma}$ we get

$$\partial_\mu\partial^\mu\chi = 0\,, \tag{2.32}$$

and substituting (2.32) in (2.31)

$$\partial_\rho\partial_\sigma\chi = 0\,.$$

This implies that $\chi(x)$ is of the form

$$\chi(x) = u + v_\mu x^\mu \qquad u, v_\mu \text{ constant}\,. \tag{2.33}$$

Let us see the consequences of (2.33) X_σ. First of all, (2.29) implies that X_σ can be, at most, quadratic in x^μ:

$$X_\sigma = a_\sigma + b_{\sigma\rho}x^\rho + c_{\sigma\rho\mu}x^\rho x^\mu\,, \qquad c_{\sigma\rho\mu} = c_{\sigma\mu\rho}\,. \tag{2.34}$$

We introduce now the expressions (2.33) and (2.34) in (2.27) and (2.29). This determines the form of the coefficients $b_{\sigma\rho}$ and $c_{\sigma\rho\mu}$ in terms of u and v_μ:

$$b_{\sigma\rho} + b_{\rho\sigma} = u\eta_{\rho\sigma}, \tag{2.35}$$

$$c_{\sigma\rho\mu} = \eta_{\rho\sigma}v_\mu + \eta_{\mu\sigma}v_\rho - \eta_{\rho\mu}v_\sigma\,, \tag{2.36}$$

while a_σ remains unconstrained.

Let us now interpret the allowed transformations.

In the first place, we have

$$x'^\mu \approx x^\mu + \epsilon a^\mu,$$

which is simply an infinitesimal translation. The vector field generating the infinitesimal transformation is $X = a^\mu\partial_\mu$. The quantities a^μ are arbitrary, so there are indeed four independent vector fields associated with translations, namely

$$P_\mu = \partial_\mu.$$

Going back to the differential equation for the flow (2.22), we find that the finite transformation is just the same

$$x'^\mu = x^\mu + \epsilon a^\mu\,, \tag{2.37}$$

where now ϵ is a finite parameter. We recover in this way the four independent spacetime translations.

Equation (2.35) says that $b_{\sigma\rho}$ has an arbitrary antisymmetric part and a symmetric part equal to $u\eta_{\sigma\rho}/2$. The symmetric part of the tensor generates the infinitesimal transformation

$$x'^\mu \approx (1 + \epsilon\frac{u}{2})x^\mu$$

with vector field $uD/2$, where

$$D = x^\mu\partial_\mu\,.$$

Integrating the differential equation (2.22) we get the finite transformation

$$x'^\mu = \mathrm{e}^{\epsilon u/2}x^\mu\,, \tag{2.38}$$

which is a dilation.

The antisymmetric part of b, $m_{\mu\nu} = (b_{\mu\nu} - b_{\nu\mu})/2$, generates an infinitesimal transformation of the form

$$x'^{\mu} \approx x^{\mu} + \epsilon m^{\mu}{}_{\nu} x^{\nu}$$

in terms of the matrices $m^{\mu}{}_{\nu} = \eta^{\mu\rho} m_{\rho\nu}$. Rewriting the antisymmetry condition

$$\eta_{\mu\rho} m^{\rho}{}_{\nu} + \eta_{\nu\sigma} m^{\sigma}{}_{\mu} = 0\,,$$

that is,

$$m^T \eta + \eta m = 0\,,$$

we have that the matrices $m^{\mu}{}_{\nu}$ form the Lie algebra of the Lorentz group, $\mathfrak{so}(1,3)$. Six independent generators are

$$L_{\mu\nu} = x_{\nu}\partial_{\mu} - x_{\mu}\partial_{\nu}\,.$$

The finite transformations are Lorentz transformations (2.10)

$$x'^{\mu} = \Lambda^{\mu}{}_{\nu} x^{\nu}.$$

The remaining transformations are generated by $c_{\sigma\rho\mu} x^{\rho} x^{\mu}$ (2.36),

$$x'^{\mu} \approx x^{\mu} + \epsilon(2b \cdot x\, x^{\mu} - x^2\, b^{\mu})\,,$$

where we have put $b^{\mu} = v^{\mu}/4$. There are four independent vector fields denoted by

$$K_{\mu} = 2x_{\mu} x^{\nu} \partial_{\nu} - x^2 \partial_{\mu}\,.$$

It is easy to check that if x'^{μ} is of the form (2.21),

$$x'^{\mu} = \frac{x^{\mu} + \epsilon b^{\mu} x^2}{1 + 2\epsilon b \cdot x + \epsilon^2 b^2 x^2}\,, \tag{2.39}$$

then

$$\left.\frac{\mathrm{d}x'^{\mu}}{\mathrm{d}\epsilon}\right|_{\epsilon=0} = 2b \cdot x\, x^{\mu} - x^2\, b^{\mu}\,,$$

which proves that $c_{\sigma\rho\mu} x^{\rho} x^{\mu}$ generates the special conformal transformations.

The vector fields of translations, dilations and Lorentz transformations are complete vector fields. The flows, then, define one-parameter groups of transformations of Minkowski space. On the contrary, the vector fields associated to the special conformal transformations are not complete, since for a given x^{μ} there are values of the parameters b^{μ} for which the denominator of (2.39) is zero. This fact was already mentioned in Section 2.2, where we pointed out that a special conformal transformation may take a point in Minkowski space to infinity. Intuitively, we see that one could 'complete' the Minkowski space with some points at infinity to make the special conformal transformations well defined everywhere for arbitrary parameters. In Section 2.7, this intuitive point of view will be made fully rigorous and the conformal space will be introduced as a compactification of the Minkowski space.

We summarize our discussion with the following proposition.

Proposition 2.3.2. *The infinitesimal conformal transformations, that is, the homothetic Killing vectors of the Minkowski metric* η *form a Lie algebra generated by:*

$$\begin{aligned}
&P_\mu = \partial_\mu\,, && \textit{(translations)},\\
&D = x^\mu\partial_\mu\,, && \textit{(dilation)},\\
&L_{\mu\nu} = x_\nu\partial_\mu - x_\mu\partial_\nu\,, && \textit{(Lorentz transformations)},\\
&K_\mu = 2x_\mu x^\nu\partial_\nu - x^2\partial_\mu\,, && \textit{(special conformal transformations)}.
\end{aligned}$$

Each of these generators corresponds to a finite transformation:

$$\begin{aligned}
&x'^\mu = x^\mu + \epsilon a^\mu\,, && \textit{(translations)},\\
&x'^\mu = \mathrm{e}^{\epsilon u/2}x^\mu\,, && \textit{(dilation)},\\
&x'^\mu = \Lambda^\mu{}_\nu x^\nu\,, && \textit{(Lorentz transformations)},\\
&x'^\mu = \frac{x^\mu + \epsilon b^\mu x^2}{1 + 2\epsilon b\cdot x + \epsilon^2 b^2 x^2}\,, && \textit{(special conformal transformations)}.
\end{aligned}$$

2.4 The conformal group and its spin group

We want now to put the conformal Lie algebra derived in Proposition 2.3.2 in a more familiar form. Let us summarize here its generators $\{P_\mu, D, L_{\mu\nu}, K_\mu\}$

$$\begin{aligned}
&P_\mu = \partial_\mu\,, && \text{(translations)},\\
&D = x^\mu\partial_\mu\,, && \text{(dilation)},\\
&L_{\mu\nu} = x_\nu\partial_\mu - x_\mu\partial_\nu\,, && \text{(Lorentz transformations)},\\
&K_\mu = 2x_\mu x^\nu\partial_\nu - x^2\partial_\mu\,, && \text{(special conformal transformations)}.
\end{aligned}$$

One can directly compute their commutation relations:

$$\begin{aligned}
&[P_\mu, D] = P_\mu\,, && [K_\mu, D] = K_\mu\,,\\
&[L_{\mu\nu}, P_\rho] = \eta_{\rho\nu}P_\mu - \eta_{\rho\mu}P_\nu\,, && [L_{\mu\nu}, K_\rho] = \eta_{\rho\nu}K_\mu - \eta_{\rho\mu}K_\nu\,,\\
&[P_\mu, K_\nu] = 2\eta_{\mu\nu}D + 2L_{\mu\nu} &&\\
&[L_{\mu\nu}, L_{\rho\sigma}] = \eta_{\mu\rho}L_{\nu\sigma} + \eta_{\nu\sigma}L_{\mu\rho} - \eta_{\nu\rho}L_{\mu\sigma} - \eta_{\mu\sigma}L_{\nu\rho}\,, &&
\end{aligned}$$

and the rest 0. Notice that K_μ and P_μ behave in the same way with respect to D and $L_{\rho\sigma}$.

We can rearrange the generators in the following way. Let us consider

$$J_{AB}\,, \qquad A, B = -1, 0, 1, 2, 3, 4\,,$$

with $J_{AB} = -J_{BA}$ and

$$\begin{aligned}
J_{\mu\nu} &= L_{\mu\nu}\,,\\
J_{4\mu} &= \frac{1}{2}(P_\mu - K_\mu)\,,\\
J_{-1\,\mu} &= \frac{1}{2}(P_\mu + K_\mu)\,,\\
J_{-1\,4} &= D\,, \qquad\qquad \mu, \nu = 0, 1, 2, 3\,.
\end{aligned}$$

It takes a moment to check that the commutation relations in terms of J_{AB} are

$$[J_{AB}, J_{CD}] = \eta_{AC}J_{BD} + \eta_{BD}J_{AC} - \eta_{BC}J_{AD} - \eta_{AD}J_{BC}\,, \tag{2.40}$$

where

$$\eta_{AB} = \mathrm{diag}(+1, +1, -1, -1, -1, -1)\,,$$

so we have proven that the conformal algebra is the orthogonal algebra $\mathfrak{so}(2, 4)$.

It is now more clear what should be the conformal group $\mathcal{C}$. On one hand, we have seen that $\mathrm{Lie}(\mathcal{C}) = \mathfrak{so}(2, 4)$; on the other hand, we have computed the finite conformal transformations (2.37, 2.38, 2.10, 2.39), especially (2.10). This suggests us that we should take as the conformal group $\mathcal{C}$ the orthogonal group $\mathrm{O}(2, 4)$, where we have included in $\mathcal{C}$ the four connected components (see Section 2.1) to allow for the most general group.

The connected components are related by multiplication by some discrete transformations. For example, the transformation

$$(x^{-1}, x^0, x^1, x^2, x^3, x^4) \to (-x^{-1}, x^0, x^1, x^2, x^3, x^4)\,, \tag{2.41}$$

changes the sign of the total determinant. The transformation

$$(x^{-1}, x^0, x^1, x^2, x^3, x^4) \to (-x^{-1}, x^0, -x^1, x^2, x^3, x^4), \tag{2.42}$$

changes the signs of the minors $\det A$ and $\det B$, which also means a change of connected component.

In many cases it is enough to consider the connected component of the identity. We shall describe the conformal space as an homogeneous space of the conformal group, and the action of the identity component will be transitive on it. Nevertheless, these discrete transformation may play a role in physical theories, as parity and time inversion of the Lorentz group do play a significant role in particle physics.

We are now going to describe a manifold that supports an action of $\mathcal{C}$ and it is a compactification of $\mathbb{R}^4$. Let us consider the following quadric in real projective space $\mathbb{P}^5(\mathbb{R}) = \{\text{lines in } \mathbb{R}^6\}$,

$$uv - \eta_{\mu\nu}x^\mu x^\nu = 0, \qquad \eta_{\mu\nu} = \text{diag}(+1, -1, -1, -1), \quad \mu, \nu = 0, 1, 2, 3, \tag{2.43}$$

where the solutions have to be taken modulo the multiplication by a factor $\rho \in \mathbb{R}$, since the quadric is defined in projective space.

Changing variables with $u = a + b$, $v = a - b$ we get

$$a^2 - b^2 - (x^0)^2 + (x^1)^2 + (x^2)^2 + (x^3)^2 = 0,$$

so the equation is invariant under $\mathrm{O}(2,4)$. The given quadric describes a manifold in $\mathbb{P}^5(\mathbb{R})$ carrying a natural action of $\mathrm{O}(2,4)$. We shall come back to this point in Section 2.11.

We now turn to examine its topological aspects.
Writing it like

$$a^2 + (x^1)^2 + (x^2)^2 + (x^3)^2 = r^2 = b^2 + (x^0)^2,$$

we can use the freedom to multiply by a factor $\rho > 0$ to bring $r^2 = 1$ and the space is seen to have the topology of $\mathrm{S}^1 \times \mathrm{S}^3$. If we further allow $\rho = -1$, we have to identify

$$(a, b, x^0, x^1, x^2, x^3) \cong (-a, -b, -x^0, -x^1, -x^2, -x^3),$$

which gives the topology of $(\mathrm{S}^1 \times \mathrm{S}^3)/\mathbb{Z}^2$. Notice that the discrete transformations (2.41, 2.42) act non trivially in this space, although they are not necessary for transitivity.

Remark 2.4.1. We have to be aware that considering only transformations with determinant +1 we are discarding, for example, the inversion

$$x^\mu \rightarrow \frac{x^\mu}{x^2},$$

which corresponds to an O(2, 4) matrix of the form

$$\begin{pmatrix} 1 & 0 & 0 \\ 0 & -1 & 0 \\ 0 & 0 & \mathrm{id}_4 \end{pmatrix}.$$

■

Let us now put local coordinates on this space. Since we are in projective space, at least one of the homogeneous coordinates has to be different from zero. We then consider the subset of points such that $u \neq 0$. We can use the dilation freedom to bring $u = 1$ and (2.43) becomes

$$v = (x^0)^2 - (x^1)^2 - (x^2)^2 - (x^3)^2,$$

so the subset of points with $u \neq 0$ is $\mathbb{R}^4$, parametrized by the four coordinates (x^0, x^1, x^2, x^3). The Lorentz group $\mathrm{SO}(1,3) \subset \mathrm{SO}(2,4)$ clearly leaves this $\mathbb{R}^4$ fixed, and it is not difficult to see that the dilations (generated by $D = J_{-1,4}$) and translations (generated by $P_\mu = J_{-1,\mu} - J_{4,\mu}$) also leave $\mathbb{R}^4$ invariant. It is natural, therefore, to identify $\mathbb{R}^4$ with the Minkowski space. The quadric defined by (2.43) may then play the role of the conformal space **C**.

The picture that we discussed in this and the previous section arises in a natural way in Sections 2.6 and 2.5.

In Section 2.5, however, we start from a different point of view, namely, introducing the spin group, that we proceed to define.

In the complex case, the orthogonal group $\mathrm{SO}(d, \mathbb{C})$, $d \geq 3$, is connected and its fundamental group is $\mathbb{Z}_2$. The *spin group* $\mathrm{Spin}(d, \mathbb{C})$ is defined as the double cover of $\mathrm{SO}(d, \mathbb{C})$, which is the universal cover.

There is a real form of $\mathrm{Spin}(d, \mathbb{C})$ which is a double cover of $\mathrm{SO}(m, n)^0$, where the superindex '0' means the connected component of the identity (CCI) of $\mathrm{SO}(m, n)$. It is connected itself (always $d \geq 3$) and we will denote it as $\mathrm{Spin}(m, n)$. It is not always simply connected: we will study some cases in Section 2.11 (Table 2.1).

In fact, one has an exact sequence

$$1 \rightarrow (\pm 1) \rightarrow \mathrm{Spin}(m, n) \rightarrow \mathrm{SO}(m, n)^0 \rightarrow 1\,.$$

In both, the real and the complex case, the orthogonal group and its spin group have the same Lie algebra (for more details see also Section 4.1) and the same dimension. The nice feature about the spin group is that all representations of the orthogonal Lie algebra exponentiate to representations of the spin group, while there are representations of the Lie algebra (the so called *spin representations*) that do not exponentiate to the orthogonal group. We will introduce these representations in Section 4.1.

In Section 2.6 we will show that the spin group of the complex orthogonal group $\mathrm{SO}(6,\mathbb{C})$ is $\mathrm{SL}_4(\mathbb{C})$ and in Section 2.7 we will find that the spin group of $\mathrm{SO}(4,\mathbb{C}) \cong \mathrm{SO}(3,\mathbb{C}) \times \mathrm{SO}(3,\mathbb{C})$, is $\mathrm{SL}_2(\mathbb{C}) \times \mathrm{SL}_2(\mathbb{C})$.

The relevant real forms will be obtained in Section 2.11. One has that the spin group of the CCI (connected component of the identity) of the Lorentz group, $\mathrm{SO}(1,3)^0$, is $\mathrm{SL}_2(\mathbb{C})_\mathbb{R}$, that is, the underlying real group of the complex special linear group. For the CCI of the conformal group, $\mathrm{SO}(2,4)^0$, the spin group is $\mathrm{SU}(2,2)$, which is the subset of 4×4 complex matrices satisfying

$$U^\dagger I_{2,2} U = I_{2,2}\,, \text{ with } I_{2,2} = \mathrm{diag}\{1,1,-1,-1\} \text{ and } \det U = 1\,. \quad (2.44)$$

Notice that while $\mathrm{SL}_2(\mathbb{C})_\mathbb{R}$ is simply connected (and hence, becomes the universal covering of the Lorentz group), the fundamental group of $\mathrm{SU}(2,2)$ is $\mathbb{Z}$, so it is not the universal covering of $\mathrm{SO}(2,4)^0$.

The idea of using the spin group to describe spacetime is not new: it is due to Penrose [124], who introduced it with his twistorial approach to spacetime. In the next section we will start from the basics of this approach.

2.5 The Grassmannian $G(2,4)$

In this section we describe the Grassmannian $G(2,4)$ of planes (i.e. 2-dimensional subspaces) in the 4-dimensional complex vector space $\mathbb{C}^4$ as an analytic manifold.

In the previous sections we described the conformal group, which turns out to be $\mathrm{O}(2,4)$, and the spin group of its CCI, $\mathrm{SU}(2,2)$. $\mathrm{SL}_4(\mathbb{C})$, the complexification of $\mathrm{SU}(2,2)$, has a natural transitive action on $G(2,4)$, which will enable us to identify $G(2,4)$ with the complex conformal space. For now we look only at the complexification of the conformal space; the reality issues need some extra care and will be discussed in Section 2.11.

Our basic idea is to realize the complex Minkowski spacetime as a certain open set, called the big cell, inside the Grassmannian manifold $G(2,4)$.

The big cell is a *Zariski open set* in $G(2,4)$. For us, a Zariski open set simply means that its complement in $G(2,4)$ is the set of zeros of a certain polynomial function. The big cell is dense in $G(2,4)$, which is also a projective variety and hence compact, so $G(2,4)$ is a compactification of the big cell.

Our goal then is to show that the complexified conformal space is $\mathbf{C}^c = G(2,4)$ and that its big cell can be identified with the complex Minkowski space, $\mathbf{M}^c$.

As we remarked above, $G(2,4)$ admits a natural action of the complex special linear group $\mathrm{SL}_4(\mathbb{C})$, since the image of a plane by an element of $\mathrm{SL}_4(\mathbb{C})$ is again a plane. Only a subgroup of $\mathrm{SL}_4(\mathbb{C})$ leaves the big cell invariant. We will prove that this subgroup is[5]

$$F_0^c = \left(\mathrm{SL}_2(\mathbb{C}) \times \mathrm{SL}_2(\mathbb{C}) \times \mathbb{C}^\times\right) \ltimes \mathbb{C}^4 .$$

The factor $\mathrm{SL}_2(\mathbb{C}) \times \mathrm{SL}_2(\mathbb{C})$ will be interpreted as the spin group of the complexified Lorentz group, $\mathbb{C}^\times$ as the dilations and $\mathbb{C}^4$ as the spacetime translations. F_0^c is a version of $\mathcal{P}$ (complexified) times dilations, where the Lorentz group has been substituted by the spin group. As we will formally establish later, the group F_0^c will be called in this book the complex Poincaré group times dilations.

Let us start by looking at the Grassmannian $G(2,4)$:

$$G(2,4) = \left\{\text{planes in } \mathbb{C}^4\right\}$$

as the set of two dimensional subspaces or planes in $\mathbb{C}^4$. A plane is given by two linearly independent vectors:

$$\pi = \mathrm{span}\{a, b\} \in G(2,4)$$

where a and $b \in \mathbb{C}^4$ are to be thought as column vectors. The pair (a,b) is not unique. In fact, given two pairs of vectors (a,b) and (a',b') describing the same plane

$$\mathrm{span}\{a, b\} = \mathrm{span}\{a', b'\} ,$$

[5] We use the superindex 'c' to indicate that it is a complex group and the subindex '0' to indicate that it is the non supersymmetric version. The usefulness of the notation will be recognized in Chapter 4, where we transport this framework from ordinary geometry to supergeometry.

then one pair can be expressed as a linear combination of the other, which means that they are related by an element of $\mathrm{GL}_2(\mathbb{C})$

$$(a', b') = (a, b) \cdot g := \begin{pmatrix} a_1 & b_1 \\ a_2 & b_2 \\ a_3 & b_3 \\ a_4 & b_4 \end{pmatrix} \begin{pmatrix} g_{11} & g_{12} \\ g_{21} & g_{22} \end{pmatrix}, \quad g = \begin{pmatrix} g_{11} & g_{12} \\ g_{21} & g_{22} \end{pmatrix} \in \mathrm{GL}_2(\mathbb{C}) .$$

Hence, $G(2, 4)$ is identified with the pairs of linearly independent vectors in $\mathbb{C}^4$ modulo the right action of $\mathrm{GL}_2(\mathbb{C})$ above, which amounts to a change of basis in the description of the plane:

$$G(2, 4) = \big\{(a, b) \,|\, a, b \in \mathbb{C}^4 \ a,\ b \text{ linearly independent}\big\} \,/\, \mathrm{GL}_2(\mathbb{C}) .$$

We now want to look at $G(2, 4)$ as an homogeneous space for the group $\mathrm{SL}_4(\mathbb{C})$. We first notice that the special linear group $\mathrm{SL}_4(\mathbb{C})$ acts transitively on $G(2, 4)$; in fact, it acts on a pair of generators of a given plane

$$g \cdot (a, b) = (ga, gb), \qquad g \in \mathrm{SL}_4(\mathbb{C})$$

and this action is well defined modulo the right action of $\mathrm{GL}_2(\mathbb{C})$ described above. Let $\{e_1, e_2, e_3, e_4\}$ be the standard basis of $\mathbb{C}^4$ and consider the plane π_0 spanned by the vectors e_1 and e_2. The isotropy group of π_0 is

$$I_0 = \left\{ \begin{pmatrix} L & M \\ 0 & R \end{pmatrix} \in \mathrm{SL}_4(\mathbb{C}) \right\} , \tag{2.45}$$

where L, M, R are 2×2 matrices, L and R invertible. Since $\mathrm{SL}_4(\mathbb{C})$ is an analytic Lie group and I_0 a closed subgroup, by the theory of homogeneous spaces, the quotient $\mathrm{SL}_4(\mathbb{C})/I_0$ acquires a complex analytic manifold structure, which is unique, provided some natural conditions are satisfied (see Chapter 2 of Ref. [146] for more details in the ordinary setting and also Chapter 1 for the super setting). In other words, any manifold on which $\mathrm{SL}_4(\mathbb{C})$ acts transitively, with isotropy group I_0, is diffeomorphic to $\mathrm{SL}_4(\mathbb{C})/I_0$, so we conclude that

$$G(2, 4) \cong \mathrm{SL}_4(\mathbb{C})/I_0 .$$

We now want to describe the analytic manifold structure of $G(2, 4)$ directly by providing a cover by open subsets each isomorphic to $\mathbb{C}^4$.

If $\pi = \mathrm{span}\{a, b\}$ is a plane, since the vectors a and b are linearly independent, the 4×2 matrix

$$(a, b) = \begin{pmatrix} a_1 & b_1 \\ a_2 & b_2 \\ a_3 & b_3 \\ a_4 & b_4 \end{pmatrix}$$

has rank 2, so at least one of the 6 minors $y_{ij} = a_i b_j - b_i a_j$, $1 \le i < j \le 4$ is different from zero. Hence we can cover $G(2,4)$ with open sets

$$U_{ij} = \{(a,b) \mid y_{ij} \neq 0\} / \mathrm{GL}_2(\mathbb{C}) \subset G(2,4), \quad i < j, \tag{2.46}$$

so

$$G(2,4) = \bigcup_{i<j} U_{ij}, \qquad i,j = 1,\ldots,4.$$

We will show now that $U_{ij} \cong \mathbb{C}^4$. For example, consider U_{12}. We can write any element in U_{12} in the form

$$U_{12} \cong \left\{ \begin{pmatrix} 1 & 0 \\ 0 & 1 \\ \alpha & \beta \\ \gamma & \delta \end{pmatrix} \right\} \cong \mathbb{C}^4$$

by multiplying on the right by an element in $\mathrm{GL}_2(\mathbb{C})$. This representative is unique and α, β, γ and δ are arbitrary complex numbers which we can take as coordinates on U_{12}. So $U_{12} \cong \mathbb{C}^4$. Similarly, for general U_{ij}, we can choose a representative (a,b) of the form

$$a = e_i + \alpha e_k + \gamma e_l, \qquad b = e_j + \beta e_k + \delta e_l,$$

$$\text{with } k \neq l \text{ and } k,l \neq i,j$$

by right multiplication by a suitable element in $\mathrm{GL}_2(\mathbb{C})$. Hence $U_{ij} \cong \mathbb{C}^4$.

Notice that each of the six open sets $\{U_{ij}\}_{i<j}$ is dense in $G(2,4)$. It is convenient to organize the coordinates of U_{ij} as a 2×2 matrix, so $U_{ij} \approx \mathrm{M}_2(\mathbb{C})$. In the case of U_{12}:

$$A = \begin{pmatrix} \alpha & \beta \\ \gamma & \delta \end{pmatrix}, \qquad \pi = \begin{pmatrix} \mathrm{id} \\ A \end{pmatrix} \in U_{12}. \tag{2.47}$$

We call U_{12} the *big cell* in $G(2,4)$. It will be our model for the complex Minkowski space.

The change of chart $\psi_{ij,kl} : U_{ij} \cap U_{kl} \longrightarrow U_{ij} \cap U_{kl}$ is indeed an analytic diffeomorphism, so we have defined a complex analytic manifold structure on $G(2,4)$. Let us see an example. The analytic diffeomorphism $\psi_{12,13}$ is realized as follows. If π is a plane in $U_{12} \cap U_{13}$, it has coordinates $(\alpha, \beta, \gamma, \delta)$

in U_{12} and coordinates $(\alpha', \beta', \gamma', \delta')$ in U_{13}, according to the identification $U_{ij} \cong \mathbb{C}^4$ described above. So there are representatives

$$\begin{pmatrix} 1 & 0 \\ 0 & 1 \\ \alpha & \beta \\ \gamma & \delta \end{pmatrix} = \begin{pmatrix} 1 & 0 \\ \alpha' & \beta' \\ 0 & 1 \\ \gamma' & \delta' \end{pmatrix} \begin{pmatrix} a & b \\ c & d \end{pmatrix}, \qquad \begin{pmatrix} a & b \\ c & d \end{pmatrix} \in \mathrm{GL}_2(\mathbb{C}),$$

which are related by a transformation of $\mathrm{GL}_2(\mathbb{C})$. From here we can compute a, b, c, d and $\alpha', \beta', \gamma', \delta'$ in terms of α, β, γ and δ,

$$\alpha' = -\frac{\alpha}{\beta}, \qquad \beta' = \frac{1}{\beta}$$
$$\delta' = \frac{\delta}{\beta}, \qquad \gamma' = \gamma - \frac{\alpha\delta}{\beta}. \tag{2.48}$$

Notice that $\beta, \beta' \neq 0$ if $\pi \in U_{12} \cap U_{13}$. This shows that the coordinate change $\psi_{12,13}$ is analytic. With a similar calculation one can check that all of the $\psi_{ij,kl}$ are analytic diffeomorphism, hence $G(2,4)$ has the structure of an analytic manifold.

Remark 2.5.1. Following the same procedure as above, we can also give to $G(2,4)$ an algebraic variety structure. In fact, we can identify each U_{ij} with the complex affine space $\mathbb{A}^4$, and then check that the change of chart $\psi_{ij,kl}$ is indeed also an algebraic isomorphism. ∎

2.6 The Plücker embedding and the Klein quadric

In this section we want to show that $G(2,4)$ is a projective variety and we shall achieve this by providing an explicit embedding into the projective space

$$\mathbb{P}^5(\mathbb{C}) = \{ \text{ lines in } \mathbb{C}^6 \}$$

called the *Plücker embedding.* We identify the image of $G(2,4)$ under such embedding with a closed subvariety of $\mathbb{P}^5(\mathbb{C})$, the *Klein quadric.*

Let us consider the vector space $\wedge^2(\mathbb{C}^4) \approx \mathbb{C}^6$, with basis $\{e_i \wedge e_j\}_{i<j}$, $i, j = 1, \ldots, 4$, and let us consider the plane:

$$\pi = \mathrm{span} \left\{ \begin{pmatrix} a_1 & b_1 \\ a_2 & b_2 \\ a_3 & b_3 \\ a_4 & b_4 \end{pmatrix} \right\}.$$

We have:

$$a \wedge b = \sum_{i<j} y_{ij} e_i \wedge e_j, \quad y_{ij} = a_i b_j - b_i a_j, \quad 1 \leq i < j \leq 4 .$$

If we choose a different basis for π, $(a', b') = (a, b)g$, for $g \in \mathrm{GL}_2(\mathbb{C})$ we obtain

$$a' \wedge b' = \det(g) a \wedge b .$$

So there is a well defined map:

$$\begin{array}{ccc} G(2,4) & \xrightarrow{\phi} & \mathbb{P}(\wedge^2(\mathbb{C}^4)) \cong \mathbb{P}^5(\mathbb{C}) \\ \pi = \mathrm{span}\{a, b\} & \longrightarrow & [a \wedge b] \cong [y_{12}, y_{13}, y_{14}, y_{23}, y_{24}, y_{34}] , \end{array}$$

where we have identified $\mathbb{P}(\wedge^2(\mathbb{C}^4)) \cong \mathbb{P}^5(\mathbb{C})$ by establishing the vector space isomorphism $\wedge(\mathbb{C}^4) \cong \mathbb{C}^6$,

$$\begin{array}{llll} e_1 \wedge e_2 \longrightarrow f_1, & & e_2 \wedge e_3 \longrightarrow f_4, \\ e_1 \wedge e_3 \longrightarrow f_2, & & e_2 \wedge e_4 \longrightarrow f_5, \\ e_1 \wedge e_4 \longrightarrow f_3, & & e_3 \wedge e_4 \longrightarrow f_6 . \end{array}$$

This map is injective and analytic. In fact, we notice that when ϕ is restricted to the open subsets U_{ij}, it is a morphism of complex analytic manifolds. For example when we restrict ϕ to the big cell U_{12}, we have:

$$\begin{array}{ccc} U_{12} & \xrightarrow{\phi|_{U_{12}}} & \mathbb{P}(\wedge^2(\mathbb{C}^4)) \cong \mathbb{P}^5(\mathbb{C}) \\ \pi = \mathrm{span}\left\{ \begin{pmatrix} 1 & 0 \\ 0 & 1 \\ \alpha & \beta \\ \gamma & \delta \end{pmatrix} \right\} & \longrightarrow & [1, \beta, \delta, -\alpha, -\gamma, \alpha\delta - \beta\gamma] . \end{array}$$

Hence, $G(2,4)$ is a complex projective manifold.

The map ϕ is called the *Plücker embedding* and y_{ij} are the *homogeneous Plücker coordinates.*

We now wish to characterize the image of the Plücker map in order to identify $G(2,4)$ with an algebraic closed subvariety of $\mathbb{P}^5(\mathbb{C})$. If $p \in \wedge^2(\mathbb{C}^4)$, $[p]$ is in the image of the Plücker map if and only if $p = a \wedge b$ for some $a, b \in \mathbb{C}^4$. Such p is written as

$$p = y_{12} e_1 \wedge e_2 + y_{13} e_1 \wedge e_3 + \cdots + y_{34} e_3 \wedge e_4 ,$$

with $y_{ij} = a_i b_j - b_i a_j$ being the determinant formed by the rows i and j in the matrix with columns the vectors a and b. One can easily check that such minors satisfy

$$y_{12}y_{34} - y_{13}y_{24} + y_{14}y_{23} = 0\,. \tag{2.49}$$

This is called the *Plücker relation* and it defines a quadric K in $\mathbb{P}^5(\mathbb{C})$ called the *Klein quadric.* The quadric appearing in (2.43), in our discussion of the real conformal space, is a real form of the Klein quadric. We will determine in Section 2.11 that it is precisely this real form that corresponds to take the real form $\mathrm{SU}(2,2)$ of the complex conformal group $\mathrm{SL}_4(\mathbb{C})$.

We now wish to show that there are no further relations among the Plücker coordinates y_{ij}. Let us consider one of the open sets, say U_{12}. On such open set $y_{12} = 1$, hence the relation reads:

$$y_{34} = y_{13}y_{24} - y_{14}y_{23}\,.$$

We can express the variable y_{34} in terms of y_{13}, y_{24}, y_{14}, y_{23} and then any further relation occurring could be expressed in terms of these four variables only. There cannot be other relations, however, since $(y_{13}, y_{14}, y_{23}, y_{24})$ give the coordinates for the isomorphism of U_{12} with $\mathbb{C}^4$. A similar argument can be repeated for all of the U_{ij}. Thus the set of solutions of the Klein quadric in projective space are in one to one correspondence with the elements of $G(2,4)$. We shall return to this point in the Section 2.9, when we discuss the Grassmannian via invariant theory.

Remark 2.6.1. There is an equivalent derivation of the *Plücker relation*, that is useful for explicit calculations. As we remarked above, given $p \in \wedge^2(\mathbb{C}^4)$, $[p]$ is in the image of the Plücker map if and only if $p = a \wedge b$. On the other hand, $p = a \wedge b$ if and only if $p \wedge p = 0$. If one writes p in the coordinates y_{ij}, a simple calculation gives

$$p \wedge p = (y_{12}y_{34} - y_{13}y_{24} + y_{14}y_{23}) e_1 \wedge e_2 \wedge e_3 \wedge e_4 = 0\,.$$

Thus we retrieve the Plücker relation described above. ∎

The variety K is smooth: this can be easily verified by computing the differential of the Plücker relation on each of the open affine subsets U_{ij} and observing that it never vanishes. Alternatively, since $K \cong G(2,4)$ is an homogeneous space, it has to be smooth.

We summarize the previous discussion with the following theorem.

Theorem 2.6.2. *The Plücker embedding $G(2,4) \to \mathbb{P}^5(\mathbb{C})$ is a morphism of analytic manifolds and realizes an isomorphism between $G(2,4)$ and the Klein quadric K in $\mathbb{P}^5(\mathbb{C})$ defined by equation (2.49). Hence, $G(2,4)$ acquires the structure of complex analytic projective manifold and the open cover $\{U_{ij}\}$ is compatible with the standard open cover $\{U_i\}$ of $\mathbb{P}^5(\mathbb{C})$:*

$$U_{ij} = G(2,4) \cap U_i, \quad U_i = \{[x_0, \dots, x_5] \,|\, x_i \neq 0\} \subset \mathbb{P}^5(\mathbb{C}).$$

Furthermore, this embedding gives to $G(2,4)$ an algebraic projective variety structure, which is compatible with the analytic structure defined in Section 2.5. ■

Let us analyze now the action of the group $\mathrm{SL}_4(\mathbb{C})$ on K. The group $\mathrm{GL}_4(\mathbb{C})$ acts on $\wedge^2(\mathbb{C}^4)$, and for any $g \in \mathrm{GL}_4(\mathbb{C})$ we have that

$$Q(gp) = \det(g) Q(p),$$

so if $g \in \mathrm{SL}_4(\mathbb{C})$, then g preserves the quadratic form Q. This action defines a map

$$\mathrm{SL}_4(\mathbb{C}) \to \mathrm{SO}(Q) \cong \mathrm{SO}(6, \mathbb{C}) \tag{2.50}$$

with kernel $\pm\mathrm{id}$, as we mentioned at the end of Section 2.4. Both groups have dimension 15 and one has the exact sequence

$$1 \longrightarrow \{\pm\mathrm{id}\} \longrightarrow \mathrm{SL}_4(\mathbb{C}) \longrightarrow \mathrm{SO}(Q) \longrightarrow 1.$$

We conclude that $\mathrm{SL}_4(\mathbb{C})$ is the double cover, which in this case is also the universal cover, of $\mathrm{SO}(6, \mathbb{C})$. $\mathrm{SL}_4(\mathbb{C})$ is called the *spin group* of $\mathrm{SO}(6, \mathbb{C})$. The fundamental representation of $\mathrm{SL}_4(\mathbb{C})$ and its dual (or *contragradient*) are called in the physics literature the *spin representations* of the orthogonal group, although they are not, technically, representations of the group but of the Lie algebra. We shall return on these very important points in Chapter 4.

2.7 The complex Minkowski space

In this section we show how the big cell U_{12} inside the Grassmannian manifold $G(2,4)$ is interpreted as the complex Minkowski space.

The big cell $U_{12} \subset G(2,4)$ is the Zariski open set corresponding to the condition $y_{12} \neq 0$ in the Klein quadric identification $K \cong G(2,4)$ described above. A plane $\pi \in U_{12}$ is uniquely represented as

$$\pi = \begin{pmatrix} \mathrm{id} \\ A \end{pmatrix}, \qquad A = \begin{pmatrix} \alpha & \beta \\ \gamma & \delta \end{pmatrix},$$

and the Plücker coordinates of π are

$$[y_{12}, y_{13}, y_{14}, y_{23}, y_{24}, y_{34}] = [1, \beta, \delta, -\alpha, -\gamma, \alpha\delta - \beta\gamma]\,. \tag{2.51}$$

Let $\{e_1, e_2, e_3, e_4\}$ be the standard basis in $\mathbb{C}^4$. We define

$$\pi_0 = \text{span}\{e_1, e_2\}\,, \qquad \pi_\infty = \text{span}\{e_3, e_4\}\,,$$

so we have $\mathbb{C}^4 = \pi_0 \oplus \pi_\infty$.

We describe now the complement of U_{12} in $G(2, 4)$. An arbitrary plane $\pi \in G(2, 4)$ is said to be *finite* if $\pi \cap \pi_\infty = \emptyset$. The big cell U_{12} is equivalently characterized as the set of finite planes. Notice that a plane that has non-empty intersection with π_∞ can always be represented by two independent vectors $a \in \pi_\infty$ and b of the form

$$a = a_3e_3 + a_4e_4\,, \qquad b = b_1e_1 + b_2e_2 + b_3e_3 + b_4e_4\,,$$

so $y_{12} = 0$. Substituting in the Plücker relation we obtain

$$y_{23}y_{14} - y_{13}y_{24} = 0\,. \tag{2.52}$$

This equation gives geometrically a cone with some extra points. In fact, we have two possibilities:

- $y_{34} \neq 0$, in which case we use the equivalence of projective space to set $y_{34} = 1$. The equation (2.52) is a (null) cone in $\mathbb{C}^4$.
- $y_{34} = 0$. Then (2.52) is a cone, but now in the projective space $\mathbb{P}^3(\mathbb{C})$. Again we have that some of the y's should be different from zero and the solutions have to be taken up to multiplication by a constant non zero factor. These points define a projective variety whose affine cone is the one obtained in the previous paragraph (see Chapter 1, pg 12 of Ref. [85] more details on the cones on projective varieties).

Our next step is to analyze the group action on U_{12}. We are going to see that the complex Poincaré group times dilations, denoted as F_0^c, sits inside $\text{SL}_4(\mathbb{C})$ as the subgroup that leaves U_{12} invariant. So, as we announced in Section 2.4, we will have a natural identification of U_{12} with the complex Minkowski spacetime, together with the natural action of the complex Poincaré group times dilations on it. Consequently, $G(2, 4)$ results to be a compactification of Minkowski spacetime. The points at infinity that we are adding form the cone described above.

Let us start by computing F_0^c, the group that leaves π_∞ invariant. In the chosen coordinates,

$$\pi_\infty = \begin{pmatrix} 0 \\ \text{id} \end{pmatrix}.$$

An easy calculation shows that

$$F_0^c = \left\{ \begin{pmatrix} L & 0 \\ N & R \end{pmatrix}, \quad L, R \text{ invertible} \right\}.$$

Proposition 2.7.1. *The subgroup of* $\mathrm{SL}_4(\mathbb{C})$ *leaving the big cell* U_{12} *invariant is* F_0^c.

Proof. Consider the action of $g = \begin{pmatrix} L & M \\ N & R \end{pmatrix} \in \mathrm{SL}_4(\mathbb{C})$ on the generic element $\begin{pmatrix} I \\ A \end{pmatrix}$ in $U_{12} \subset G(2,4) \cong \mathrm{SL}_4(\mathbb{C})/I_0$:

$$\begin{pmatrix} L & M \\ N & R \end{pmatrix} \begin{pmatrix} I \\ A \end{pmatrix} = \begin{pmatrix} L + MA \\ N + RA \end{pmatrix}$$

g leaves U_{12} invariant if and only if $L + MA$ is invertible for all 2×2 matrices A. In particular, when $A = 0$ this amounts to say that L is invertible. $L + MA = L(I + L^{-1}MA)$ is invertible if and only if $I + XA$ is invertible, for $X = L^{-1}M$. If $M \neq 0$, also X and X^* are non zero and XX^* has an eigenvalue $\gamma > 0$. Hence $I + XA$ is not invertible with $A = -\gamma^{-1}X^*$. Hence in order to leave U_{12} invariant, g must be of the prescribed form, i.e. $M = 0$. ∎

We proceed now to define formally the complex Poincaré group. We will presently justify the definition in terms of the standard action of the Poincaré group on Minkowski space.

Definition 2.7.2. We define the *complex Poincaré group times dilations* as the subgroup of $\mathrm{SL}_4(\mathbb{C})$ leaving the big cell invariant. By the previous discussion we have that such a subgroup is

$$F_0^c = \left\{ \begin{pmatrix} L & 0 \\ NL & R \end{pmatrix}, \quad L, R \text{ invertible} \right\}.$$

The *complex Poincaré group* is the subgroup of F_0^c with $\det L = \det R = 1$. It will be denoted as P_0^c. ∎

Notice that in the proof of Proposition 2.7.1 we proved that F_0^c consists of matrices in $\mathrm{SL}_4(\mathbb{C})$ such that the upper right entry is zero and the bottom left entry is an arbitrary 2×2 matrix. We have decided to write it as NL, with N arbitrary, since L is invertible. We shall see the importance of such a notation in a moment.

Also, one can prove that $F_0^c \cong I_0$, the isotropy group in (2.45). We will see more of this in Section 4.5.

The action of F_0^c on the big cell

$$U_{12} = \left\{ \begin{pmatrix} I \\ A \end{pmatrix} I_0 \right\} \subset G(2,4) \cong \mathrm{SL}_4(\mathbb{C})/I_0$$

is as follows:

$$\begin{pmatrix} L & 0 \\ NL & R \end{pmatrix} \begin{pmatrix} I \\ A \end{pmatrix} I_0 = \begin{pmatrix} L \\ NL + RA \end{pmatrix} I_0 = \begin{pmatrix} I \\ N + RAL^{-1} \end{pmatrix} I_0 \,.$$

More concisely,

$$A \longrightarrow N + RAL^{-1} \,. \tag{2.53}$$

F_0^c has the structure of a semidirect product, $F_0^c = H \ltimes \mathrm{M}_2(\mathbb{C})$, where the elements in $\mathrm{M}_2(\mathbb{C})$, the set of 2×2 matrices N, act as translations in (2.53) and

$$H = \left\{ \begin{pmatrix} L & 0 \\ 0 & R \end{pmatrix}, \quad L, R \in \mathrm{SL}_2(\mathbb{C}), \quad \det L \cdot \det R = 1 \right\}.$$

The subgroup of H is the direct product $H = \mathrm{SL}_2(\mathbb{C}) \times \mathrm{SL}_2(\mathbb{C}) \times \mathbb{C}^\times$.

To see that Definition 2.7.2 is fully justified, let us put this in a more standard form. We consider the following basis of the complex 2×2 matrices

$$\sigma_0 = \begin{pmatrix} 1 & 0 \\ 0 & 1 \end{pmatrix}, \quad \sigma_1 = \begin{pmatrix} 0 & 1 \\ 1 & 0 \end{pmatrix}, \quad \sigma_2 = \begin{pmatrix} 0 & -i \\ i & 0 \end{pmatrix}, \quad \sigma_3 = \begin{pmatrix} 1 & 0 \\ 0 & -1 \end{pmatrix}. \tag{2.54}$$

The matrices σ_1, σ_2 and σ_3 are the *Pauli matrices*. We expand A and N in this basis

$$A = x^\mu \sigma_\mu = \begin{pmatrix} x^0 + x^3 & x^1 - \mathrm{i}x^2 \\ x^1 + \mathrm{i}x^2 & x^0 - x^3 \end{pmatrix},$$

$$N = n^\mu \sigma_\mu = \begin{pmatrix} n^0 + n^3 & n^1 - \mathrm{i}n^2 \\ n^1 + \mathrm{i}n^2 & n^0 - n^3 \end{pmatrix}.$$

We note that

$$\det A = (x^0)^2 - (x^1)^2 - (x^2)^2 - (x^3)^2 \,, \tag{2.55}$$

so the big cell U_{12} can be effectively identified with the complexified Minkowski space $\mathbf{M}^c$.

The transformation

$$A \longrightarrow RAL^{-1} \,, \qquad R, L \in \mathrm{SL}_2(\mathbb{C}) \,, \tag{2.56}$$

preserves the non degenerate quadratic form (2.55). This establishes an homomorphism

$$\mathrm{SL}_2(\mathbb{C}) \times \mathrm{SL}_2(\mathbb{C}) \longrightarrow \mathrm{SO}(4, \mathbb{C}) \cong \mathrm{SO}(3, \mathbb{C}) \times \mathrm{SO}(3, \mathbb{C}) \,,$$

that, as in the six dimensional case, has kernel $\pm$id. $\mathrm{SL}_2(\mathbb{C}) \times \mathrm{SL}_2(\mathbb{C})$ is in fact the double (universal) cover or spin group of $\mathrm{SO}(4, \mathbb{C})$, the CCI of the complexified Lorentz group $\mathrm{SO}(1,3)^0$. If we denote by $\mathbf{2}$ the fundamental representation of SL_2 and by $\bar{\mathbf{2}}$ its contragredient one[6], then the fundamental representation of $\mathrm{SO}(4, \mathbb{C})$ is constructed as $\mathbf{2} \oplus \bar{\mathbf{2}}$ of $\mathrm{SL}_2(\mathbb{C}) \times \mathrm{SL}_2(\mathbb{C})$. The matrices Λ of this representation can be computed from (5.8) as

$$\Lambda^{\nu}{}_{\mu}\sigma_{\nu} = R\sigma_{\mu}L^{-1}\,. \tag{2.57}$$

Finally, it is clear that $\mathbb{C}^{\times}$ acts as a dilation. So we have that the complex Poincaré group times dilations, according to Definition 2.7.2 is

$$F_0^c = \left(\mathrm{SL}_2(\mathbb{C}) \times \mathrm{SL}_2(\mathbb{C}) \times \mathbb{C}^{\times}\right) \ltimes \mathbb{C}^4.$$

The complex Poincaré group is

$$P_0^c = \left(\mathrm{SL}_2(\mathbb{C}) \times \mathrm{SL}_2(\mathbb{C})\right) \ltimes \mathbb{C}^4\,,$$

and it is not exactly the complexification of $\mathcal{P}$ in (2.13), since we have substituted the Lorentz group $\mathcal{L}$ there by the spin group. This will be our standard terminology for the rest of the book (unless otherwise specified).

We have then succeeded in embedding the complex Minkowski space $\mathbf{M}^c \cong U_{12}$ into one of its compactifications, $G(2,4)$, in such a way that the action of the complex Poincaré group times dilations is coming as the action of the subgroup F_0^c of $\mathrm{SL}_4(\mathbb{C})$ leaving U_{12} invariant. The action of $\mathrm{SL}_4(\mathbb{C})$ on its quotient $G(2,4) = \mathrm{SL}_4(\mathbb{C})/I_0$ is the natural one.

2.8 $G(2, 4)$ as the conformal space

In the previous sections we examined the action of the special linear group $\mathrm{SL}_4(\mathbb{C})$ on the Grassmannian manifold of planes in $\mathbb{C}^4$, that is, $G(2,4)$. We have seen that if we identify the big cell in $G(2,4)$ with the complexified Minkowski space, then the Poincaré group becomes identified with the subgroup of $\mathrm{SL}_4(\mathbb{C})$ that leaves the Minkowski space invariant.

The Minkowski space is characterized by its pseudoriemannian metric (2.7, 2.16). To extend it to the whole conformal space, one needs to go to the more general concept of *conformal metric*, that we will presently define. Then, one can show (Proposition 2.8.4) that such extension exists and it is unique under very natural requirements.

[6]Note that for $\mathrm{SL}_2(\mathbb{C})$ the fundamental representation and its contragradient are equivalent.

Since we are dealing with the complexification of Minkowski and conformal spaces, we will be interested in *holomorphic metrics*. Let us briefly recall the definition.

Definition 2.8.1. Let X be a complex manifold and let T^0X be the holomorphic tangent bundle, with $T^{0*}(X)$ its dual bundle. An *holomorphic* metric on X is a symmetric, non degenerate, holomorphic section of the bundle $T^{0*}X \otimes T^{0*}X$. If $(z^1, \dots, z^n)$ are complex coordinates in an open set of X, then an holomorphic metric has a local representation of the form

$$\mathrm{d}s^2 = g_{ij}(z)\mathrm{d}z^i\mathrm{d}z^j\,, \qquad \text{where} \quad \frac{\partial g_{ij}}{\partial \bar{z}^i} = 0\,.$$

■

Notice that an holomorphic metric does not define a metric on $X_{\mathbb{R}}$, that is, on X seen as a real manifold, since in the full tangent bundle TX, the field of quadratic forms $\mathrm{d}s^2$ is degenerate. But given a real form of X, the holomorphic metric may give a well defined metric. For example, the quadratic form (2.55) defines a holomorphic metric in the complexification of Minkowski space that descends to the standard metric in the real Minkowski space.

We are now ready for the definition of conformal metric both, in the real differentiable and complex analytic case.

Definition 2.8.2. A *conformal metric* on a real (resp. complex) manifold X is an assignment that associates to each point x of X a set of nonsingular quadratic forms on the tangent space at x, T_xX, (resp. the holomorphic tangent space T_x^0X) with the following properties:

- Any two quadratic forms are non zero scalar multiples of each other.
- On a neighborhood of each point we can choose a smooth metric or an holomorphic metric respectively, whose quadratic form belongs to this assignment.

When these conditions are satisfied, we say that the metric on the neighborhood of a point defines the *conformal structure* on that neighborhood.

Let $\alpha : X \longrightarrow X$ be an automorphism of X. We say that a conformal metric is *invariant* under α if for $\alpha(x) = y$, the pull back by α takes the set of metrics at y over the set of metrics at x. Such an α is then said to be *conformal*. If $\mathrm{d}s^2$ denotes one of the quadratic forms, the condition reads

$$(\alpha^*\mathrm{d}s^2)_x(X,Y) =_{\text{def}} \mathrm{d}s^2_y(\mathrm{d}\alpha X, \mathrm{d}\alpha Y) = \lambda(x)\mathrm{d}s^2_x(X,Y)\,.$$

■

On a Lie group G we can define a metric invariant under left translations just by taking a symmetric non degenerate bilinear form on the tangent space at the identity and then transporting it with the left translations. In fact, let $\langle\ ,\ \rangle : \mathfrak{g} \times \mathfrak{g} \to \mathbb{C}$ be a symmetric, non degenerate, bilinear form on $\mathfrak{g} = \mathrm{Lie}(G)$. Then, we define

$$\begin{array}{ccc} T_gG \times T_gG & \xrightarrow{\mu_g} & \mathbb{C} \\ (X_g, Y_g) & \longrightarrow & \mu_g(X_g, Y_g) =_{\mathrm{def}} \langle X, Y \rangle \,, \end{array}$$

where $X, Y \in \mathfrak{g}$ and $X_g = \mathrm{d}\ell_g X$, $Y_g = \mathrm{d}\ell_g Y$. Clearly, μ is a metric and it is invariant under left translations.

We consider now the quotient space G/H, where H is a closed Lie subgroup of G. We want to outline how we can define a metric G/H that is also invariant under left translations. We start by regarding a few facts on the quotients of Lie groups that hold for both the real differential and the complex analytic category. For more details we invite the reader to go to Section 2.9 of Ref. [146].

Let G be a real Lie group or a complex analytic group. Let $X = G/H$, with H a closed subgroup of G. G/H is a real differentiable or complex analytic manifold with a unique differentiable (or analytic) structure for which the action of G on G/H is differentiable (resp. analytic). Let $\mathcal{O}_G$ denote the sheaf of real differentiable (or complex analytic) functions on G and let $\mathfrak{g} = \mathrm{Lie}(G)$ and $\mathfrak{h} = \mathrm{Lie}(H)$. The key step in the construction of the quotient G/H is the realization of a local section of the natural projection $\pi : G \longrightarrow G/H$. In fact, there exists a neighbourhood V of $[e] \in G/H$ such that $\pi^{-1}(V) = WH \cong W \times H$, where W is a closed submanifold of G, $\pi(W) = V$. We can then define a local section $V \mapsto W \times H \subset G$. The existence of the local section comes through the application of the Frobenius theorem to the integral distribution:

$$\mathcal{D}_{\mathfrak{h}} = \{\mathrm{d}\ell_g X \ \mid \ X \in T_eH\} \,,$$

where as usual, ℓ_g is the left translation in G by the element $g \in G$. This allows us to define a real differentiable or complex analytic structure on G/H, and the structural sheaf $\mathcal{O}_{G/H}$ consists of the right invariant functions on G:

$$\mathcal{O}_{G/H}(U) = \{f \in \mathcal{O}_G(U) \ \mid \ f(gh) = f(g), \forall g \in G, \forall h \in H\} \,.$$

By the very construction, at each point $[g] \in G/H$ we have that

$$T_{[g]}(G/H) \cong T_gG/T_ggH \,,$$

where g is any representative in the class $[g] = gH$ (see Ref. [147] for more details). We are going to compute the tangent space $T_{[g]}(G/H)$ explicitly in the differentiable setting. In the analytic setting the calculation is similar, but one must exert some care since the sheaf of analytic functions is not recovered through its global sections.

Let us consider the tangent bundle TG and let $\mathcal{X}(G)$ denote the set of vector fields on G. TG is a trivial bundle, and it is in fact spanned by the left invariant vector fields

$$X_g = \mathrm{d}\ell_g X, \qquad X \in T_e G \cong \mathfrak{g} . \tag{2.58}$$

Let $X_g = \mathrm{d}\ell_g X \in \mathcal{D}_{\mathfrak{h}}$. We compute its action on a function $f \in C^\infty(G/H)$ which, we recall, is identified with the right H-invariant functions in $C^\infty(G)$.

$$X_g(f) = \mathrm{d}f(X_g) = \mathrm{d}f \circ \mathrm{d}\ell_g|_e X = \mathrm{d}(f \circ \ell_g)\Big|_e X .$$

If $c(t)$ is the integral curve of X_g passing through e we have

$$\frac{\mathrm{d}}{\mathrm{d}t} c(t)\Big|_{t=0} = X , \qquad c(t) \in H ,$$

and

$$X_g(f) = \mathrm{d}f(gc(t))|_{t=0} = 0$$

by the property of $f \in C^\infty(G/H)$. So the elements in $\mathcal{D}_{\mathfrak{h}}$ have a null action on $C^\infty(G/H)$ and the action of the equivalence classes in the quotient bundle $TG/\mathcal{D}_{\mathfrak{h}}$ is well defined. We now want to show that an arbitrary vector and its right translate

$$X_g , \qquad Y_{gh} = \mathrm{d}r_h X_g ,$$

give the same result when acting on $f \in C^\infty(G/H)$:

$$Y_{gh}(f) = \mathrm{d}r_h X_g(f) = \mathrm{d}f|_{gh}(\mathrm{d}r_h X_g) = \mathrm{d}(f \circ r_h)|_g X_g = \mathrm{d}f|_g X_g = X_g(f) ,$$

as we wanted to show. This means that we can identify tangent vectors at different points of the coset when acting on $C^\infty(G/H)$

$$X_g \sim \mathrm{d}r_h X_g \qquad \text{for some } h \in H .$$

The quotient spaces of TG and $\mathcal{D}_{\mathfrak{h}}$ are, by this equivalence relation, two bundles over G/H that we denote as $\widetilde{TG}$ (with standard fiber $\mathfrak{g}$) and $\widetilde{\mathcal{D}_{\mathfrak{h}}}$ (with standard fiber $\mathfrak{h}$) respectively. The quotient

$$\widetilde{TG}/\widetilde{\mathcal{D}_{\mathfrak{h}}} \cong T(G/H)$$

is a bundle over G/H with fiber $\mathfrak{g}/\mathfrak{h}$, which can be identified with the tangent space of the manifold G/H. In particular, we have that

$$T_{[e]}(G/H) \cong \mathfrak{g}/\mathfrak{h}\,.$$

In the following, we freely use this description of $T(G/H)$.

To construct a metric on G/H we shall follow the same procedure as for G, with suitable changes. We start with a symmetric, non degenerate bilinear form on $\mathfrak{g}/\mathfrak{h}$,

$$\langle\ ,\ \rangle : \mathfrak{g}/\mathfrak{h} \times \mathfrak{g}/\mathfrak{h} \to \mathbb{C}$$

and we assume that $\langle\ ,\ \rangle$ is invariant under the adjoint action of H, that is

$$\langle \mathrm{Ad}_h[X],\ \mathrm{Ad}_h[Y]\rangle = \langle [X], [Y]\rangle \qquad \forall h \in H\,,$$

where $[X]$ denotes the equivalence class of $X \in \mathfrak{g}$ in $\mathfrak{g}/\mathfrak{h}$. Notice that the adjoint action of H is induced by the adjoint action of G, since

$$\mathrm{Ad}_h(X) - \mathrm{Ad}_h(X') \in \mathfrak{h} \qquad \text{if} \qquad X - X' \in \mathfrak{h}\,.$$

We use the same notation as for the group, since there is no risk of confusion. The invariance of $\langle\ ,\ \rangle$ under the adjoint action of H will turn out to be a crucial requirement for our metric to exist, as we shall presently see.

We proceed as before and define the metric on G/H as

$$\begin{array}{ccc} T_{[g]}(G/H) \times T_{[g]}(G/H) & \xrightarrow{\mu_{[g]}} & \mathbb{C} \\ ([X_g], [Y_g]) & \longrightarrow & \mu_{[g]}([X_g], [Y_g]) =_{\text{def}} \langle [X], [Y]\rangle\,, \end{array}$$

where $X, Y \in \mathfrak{g}$, and $X_g = \mathrm{d}\ell_g X$ and $Y_g = \mathrm{d}\ell_g Y$ are arbitrary representatives of the corresponding equivalence classes. This is well defined since for other representatives, say $g' = gh$, $X'_{g'} = \mathrm{d}\ell_{g'} X'$ and $Y'_{g'} = \mathrm{d}\ell_{g'} Y'$ we have

$$X_g \sim X'_{g'} \ \Rightarrow\ X_g - \mathrm{d}r_{h^{-1}} X'_{g'} \in (\mathcal{D}_{\mathfrak{h}})_g\,,$$

and taking $g = e$ we get

$$X - \mathrm{d}r_{h^{-1}} X'_h = X - \mathrm{d}r_{h^{-1}} \circ \mathrm{d}\ell_h X' = X - \mathrm{Ad}_h X' \in \mathfrak{h}\,.$$

The same is true for Y. So we have

$$\mu_{[g']}([X'_{g'}], [Y'_{g'}]) \equiv \langle [X'], [Y']\rangle = \langle [\mathrm{Ad}_{h^{-1}} X], [\mathrm{Ad}_{h^{-1}} Y]\rangle = \langle [X], [Y]\rangle\,,$$

and the metric is well defined. It is easy to prove that it is left invariant,

$$\mu_{\tilde{g}g}(\mathrm{d}\ell_{\tilde{g}} X_g, \mathrm{d}\ell_{\tilde{g}} Y_g) = \langle [X], [Y]\rangle\,.$$

Let us now consider the case of a conformal metric μ on G/H. If the bilinear form on $\mathfrak{g}/\mathfrak{h}$ is invariant under H only up to a factor,

$$\langle [\mathrm{Ad}_h X], [\mathrm{Ad}_h Y] \rangle = c \langle [X], [Y] \rangle \,, \qquad c \in \mathbb{C} \,,$$

then μ assigns to each point in G/H a set of proportional quadratic forms in the tangent space. Moreover, in a certain neighborhood U of $[g]$, we can always choose a section

$$U \subset G/H \xrightarrow{\ \rho\ } G$$

and then there is a metric on U given by

$$\mathrm{d}s^2_{[g]}([X_g], [Y_g]) = \langle [X], [Y] \rangle \,,$$

where $X_{\rho([g])} = \mathrm{d}\ell_{\rho([g])} X$ and $Y_{\rho([g])} = \mathrm{d}\ell_{\rho([g])} Y$.

The quadratic form $\mathrm{d}s^2_{[g]}$ belongs to the set $\mu_{[g]}$, so we conclude that $\mu_{[g]}$ is a conformal metric.

Let us apply this procedure to $G(2,4) = \mathrm{SL}_4(\mathbb{C})/I_0$. We recall that

$$[e] = \pi_0 = \begin{pmatrix} \mathrm{id} \\ 0 \end{pmatrix}, \qquad I_0 = \left\{ \begin{pmatrix} L & M \\ 0 & R \end{pmatrix} \in \mathrm{SL}_4(\mathbb{C}) \right\} .$$

Let $\mathfrak{i}_0$ denote the Lie algebra of I_0, with

$$\mathfrak{i}_0 = \left\{ \begin{pmatrix} l & m \\ 0 & r \end{pmatrix} \mid \mathrm{tr}\, l + \mathrm{tr}\, r = 0 \right\}, \qquad \mathfrak{sl}_4(\mathbb{C}) = \left\{ \begin{pmatrix} l & m \\ n & r \end{pmatrix} \mid \mathrm{tr}\, l + \mathrm{tr}\, r = 0 \right\}.$$

(All the matrix entries here are 2×2 complex matrices.) The tangent space at π_0 is identified with $\mathfrak{sl}_4(\mathbb{C})/\mathfrak{i}_0$,

$$\mathfrak{sl}_4(\mathbb{C})/\mathfrak{i}_0 \cong \left\{ \begin{pmatrix} 0 & 0 \\ n & 0 \end{pmatrix} \right\},$$

and the adjoint action of I_0 on this space is

$$n \to R n L^{-1} \,.$$

This action is similar to the action of the Lorentz group that we have seen in (2.56). The action, then, preserves the quadratic form q defined on $\mathfrak{sl}_4(\mathbb{C})/\mathfrak{i}_0$:

$$q \begin{pmatrix} 0 & 0 \\ n & 0 \end{pmatrix} = \alpha\delta - \beta\gamma, \qquad n = \begin{pmatrix} \alpha & \beta \\ \gamma & \delta \end{pmatrix}$$

up to a factor and it also preserves its associated symmetric bilinear form. Then, according to our previous discussion, we can transport it with the left action to any other point in $G(2,4)$ and obtain a conformal, holomorphic metric μ proportional, in an open set, to $\mathrm{d}s^2$. The metric is invariant under $\mathrm{SL}_4(\mathbb{C})$.

More explicitly, consider the local section $\rho : U_{12} \to \mathrm{SL}_4(\mathbb{C})$

$$\rho(\pi) = \begin{pmatrix} \mathrm{id} & 0 \\ A & \mathrm{id} \end{pmatrix} \qquad \text{for } \pi = \begin{pmatrix} \mathrm{id} \\ A \end{pmatrix} \in U_{12} \qquad \text{and } A = \begin{pmatrix} \alpha & \beta \\ \gamma & \delta \end{pmatrix}.$$

Then the left action is just

$$\ell_{\rho(\pi)} \begin{pmatrix} \mathrm{id} & 0 \\ 0 & \mathrm{id} \end{pmatrix} = \begin{pmatrix} \mathrm{id} & 0 \\ A & \mathrm{id} \end{pmatrix}, \qquad \mathrm{d}\ell_{\rho(\pi)} n = n\,,$$

where the matrix n in the right hand side of the last equation has to be understood as a tangent vector at the point π. Finally, the metric

$$\mathrm{d}s^2 = \mathrm{d}\alpha\mathrm{d}\delta - \mathrm{d}\beta\mathrm{d}\gamma\,, \tag{2.59}$$

is defined on all U_{12} and it is proportional to the conformal structure.

Remark 2.8.3. It is instructive to compute the metric in another open set, say U_{13}, and study the effect of the change of coordinates (2.48) in the intersection $\pi \in U_{12} \cap U_{13}$. We have

$$\mathrm{d}s^2 = (\mathrm{d}\alpha\mathrm{d}\delta - \mathrm{d}\beta\mathrm{d}\gamma) = -\frac{1}{\beta'^2}(\mathrm{d}\alpha'\mathrm{d}\delta' - \mathrm{d}\beta'\mathrm{d}\gamma') = -\frac{1}{\beta'^2}\mathrm{d}s'^2\,. \tag{2.60}$$

So $\mathrm{d}s^2$ and $\mathrm{d}s'^2$ are proportional and they define the same family of quadratic forms.

Let us summarize our observations in a proposition.

Proposition 2.8.4. *The assignment:*

$$\pi \longmapsto F_\pi = \{k \cdot \mathrm{d}s^2,\, k \in \mathbb{C}\}$$

described above corresponds to the unique conformal, holomorphic metric on $G(2,4)$ *which is invariant under the action of* $\mathrm{SL}_4(\mathbb{C})$*. On the Minkowski spacetime* $\mathbf{M}^c = U_{12}$ *such a metric takes the form (up to a constant)*

$$\mathrm{d}s^2 = \mathrm{d}\alpha\mathrm{d}\delta - \mathrm{d}\beta\mathrm{d}\gamma = (\mathrm{d}x^0)^2 - (\mathrm{d}x^1)^2 - (\mathrm{d}x^2)^2 - (\mathrm{d}x^3)^2$$

under the identification

$$\begin{pmatrix} \alpha & \beta \\ \gamma & \delta \end{pmatrix} \cong \begin{pmatrix} x^0 + x^3 & x^1 - ix^2 \\ x^1 + ix^2 & x^0 - x^3 \end{pmatrix}.$$

Proof. For the invariance under the $\mathrm{SL}_4(\mathbb{C})$ action, it is enough to show that a metric in F_π is changed into a multiple of itself under the action of the stabilizer I_0 of an element π in $G(2,4)$. This calculation was done above.

We now come to the uniqueness. As the subgroup of the stabilizer defined by the condition $M = 0$, $R, L \in \mathrm{SL}_2(\mathbb{C})$ has no non trivial characters, any quadratic form that is changed into a multiple of itself by elements of this subgroup will have to be invariant under it. Since the action of this subgroup is already irreducible, such a form has to be a multiple of $\mathrm{d}s^2$. ■

Up to now we have studied the geometric properties of the complex conformal space $G(2,4)$. In the next section we deepen our understanding of the conformal space as an algebraic variety by using invariant theory.

2.9 The Grassmannian via invariant theory

In this section we want to give another characterization of the Grassmannian $G(2,4)$ which was implicitly introduced in Section 2.5. We will show that the ring of regular functions of $G(2,4)$ with respect to the Plücker embedding coincides with a certain ring of invariants and, furthermore, we will give this ring explicitly in terms of generators and relations. We will also give a basis, using the terminology of *Young tableau.*

This section requires some elementary prerequisites on algebraic projective varieties; the reader can consult, for example, Chapter 1 of Ref. [85]. Also, this section is not strictly needed in the sequel and the reader can omit it without any trouble.

Consider the set

$$S = \{(v, w) \in \mathbb{C}^4 \oplus \mathbb{C}^4 \mid \mathrm{rank}(v, w) = 2\},$$

together with the equivalence relation

$$(v, w) \sim (v', w') \quad \Leftrightarrow \quad \mathrm{span}\{v, w\} = \mathrm{span}\{v', w'\},$$

or equivalently

$$(v, w) \sim (v', w') \quad \Leftrightarrow \quad \exists g \in \mathrm{GL}_2(\mathbb{C}) \text{ such that } (v', w') = (v, w)g\,.$$

We can restrict ourselves to consider the group $\mathrm{SL}_2(\mathbb{C})$, since its action on S is transitive. Then we have that $G(2,4) = S/\sim$.

Let $\mathcal{O}(S)$ be the ring of homogeneous polynomials on S. This is just the ring of polynomials on S with the usual polynomial grading. The ring $\mathcal{O}(S)$ admits a left action of $\mathrm{SL}_2(\mathbb{C})$

$$(g \cdot f)(x) = f(xg)\,.$$

A regular function on S that is invariant under the action of G, that is, $gf = f$, defines a function on $G(2,4)$. It is quite immediate to verify that the polynomial functions

$$y_{ij} = v^i w^j - v^j w^i, \quad \text{with } i < j \text{ and } v = \begin{pmatrix} v^1 \\ v^2 \\ v^3 \\ v^4 \end{pmatrix}, \ w = \begin{pmatrix} w^1 \\ w^2 \\ w^3 \\ w^4 \end{pmatrix} \in \mathbb{C}^4$$

are invariant under the given $\mathrm{SL}_2(\mathbb{C})$ action and that they satisfy the Plücker relation

$$y_{12}y_{34} - y_{13}y_{24} + y_{14}y_{23} = 0\,.$$

We now recall the first and second fundamental theorems of invariant theory in our particular case, which establish that the y_{ij}'s are actually the generators of the ring of invariants and that the Plücker relation generates the ideal of all of the relations among the y_{ij}.

Theorem 2.9.1. *First Fundamental Theorem of invariant theory. Let us consider the action of* $\mathrm{SL}_2(\mathbb{C})$ *on* $\mathcal{O}(S)$ *as above. Then, the elements* y_{ij}*'s generate the ring of invariants* $\mathcal{O}(S)^{\mathrm{SL}_2(\mathbb{C})}$. ■

Theorem 2.9.2. *Second Fundamental Theorem of invariant theory. Let us consider the action of* $\mathrm{SL}_2(\mathbb{C})$ *on* $\mathcal{O}(S)$ *as above. Then, the ring of invariants is given in terms of generators and relations as*

$$\mathbb{C}[y_{12}, y_{13}, y_{14}, y_{23}, y_{24}, y_{23}] \,/\, (y_{12}y_{34} - y_{13}y_{24} + y_{14}y_{23})\,.$$

■

The proof of these theorems is very interesting, but it would lead us too far from the purpose of the present book. We send the reader to the Appendix F of Ref. [72] for a combinatorial treatment or to Chapter 9 of Ref. [71] for a more geometric one.

The Plücker embedding described in Section 2.6, through the coordinates y_{ij}'s allows us to realize $G(2,4)$ also as complex algebraic projective variety. In fact $G(2,4)$ is realized as the set of zeros of an homogeneous algebraic equation in $\mathbb{P}^5(\mathbb{C})$: the equation of the Klein quadric. One can prove that the coordinate ring of $G(2,4)$ with respect to this projective embedding is precisely the homogeneous subring of $\mathcal{O}(S)$ consisting of the invariants with respect to the given action (see Chapter 9 of Ref. [72]). It is important to remark at this point that such ring is not intrinsically associated to the algebraic variety $G(2,4)$. Instead, it is the result of some choices that we are making, which effectively correspond to the Plücker embedding of $G(2,4)$ into the projective space $\mathbb{P}^5(\mathbb{C})$. We will denote this ring as $\mathbb{C}[G(2,4)]$.

We end this section by computing a basis for $\mathbb{C}[G(2,4)]$. In order to do this we need the notion of *tableau*. We send the reader to Ref. [72] for a complete and exhaustive treatment.

Definition 2.9.3. A *Young diagram* is a finite collection of boxes or cells, arranged in left-justified rows, with the row lengths weakly decreasing (each row has the same or shorter length than its predecessor). Listing the number of boxes in each row gives a partition of a non-negative integer n, the total number of boxes of the diagram. A *Young standard tableau*[7] (or *standard tableau*) is obtained by filling in the boxes of the Young diagram with the first m natural numbers in such a way that

- Each row consists of strictly increasing integers.
- Each column consists of weakly increasing integers.

■

In our special case, we are interested in Young diagrams with all of the rows of the same length equal to 2 and filled with the numbers 1, 2, 3, 4.

Let $I_1, \dots, I_r$ be r pairs of non negative integers between 1 and 4. We say the indices $I_1 = (i_i, j_1), \dots, I_r = (i_r, j_r)$ form a *standard tableau* if they are the fillings of a Young standard tableau.

For example, $I_1 = (1,2)$, $I_2 = (1,3)$, $I_3 = (3,4)$ form a tableau, while $J_1 = (1,4)$ and $J_2 = (2,3)$ do not form a tableau:

1	2
1	3
3	4

✓

1	4
2	3

×

Proposition 2.9.4. *Let the notation be as above. Then the set:*

$$1, \quad y_{I_1} \cdots y_{I_r},$$

[7] Some authors will call these *semistandard* or simply tableau without any adjective (see for example Ref. [71]).

with each sequence $I_1, \dots, I_r$ forming a standard tableau, is a basis for $\mathbb{C}[G(2,4)]$.

■

Again, a complete proof of this statement is beyond the scope of the present work and we send the reader to Ref. [72].

Intuitively, however, the statement seems clear. In fact, the set

$$\{1, y_{I_1} \cdots y_{I_r}\},$$

without any condition on the I_j's, certainly generates the ring $\mathbb{C}[G(2,4)]$ as a vector space. Let us consider one monomial $y_{I_1} \cdots y_{I_r}$. If we assume, without loss of generality, the indices $I_1, \dots, I_r$ to be ordered lexicographically, they form a standard tableau if and only if there are no instances of the pair $y_{14}y_{23}$ in $y_{I_1} \cdots y_{I_r}$. Using the Plücker relation:

$$y_{14}y_{23} = y_{13}y_{24} - y_{12}y_{34}$$

we can always elimine every such pair and one can show that there are no other relations among the monomials $y_{J_1} \cdots y_{J_r}$ when such pair does not occur. Hence, the remaining monomials $y_{J_1} \cdots y_{J_r}$ are independent.

2.10 The functor of points of the Grassmannian manifold $G(2,4)$

In this section we want to describe the functor of points of the Grassmannian $G(2,4)$ viewed as an analytic manifold. In Chapter 1 we have defined the functor of points for supermanifolds, but let us briefly recall here the corresponding ordinary notions.

Definition 2.10.1. Let X be a complex analytic manifold. We call a T-*point* of X any morphism $f : T \longrightarrow X$, where T is a complex analytic manifold. We denote with $X(T)$ the set of all T-points of X:

$$X(T) = \{f : T \longrightarrow X\}.$$

■

Example 2.10.2. We want to examine the T-points of two complex analytic manifolds, which are very relevant to us.

(1) *The T-points of $\mathbb{C}^4$.* $\mathbb{C}(T)$ consists by definition of all morphisms $T \longrightarrow \mathbb{C}^4$. In practice giving one such morphism f amounts to choose

four holomorphic functions on T, that is, four global sections of the holomorphic sheaf on T, t_1, t_2, t_3, $t_4 \in \mathcal{O}(T)$:

$$\begin{array}{ccc} T & \xrightarrow{f} & \mathbb{C}^4 \\ t & \longrightarrow & (t_1(t),\, t_2(t),\, t_3(t),\, t_4(t)) \,. \end{array}$$

Hence we can identify $\mathbb{C}^4(T)$ with $\mathcal{O}(T)^4 := \mathcal{O}(T) \otimes \mathbb{C}$.

(2) *The T-points of* $\mathrm{SL}_n(\mathbb{C})$. We first determine the T-points of the analytic complex group $\mathrm{SL}_2(\mathbb{C})$. A morphism $f : T \longrightarrow \mathrm{SL}_2(\mathbb{C})$ consists of the choice of four holomorphic functions t_{11}, t_{12}, t_{21}, t_{22} on T, satisfying the relation: $t_{11}t_{22} - t_{12}t_{21} = 1$. Hence, we can identify the T-points of $\mathrm{SL}_2(\mathbb{C})$ with the 2×2 matrices with coefficients in $\mathcal{O}(T)$, the global, holomorphic sections on T,

$$\mathrm{SL}_2(T) = \left\{ \begin{pmatrix} t_{11} & t_{12} \\ t_{21} & t_{22} \end{pmatrix} \,\middle|\, t_{ij} \in \mathcal{O}(T),\, t_{11}t_{22} - t_{12}t_{21} = 1 \right\}.$$

The generalization to $\mathrm{SL}_n(\mathbb{C})$ is now obvious.

■

We now turn to the problem of describing the T-points of the Grassmannian manifold $G(2,4)$.

Let $f : T \longrightarrow G(2,4)$ be a T-point. Since $G(2,4)$ is a projective manifold, we cannot expect a global description of a T-point as the one we found in the previous examples. Let us consider the open cover of $G(2,4)$ by the six (Zariski) open subsets U_{ij}, $1 \leq i < j \leq 4$, described in the previous sections. Then our morphism f is completely determined by the family of the six morphisms

$$V_{ij} \xrightarrow{f_{ij}} U_{ij} \cong \mathbb{C}^4 \qquad \text{with} \qquad V_{ij} = f^{-1}(U_{ij}),\ f_{ij} = f|_{V_{ij}}\,,$$

and $1 \leq i < j \leq 4$. If $\mathcal{O}_T$ is the sheaf of holomorphic functions on T, each f_{ij} is a V_{ij}-point of $U_{ij} \cong \mathbb{C}^4$ and it corresponds to the choice of four global sections on V_{ij}, namely, four elements in $\mathcal{O}_T(V_{ij})$. We have then the following identification (for $i, j = 1, 2$ and similarly for the other values):

$$f_{12} \cong \begin{pmatrix} 1 & 0 \\ 0 & 1 \\ t_{11} & t_{12} \\ t_{21} & t_{22} \end{pmatrix}, \qquad t_{kl} \in \mathcal{O}_T(V_{ij})\,.$$

In other words, the morphism $f_{12} : V_{12} \to U_{12} \cong \mathbb{C}^4$ corresponds to the choice of the four sections t_{ij}, $i, j = 1, 2$ in $\mathcal{O}(V_{12})$ (see Example 2.10.2).

Let us define

$$\mathcal{F}_T(V_{12}) := \text{span}\left\{ \begin{pmatrix} 1 \\ 0 \\ t_{11} \\ t_{21} \end{pmatrix}, \begin{pmatrix} 0 \\ 1 \\ t_{12} \\ t_{22} \end{pmatrix} \right\} \subset \mathcal{O}_T(V_{12})^4 := \mathcal{O}_T(V_{12}) \otimes \mathbb{C}^4 ,$$

and similarly we can define $\mathcal{F}_T(V_{ij})$ for all i, j, $1 \leq i < j \leq 4$. For example,

$$\mathcal{F}_T(V_{13}) := \text{span}\left\{ \begin{pmatrix} 1 \\ s_{11} \\ 0 \\ s_{21} \end{pmatrix} \begin{pmatrix} 0 \\ s_{12} \\ 0 \\ s_{22} \end{pmatrix} \right\} \subset \mathcal{O}_T(V_{13})^4 .$$

The s_{ij}'s are not arbitrary sections in $\mathcal{O}(V_{13})$; they are related to the t_{ij} on $V_{12} \cap V_{13}$ in the following way:

$$\begin{pmatrix} 1 & 0 \\ 0 & 1 \\ t_{11} & t_{12} \\ t_{21} & t_{22} \end{pmatrix} = \begin{pmatrix} 1 & 0 \\ s_{11} & s_{12} \\ 0 & 1 \\ s_{21} & s_{22} \end{pmatrix} \begin{pmatrix} 1 & 0 \\ s_{11} & s_{12} \end{pmatrix}^{-1}$$

(as customary, one should consider the four equalities in the t_{ij}'s and s_{kl}'s obtained by equating the two matrices). We invite the reader to compare this equation to (2.48) in Section 2.5: they are formally the same, though the meaning of the two sets of variables appearing is quite different.

The assignment $\mathcal{F}_T^{12} : V \mapsto \mathcal{F}_T(V_{12})|_V$, where $V \subset V_{12}$ is open and

$$\mathcal{F}_T(V_{12})|_V := \text{span}\left\{ \begin{pmatrix} 1 \\ 0 \\ t_{11}|_V \\ t_{21}|_V \end{pmatrix} \begin{pmatrix} 0 \\ 1 \\ t_{12}|_V \\ t_{22}|_V \end{pmatrix} \right\} \subset \mathcal{O}_T(V)^4 (:= \mathcal{O}_T(V) \otimes \mathbb{C}^4) .$$

Similarly, for generic i, j, one defines six sheaves $\mathcal{F}_T^{ij}$, one for each of the open sets V_{ij} which cover T. It is immediate to check that such sheaves agree on the intersections $V_{ij} \cap V_{kl}$, by their very definition, and satisfy the compatibility conditions as in Proposition 1.3.6, thus giving us a sheaf $\mathcal{F}_T$ on the whole T. Notice that $\mathcal{F}_T$ is locally free and it is a direct summand[8] of rank 2 of $\mathcal{O}(T)^4$. Hence we have associated to a morphism $f : T \longrightarrow G(2,4)$ a locally free sheaf $\mathcal{F}_T$ of rank 2 on T.

Our next proposition establishes a one to one correspondence between such sheaves and the T-points of $G(2,4)$, providing a useful characterization that we shall exploit fully in the supergeometric setting.

[8] N is a locally free submodule of M if and only if it is a direct summand of M, in other words, if $M = N \oplus N'$, where N' is another submodule of M. Such modules are also called *projective modules*. Refer to [44] Appendix A.3 for more details.

Proposition 2.10.3. *There is a one to one correspondence between the T-points of $G(2,4)$ and the subsheaves which are direct summands of $\mathcal{O}_T^4$ of rank 2:*

$$G(2,4)(T) = \left\{ \mathcal{F}_T \subset \mathcal{O}_T^4, \quad \mathcal{F}_T \textit{ locally free of rank } 2 \right\}.$$

Proof. We already showed how to associate to each T-point of $G(2,4)$ a unique $\mathcal{F}_T$ with the required properties. In fact, it is quite immediate to verify, in the above discussion, that $\mathcal{F}_T$ is independent from the choice of the open cover V_{ij} of T.

We now show the other direction. Let $\mathcal{F}_T$ be one such sheaves. We are going to build a morphism $f : T \to G(2,4)$. Let us first cover T with open subsets V_{ij}, such that $\mathcal{F}_T|_{V_{ij}}$ is free of rank 2. Hence

$$\mathcal{F}_T(V_{ij}) = \text{span}\left\{ \begin{pmatrix} a_1 \\ a_2 \\ a_3 \\ a_4 \end{pmatrix}, \begin{pmatrix} b_1 \\ b_2 \\ b_3 \\ b_4 \end{pmatrix} \right\} \subset \mathcal{O}_T(V_{ij})^4 .$$

Since the local rank of $\mathcal{F}_T(V_{ij})$ is 2, after suitably refining the cover if needed, we can assume that one of the determinants, say $a_i b_j - b_i a_j$, is invertible. Assume now, for concreteness, that $i = 1$, $j = 2$. After a suitable change of basis we can write:

$$\mathcal{F}_T(V_{12}) := \text{span}\left\{ \begin{pmatrix} 1 \\ 0 \\ t_{11} \\ t_{21} \end{pmatrix}, \begin{pmatrix} 0 \\ 1 \\ t_{12} \\ t_{22} \end{pmatrix} \right\} \subset \mathcal{O}_T(V_{12})^4 .$$

Such change of basis effectively corresponds to the right multiplication of the generators by the matrix

$$\begin{pmatrix} a_1 & b_1 \\ a_2 & b_2 \end{pmatrix}^{-1} \in \mathrm{GL}(\mathcal{O}_T(V_{ij})^2) .$$

We can then define immediately the morphism:

$$V_{12} \xrightarrow{f_{12}} U_{12} \subset G(2,4)$$

$$x \longrightarrow \begin{pmatrix} 1 & 0 \\ 0 & 1 \\ t_{11}(x) & t_{12}(x) \\ t_{21}(x) & t_{22}(x) \end{pmatrix} .$$

In the same way we can define $f_{ij} : V_{ij} \longrightarrow U_{ij} \subset G(2,4)$. It is easy to check

$$f_{ij}|_{V_{ij}\cap V_{kl}} = f_{kl}|_{V_{ij}\cap V_{kl}}\,,$$

so we obtain a morphism $f : T \longrightarrow G(2,4)$. ■

In the next remark we see how we retrieve the topological points of the complex manifold $G(2,4)$ as the T-points, when T is the space consisting only of one point (refer also to Example 1.7.5).

Remark 2.10.4. We want to determine $G(2,4)(T)$, when T consist only of one point. Clearly, the holomorphic sections on $T = \{p\}$ are just the constants so $\mathcal{O}(T) = \mathbb{C}$, hence $\mathcal{O}(T)^4 = \mathbb{C} \otimes \mathbb{C}^4 \cong \mathbb{C}^4$. The above theorem states that $G(2,4)(T)$ is the set of locally free subsheaves of $\mathbb{C}^4$ of rank 2. In this particular situation, locally free means free over $\mathbb{C}$, hence we have that $G(2,4)(T)$ are the dimension 2 subspaces of $\mathbb{C}^4$, thus retrieving the topological space underlying $G(2,4)$ described in Section 2.5. ■

We now recall the definition of functor of points in our setting, sending the reader to Chapter 1 for the more general situation.

Definition 2.10.5. Let X be a complex analytic manifold and let (mflds) denote the category of such manifolds. We define the *functor of points* of X the following functor:

$$X : (\text{mflds})^o \longrightarrow (\text{sets}), \qquad T \longmapsto X(T)$$

which associates to each manifold T, the T-points of X. On the morphisms $\phi : T \to T'$,

$$X(\phi)(f) = f \circ \phi, \qquad \text{for } f \in X(T'), \quad f : T' \to X\,.$$

■

The importance of the functor of points stems from Yoneda's Lemma (see Appendix A for more details).

Theorem 2.10.6. *Yoneda's Lemma. Let X and Y be two complex analytic manifolds. There is a bijection from the set of morphisms $\varphi : X \longrightarrow Y$ to the set of maps $\varphi_T : X(T) \longrightarrow Y(T)$, functorial in T. In particular, X and Y are isomorphic if and only if their functors of points are isomorphic.* ■

We now want to give a characterization of the functor of points of the Grassmannian manifold.

Consider the functor $\mathcal{G} : (\text{smflds})^o \longrightarrow (\text{sets})$

$$\mathcal{G}(T) = \left\{\mathcal{F}_T \subset \mathcal{O}_T^4,\ \mathcal{F}_T \text{ locally free of rank } 2\right\} .$$

On the morphisms, $\mathcal{G}$ is defined as follows. If $\psi : T \longrightarrow S$ is a morphism of complex analytic manifolds, then $\psi_U^* : \mathcal{O}_S(U) \longrightarrow \mathcal{O}_T(\psi^{-1}(U))$ induces a morphism on the sheaves, from which we get a morphism $\mathcal{G}(\psi) : \mathcal{G}(S) \longrightarrow \mathcal{G}(T)$.

Proposition 2.10.7. *The functor $\mathcal{G} : (\text{smflds})^o \longrightarrow (\text{sets})$ defined above is the functor of points of the Grassmannian manifold $G(2,4)$.*

Proof. In Proposition 2.10.3 we gave a bijective correspondence between $G(2,4)(T)$ and $\mathcal{G}(T)$, and as one can readily check the correspondence is functorial. We leave to the reader the corresponding check for the morphisms. ∎

We end this section with a discussion of the functor of points of the Grassmannian manifold as an *homogeneous space*. Let us consider the functor $\mathcal{G}$ described above, which gives an equivalent way to view the functor of points of the Grassmanniann manifold. $\mathrm{SL}_4(T)$ acts in a natural way on $\mathcal{G}(T)$ for all analytic manifolds T. In fact, let $\mathcal{F}_T \in \mathcal{G}(T)$ and let us choose an open cover V_{ij} of T as above, such that $\mathcal{F}_T(V_{ij})$ is free of rank 2 in $\mathcal{O}_T(V_{ij})^4$:

$$\mathcal{F}_T(V_{ij}) = \mathrm{span}_{\mathcal{O}_T(V_{ij})}\{a, b\} \subset \mathcal{O}_T(V_{ij})^4 .$$

Since $\mathrm{SL}_4(\mathcal{O}_T(V_{ij}))$ acts naturally on $\mathcal{O}_T(V_{ij})^4$ we can define

$$(g \cdot \mathcal{F}_T)(V_{ij}) = \mathrm{span}_{\mathcal{O}_T(V_{ij})}\{g \cdot a, g \cdot b\} .$$

Patching the $(g \cdot \mathcal{F}_T)(V_{ij})$ for $1 \leq i, j \leq 4$ we obtain a locally free rank 2 subsheaf, denoted with $g \cdot \mathcal{F}_T$, of $\mathcal{O}_T^4$, hence an element of $\mathcal{G}(T)$.

In view of this result, we can then state the following proposition. It is of great importance to us because its generalization to the super setting will give the action of the conformal supergroup on the conformal superspace.

Proposition 2.10.8.

(1) The natural transformation

$$\begin{array}{ccc} \mathrm{SL}_4(T) \times \mathcal{G}(T) & \xrightarrow{\rho_T} & \mathcal{G}(T) \\ g, \mathcal{F}_T & \longrightarrow & g \cdot \mathcal{F}_T \end{array},$$

gives a well defined action of $\mathrm{SL}_4(\mathbb{C})$ *on* $G(2,4)$ *in terms on their functor of points.*

(2) *The stabilizer of the topological point* $\mathcal{I} = \text{span}\{e_1, e_2\}$ *in* $\text{SL}_4(T)$ *is:*

$$I_0(T) = \left\{ \begin{pmatrix} L & M \\ 0 & R \end{pmatrix} \in \text{SL}_4(T) \right\},$$

where L, M, R *are all matrices with coefficient in* $\mathcal{O}(T)$. *Hence*

$$G(2,4) \cong \text{SL}_4(\mathbb{C})/I_0 .$$

■

We end this section by giving an answer to a very natural question at this point: what is the functor of points of the Grassmannian manifold viewed as the quotient $\text{SL}_4(\mathbb{C})/I_0$? One may think, in the light of the previous proposition, that this functor, on the objects, is $T \mapsto \text{SL}_4(T)/I_0(T)$. However one can see after a small calculation that $\text{SL}_4(T)/I_0(T)$ consists of the *free subsheaves* of $\mathcal{O}_T^4$ of rank 2. Instead, the functor of points of $G(2,4)$ is $\mathcal{G}$ and $\mathcal{G}(T)$ consists of the *locally free* sheaves of rank 2, and not just the free ones. However, for each T and $\mathcal{F}_T \in \mathcal{G}(T) \cong (\text{SL}_4/I_0)(T)$, if we choose a suitable cover of T, $\{V_i\}$, $\mathcal{F}_T$ becomes free, so it is an element in $\text{SL}_4(V_i)/I_0(V_i)$. In other words, the two functors:

$$T \to \text{SL}_4(T)/I_0(T), \qquad T \to (\text{SL}_4/I_0)(T),$$

coincide *locally*. This is an important property that we shall exploit in our supergeometric treatment of quotients. In fact, while the functor that we want to study is $T \to (\text{SL}_4/I_0)(T)$, the functor $T \to \text{SL}_4(T)/I_0(T)$ is much more treatable from a computational point of view. One must be aware of the fact that it coincides with the functor of points of the Grassmannian manifold only locally.

2.11 The real Minkowski and conformal spaces

In Section 2.5 we introduced the complex Minkowski and conformal spaces. We then studied their properties and functorial interpretation. In this Section we want to obtain the corresponding real forms that were at the heart of the physical description in Sections 2.1 and 2.4. It is then clear what is the real form that we have to choose for the complex Minkowski space: physics requires that the quadratic form (2.7) must have signature $(1,3)$. Hence, the real form has to be chosen accordingly. This will be our starting point.

We will show that there is a unique real form of the conformal space, together with real forms of the conformal group, the Poincaré group and its spin groups, which are consistent with this choice for the real Minkowski space and with our construction. We shall also explicitly compute them. It will then become evident that the real conformal space does not arise as a real Grassmannian, hence justifying the complexification introduced in our treatment.

For brevity, all through this section we will denote $T \cong \mathbb{C}^4$ and $E \cong \Lambda\mathbb{C}^4 \cong \mathbb{C}^6$.

In Sections 2.6 and 2.7 we identified the complex Minkowski space with the big cell U_{12} of the Klein quadric K in the projective space $\mathbb{P}(E) \cong \mathbb{P}^5(\mathbb{C})$ (see (2.49)). Furthermore, we wrote an element of U_{12} as a matrix

$$A = \begin{pmatrix} \alpha & \beta \\ \gamma & \delta \end{pmatrix} \cong \begin{pmatrix} x^0 + x^3 & x^1 - \mathrm{i}x^2 \\ x^1 + \mathrm{i}x^2 & x^0 - x^3 \end{pmatrix},$$

in terms of which the Minkowski metric is just

$$\det A = (x^0)^2 - (x^1)^2 - (x^2)^2 - (x^3)^2 . \tag{2.61}$$

Comparing with (2.7), it is clear that the real form of the Minkowski space that we wish to obtain is the one in which (x^0, x^1, x^2, x^3) are all real, which means that the matrix A is *hermitian*. In other words, given the involutive, antilinear map or *conjugation*

$$\begin{array}{ccc} \mathrm{M}_2(\mathbb{C}) & \xrightarrow{\;s\;} & \mathrm{M}_2(\mathbb{C}) \\ & & \\ A = \begin{pmatrix} \alpha & \beta \\ \gamma & \delta \end{pmatrix} & \longrightarrow & A^s := A^\dagger = \begin{pmatrix} \bar\alpha & \bar\gamma \\ \bar\beta & \bar\delta \end{pmatrix}, \end{array}$$

the set of fixed points of s is precisely the set of hermitian matrices

$$\mathrm{M}_2(\mathbb{C})^s := \{A \in \mathrm{M}_2(\mathbb{C}) \mid A^s = A\} .$$

$\mathrm{M}_2(\mathbb{C})^s$ is a real vector space and we identify it with the real Minkowski space: $\mathbf{M} \cong \mathrm{M}_2(\mathbb{C})^s$. The quadratic form (2.61), restricted to $\mathbf{M}$, has the correct signature $(1, 3)$.

The question now is to understand what is the real form of E (and $\mathbb{P}(E)$) that, restricted to U_{12}, gives rise to this real form of the Minkowski space.

The restriction of the Plücker map (2.51) to the big cell suggests readily what the conjugation in E should be

$$\begin{array}{ccc} E & \xrightarrow{\theta} & E \\ \begin{pmatrix} y_{12} \\ y_{13} \\ y_{14} \\ y_{23} \\ y_{24} \\ y_{34} \end{pmatrix} & \longrightarrow & \begin{pmatrix} \bar{y}_{12} \\ -\bar{y}_{24} \\ \bar{y}_{14} \\ \bar{y}_{23} \\ -\bar{y}_{13} \\ \bar{y}_{34} \end{pmatrix} \end{array} .$$

Furthermore, it is clear that θ also descends to a conjugation on $\mathbb{P}(E)$. We shall denote it also as θ.

We now replace the Plücker coordinate y_{13} with $y_{31} := -y_{13}$, in order to avoid a cumbersome minus sign and we reorder the coordinates. With this linear change of coordinates, θ reads

$$\begin{array}{ccc} E & \xrightarrow{\theta} & E \\ \begin{pmatrix} y_{12} \\ y_{23} \\ y_{14} \\ y_{34} \\ y_{31} \\ y_{24} \end{pmatrix} & \longrightarrow & \Theta \begin{pmatrix} \bar{y}_{12} \\ \bar{y}_{23} \\ \bar{y}_{14} \\ \bar{y}_{34} \\ \bar{y}_{31} \\ \bar{y}_{24} \end{pmatrix} \end{array} ,$$

where Θ is the matrix

$$\Theta = \begin{pmatrix} \mathrm{id}_4 & 0 \\ 0 & J_2 \end{pmatrix} \quad \text{and} \quad J_2 = \begin{pmatrix} 0 & 1 \\ 1 & 0 \end{pmatrix} . \tag{2.62}$$

The quadratic form Q defining the Klein quadric becomes

$$Q(y) = y_{12}y_{34} + y_{31}y_{24} + y_{14}y_{23} ,$$

and it satisfies

$$\overline{Q(y)} = Q\left(\theta(y)\right) , \tag{2.63}$$

so its zero set is preserved under conjugation,

$$Q(y) = 0 \quad \Rightarrow \quad Q\left(\theta(y)\right) = 0 .$$

On the real vector space

$$\begin{aligned} E^\theta &= \{y = (y_{12}, y_{31}, y_{14}, y_{23}, y_{24}, y_{34}) \in E \mid \theta(y) = y\} \\ &= \{y \in E \mid \ y_{12}, y_{23}, y_{34}, y_{14} \text{ are real and } y_{31} = \bar{y}_{24}\} , \end{aligned}$$

the quadratic form Q restricts to

$$Q_{\mathbb{R}}(y) = y_{12}y_{34} + y_{31}\bar{y}_{31} + y_{14}y_{23}\,.$$

One can perform the linear change of variables

$$\begin{aligned} y_{12} &= (x^1 + x^0), & y_{34} &= (x^1 - x^0), & y_{31} &= x^2 + \mathrm{i}x^3, \\ y_{14} &= a + b, & y_{23} &= a - b\,, \end{aligned}$$

in which E^θ is simply

$$E^\theta = \{(b, x^0, x^1, x^2, x^3, a) \in \mathbb{R}^6\} \cong \mathbb{R}^6\,,$$

and the quadratic form is

$$Q_{\mathbb{R}}(y) = -(x^0)^2 + (x^1)^2 + (x^2)^2 + (x^3)^2 + a^2 - b^2\,. \tag{2.64}$$

We can compare now with Eq. (2.43). The quadratic form $Q_{\mathbb{R}}$ has signature $(4, 2)$.

Let us focus now on the group action. We consider the group of linear transformations preserving Q:

$$\mathrm{SO}(Q) = \{g \in \mathrm{SL}_6(\mathbb{C}) \mid Q(gy) = Q(y)\} \;\cong\; \mathrm{SO}(6, \mathbb{C})\,,$$

(we restrict here to the transformations with positive determinant).

Due to the property (2.63), there is an involution in $\mathrm{SO}(Q)$ defined as

$$\begin{aligned} \mathrm{SO}(Q) &\longrightarrow \mathrm{SO}(Q) \\ g &\longrightarrow g^\Theta = \Theta\bar{g}\Theta\,, \end{aligned} \tag{2.65}$$

which defines the same conjugation on the Lie algebra $\mathfrak{so}(Q)$

$$\begin{aligned} \mathfrak{so}(Q) &\longrightarrow \mathfrak{so}(Q) \\ \lambda &\longrightarrow \lambda^\Theta = \Theta\bar{\lambda}\Theta\,. \end{aligned} \tag{2.66}$$

The set of fixed points of the involution is the set of the $g \in \mathrm{SO}(Q)$ such that

$$g = \Theta g \Theta\,, \quad \text{or, equivalently,} \quad g^{-1}\Theta g = \Theta\,.$$

It is easy to check that the set of fixed points of the involution $g \to g^\Theta$ coincides with the set of $g \in \mathrm{SO}(Q)$ leaving invariant the real vector space E^θ. We can denote this group as $\mathrm{SO}(Q_{\mathbb{R}})$: it is a real form of $\mathrm{SO}(Q)$. From (2.64) we have that $\mathrm{SO}(Q_{\mathbb{R}}) \cong \mathrm{SO}(4, 2)$.

We now want to introduce the *real Klein quadric*. However, this must be done with some care and we need the following proposition.

Proposition 2.11.1. *Let us define the sets*

$$K^\theta = \{[y] \in \mathbb{P}(E) \mid y \in E_\mathbb{R},\, Q_\mathbb{R}(y) = 0\}\,,$$
$$K^{\theta'} = \{[y] \in K \mid \theta([y]) = [y]\}\,.$$

Then $K^\theta = K^{\theta'}$. Furthermore, the connected component of the identity of the group $\mathrm{SO}(Q_\mathbb{R})$ acts transitively on K^θ.

Proof. It is obvious that $K^\theta \subset K^{\theta'}$. Let us prove now that $K^{\theta'} \subset K^\theta$.

Let $[y] \in K^{\theta'}$. Then, for an arbitrary representative y, we have that

$$\theta(y) = \lambda y, \quad \text{with } \lambda \in \mathbb{C}^\times, \text{and } |\lambda| = 1\,.$$

The last condition arises because θ is involutive. We want to show that there is a representative of $[y]$, say y'

$$y' = \alpha y, \quad \alpha \in \mathbb{C}^\times, \text{ such that } \theta(y') = y\,.$$

It is not difficult to see that one can compute such α from the equation

$$\lambda = \frac{\alpha}{\bar{\alpha}}\,, \qquad |\alpha| = 1\,,$$

which always has solution. So $K^{\theta'} \subset K^\theta$ and then $K^{\theta'} = K^\theta$.

Let us go now to the group action. We have seen that $\mathrm{SO}(Q_\mathbb{R}) \cong \mathrm{SO}(4,2)$. At the same time, for a point on the cone,

$$(x^1)^2 + (x^2)^2 + (x^3)^2 + a^2 = (x^0)^2 + b^2\,,$$

and we can set

$$(x^1)^2 + (x^2)^2 + (x^3)^2 + a^2 = (x^0)^2 + b^2 = 1\,,$$

using the transformation in projective space. Then, using a transformation of the maximal compact subgroup $\mathrm{SO}(4)\times\mathrm{SO}(2)$ we can bring this arbitrary point to

$$(x^1, x^2, x^3, a) \longrightarrow (1,0,0,0), \qquad (b, x^0) \longrightarrow (1,0)\,,$$

which shows the transitivity. ■

We can then define the *real Klein quadric* as the set $K^\theta = K^{\theta'}$.

We now want to understand which are the real forms of the complex conformal and Poincaré groups that we introduced in Section 2.7 that are consistent with the description above.

The spin group of the complex group $\mathrm{SO}(Q)$ is $\mathrm{SL}(T) \cong \mathrm{SL}_4(\mathbb{C})$, which is the complex conformal group (see Section 2.7). It is simply connected, so it is the universal covering of $\mathrm{SO}(6, \mathbb{C})$. From the properties of the universal covering, any continuous map $\mathrm{SO}(Q) \to \mathrm{SO}(Q)$ lifts to a unique map $\mathrm{SL}(T) \to \mathrm{SL}(T)$, so in particular, the involution Θ on $\mathrm{SO}(Q)$ lifts to a unique involution on $\mathrm{SL}(T)$. We now want to determine the fixed point set of such involution, $G = \mathrm{SL}(T)^{\Theta}$; this set will be the right real form of $\mathrm{SL}(T)$, that is, the *real conformal group*.

The real forms of semisimple Lie groups are all classified (see [87]) up to conjugation[9]. For $\mathrm{SL}_4(\mathbb{C})$ and $\mathrm{SO}(6, \mathbb{C})$, the real forms and their properties are listed in in Table 2.1. The labels in that table mean:

con com:	number of connected components,
fund group:	fundamental group,
MCS:	Maximal Compact Subgroup,
dim MCS:	real dimension of the Maximal Compact Subgroup.

Table 2.1 Properties of $\mathrm{SL}_4(\mathbb{C})$, $\mathrm{SO}(6, \mathbb{C})$ and their real forms.

	group	*con comp*	*fund group*	*MCS*	*dim MCS*
	$\mathrm{SL}_4(\mathbb{C})$	1	0	$\mathrm{SU}(4)$	15
1	$\mathrm{SU}(4)$	1	0	$\mathrm{SU}(4)$	15
2	$\mathrm{SL}_4(\mathbb{R})$	1	$\mathbb{Z}_2$	$\mathrm{SO}(4)$	6
3	$\mathrm{SU}(1,3)$	1	$\mathbb{Z}$	$\mathrm{SU}(3) \times \mathrm{U}(1)$	9
4	$\mathrm{SU}(2,2)$	1	$\mathbb{Z}$	$\mathrm{SU}(2) \times \mathrm{SU}(2) \times \mathrm{U}(1)$	7
5	$\mathrm{SU}^*(4)$	1	0	$\mathrm{USp}(4)$	10
	$\mathrm{SO}(6, \mathbb{C})$	1	$\mathbb{Z}_2$	$\mathrm{SO}(6)$	15
1	$\mathrm{SO}(6)$	1	$\mathbb{Z}_2$	$\mathrm{SO}(6)$	15
2	$\mathrm{SO}(3,3)$	2	$\mathbb{Z}_2 \times \mathbb{Z}_2$	$\mathrm{SO}(3) \times \mathrm{SO}(3)$	6
3	$\mathrm{SO}^*(6)$	1	$\mathbb{Z}$	$\mathrm{SU}(3) \times \mathrm{U}(1)$	9
4	$\mathrm{SO}(2,4)$	2	$\mathbb{Z} \times \mathbb{Z}_2$	$\mathrm{SO}(2) \times \mathrm{SO}(4)$	7
5	$\mathrm{SO}(1,5)$	2	$\mathbb{Z}_2$	$\mathrm{SO}(5)$	10

We briefly explain the terminology used in Table 2.1.

In Section 2.4 we already introduced the $\mathrm{SU}(2,2)$ as the group of elements $U \in \mathrm{SL}_4(\mathbb{C})$ which leave invariant the hermitian quadratic form in

[9] Two real forms H_1 and H_2 of a complex group H are *conjugated* if and only if there exists $h \in H$ such that $hH_1h^{-1} = H_2$.

$\mathbb{C}^4$:

$$H(z) = +\bar{z}^1 z^1 + \bar{z}^2 z^2 - \bar{z}^3 z^3 - \bar{z}^4 z^4, \qquad z = \begin{pmatrix} z^1 \\ z^2 \\ z^3 \\ z^4 \end{pmatrix} \in \mathbb{C}^4\,, \tag{2.67}$$

so $U \in \mathrm{SU}(2,2)$ if and only if

$$H(Uz) = H(z), \quad \text{which is equivalent to} \quad U^\dagger I_{2,2} U = I_{2,2}$$

$$\text{with} \quad I_{2,2} = \begin{pmatrix} \mathrm{id}_2 & 0 \\ 0 & -\mathrm{id}_2 \end{pmatrix}.$$

The group $\mathrm{USp}(4)$ is the compact real form of the symplectic group $\mathrm{Sp}(4,\mathbb{C})$. It consists of the symplectic, unitary matrices,

$$\mathrm{USp}(4) = \mathrm{Sp}(4,\mathbb{C}) \cap \mathrm{U}(4)\,,$$

and it is the spin group of $\mathrm{SO}(5)$.

The group $\mathrm{SO}^*(6)$ is the group of complex orthogonal matrices that leave invariant a skew Hermitian form in $\mathbb{C}^6$

$$H(z) = \bar{z}_1 z_4 - \bar{z}_4 z_1 + \bar{z}_2 z_5 - \bar{z}_5 z_2 + \bar{z}_3 z_6 - \bar{z}_6 z_3\,.$$

The group $\mathrm{SU}^*(4)$ is the group of *quaternionic unitary matrices* in $\mathrm{SL}_4(\mathbb{C})$, that is, the group of matrices that commute with the transformation in $\mathbb{C}^4$:

$$\psi \begin{pmatrix} z^1 \\ z^2 \\ z^3 \\ z^4 \end{pmatrix} = \Omega \begin{pmatrix} \bar{z}^1 \\ \bar{z}^2 \\ \bar{z}^3 \\ \bar{z}^4 \end{pmatrix}, \qquad \Omega = \begin{pmatrix} 0 & \mathrm{id}_2 \\ -\mathrm{id}_2 & 0 \end{pmatrix}.$$

Finally, notice that $\mathrm{SO}(4)/\mathbb{Z}_2 \cong \mathrm{SO}(3) \times \mathrm{SO}(3)$.

Proposition 2.11.2. *Let the notation be as above. Then*

$$G = \mathrm{SL}(T)^\Theta = \mathrm{SU}(2,2)\,.$$

Proof. The proof of this proposition consists in a careful exam of the real forms of both $\mathrm{SO}(Q)$ and its spin group, $\mathrm{SL}(T)$, whose properties are detailed in Table 2.1.

In fact, observing the Table 2.1, we see that the different real forms of each complex Lie group have different maximal compact groups, and these have all different dimensions. The spin map from the spin group to the orthogonal group sends the maximal compact subgroup of the spin group onto the maximal compact subgroup of the orthogonal group, so it is immediate to identify the spin groups of the different real forms. This is indicated in Table 2.1 with the number to the right, so, in particular, we have that the spin group of SO(2, 4) is SU(2, 2).

More precisely, all spin groups that appear here are connected, while the orthogonal groups may have one or two connected components. Then, the spin map maps the spin group onto the connected component of the identity of the corresponding orthogonal group.

We then conclude that the spin group of $SO(Q_{\mathbb{R}})$ is isomorphic to SU(2, 2). ■

It is now instructive to compute explicitly the hermitian quadratic form in $SL_4(\mathbb{C})$ starting from the involution defined in (2.65), or the corresponding conjugation in the Lie algebra (2.66). Let G denote the spin group of $SO(Q_{\mathbb{R}})$ and $\mathfrak{g}$ its Lie algebra. $\mathfrak{g}$ is a real form of $\mathfrak{sl}(T)$. Let $U \in G$. In the basis $\{e_1, e_2, e_3, e_4\}$ of T we have the fundamental representation

$$U(e_i) = U^j_{\ i} e_j \,, \quad \det U = 1\,.$$

In the basis $\{e_{12}, e_{23}, e_{14}, e_{34}, e_{31}, e_{24}\}$ of E we have the matrix representation

$$\rho(U)(e_{ij}) = U^k_{\ i} U^l_{\ j} e_{kl} = \rho(U)^{kl}_{\ \ ij} e_{kl} \qquad i<j,\ k<l\,.$$

The matrix $\rho(U)$ in the space E satisfies the reality condition $\rho(U)^\theta = \rho(U)$ (see (2.65)), and the idea is to search for the implications that this reality condition has on the original matrix $U^j_{\ i}$.

It is easier to work with the Lie algebra. Let $Z \in \mathfrak{g}$. Its representation in T is a matrix

$$Z = \begin{pmatrix} Z_{11} & Z_{12} & Z_{13} & Z_{14} \\ Z_{21} & Z_{22} & Z_{23} & Z_{24} \\ Z_{31} & Z_{32} & Z_{33} & Z_{34} \\ Z_{41} & Z_{42} & Z_{43} & Z_{44} \end{pmatrix}$$

with the property

$$\mathrm{tr} Z = Z_{11} + Z_{22} + Z_{33} + Z_{44} = 0\,. \tag{2.68}$$

$$\rho(Z)(e_i \wedge e_j) = Z^k_{\ i} e_{kj} + Z^l_{\ j} e_{il} = (Z^k_{\ i}\delta^l_{\ j} - Z^k_{\ j}\delta^l_{\ i}) e_k \wedge e_l\,.$$

This results in a matrix

$$\rho(Z) = \left(\begin{array}{cccc|cc} Z_{11}+Z_{22} & -Z_{13} & Z_{24} & 0 & -Z_{23} & -Z_{14} \\ -Z_{31} & Z_{22}+Z_{33} & 0 & -Z_{24} & -Z_{21} & Z_{34} \\ Z_{42} & 0 & Z_{11}+Z_{44} & Z_{13} & -Z_{43} & Z_{12} \\ 0 & -Z_{42} & Z_{31} & Z_{33}+Z_{44} & Z_{41} & Z_{32} \\ \hline -Z_{32} & -Z_{12} & -Z_{34} & Z_{14} & Z_{33}+Z_{11} & 0 \\ -Z_{41} & Z_{43} & Z_{21} & Z_{23} & 0 & Z_{22}+Z_{44} \end{array}\right).$$

We have evidenced conveniently four blocks forming the matrix that we will denote as

$$\rho(Z) = \begin{pmatrix} A & B \\ C & D \end{pmatrix}.$$

The reality condition (see (2.62, 2.66))

$$\Theta\overline{\rho(Z)}\Theta = \rho(Z)$$

then implies

$$A = \bar{A}, \quad DJ_2 = J_2\bar{D}, \quad BJ_2 = \bar{B}, \quad J_2\bar{C} = C.$$

It is an easy calculation to show that these conditions, together with (2.68), imply that the matrix Z is of the form

$$Z = \begin{pmatrix} \alpha & \beta \\ \gamma & -\alpha^\dagger \end{pmatrix}, \quad \text{with } \beta \text{ and } \gamma \text{ hermitian and } \alpha \text{ arbitrary,} \tag{2.69}$$

being α, β and γ 2×2 matrices. If we define

$$F = \mathrm{i}\begin{pmatrix} 0 & -\mathrm{id}_2 \\ \mathrm{id}_2 & 0 \end{pmatrix}, \tag{2.70}$$

then the reality condition (2.69) is expressed as

$$FZ^\dagger + ZF = 0.$$

This means that the Lie algebra $\mathfrak{g}$ is the fixed point set of the conjugation

$$\begin{array}{ccc} \mathfrak{sl}(T) & \xrightarrow{\theta} & \mathfrak{sl}(T) \\ Z & \longrightarrow & Z^\theta = -FZ^\dagger F. \end{array}$$

Integrating it to the group $\mathrm{SL}(T)$, we have the involution

$$\begin{array}{ccc} \mathrm{SL}(T) & \xrightarrow{\theta} & \mathrm{SL}(T) \\ U & \longrightarrow & U^\theta = F{U^\dagger}^{-1}F. \end{array} \tag{2.71}$$

Let $u, v \in \mathbb{C}^4$. We define the hermitian form in $\mathbb{C}^4$

$$(u, v) = u^\dagger F v. \tag{2.72}$$

Diagonalizing F, the hermitian form becomes the one in (2.67), so it has signature $(2, 2)$. Moreover, the fixed point set of the involution (2.71) coincides with the subgroup of $\mathrm{SL}(T)$ that leaves invariant this hermitian form. We will denote this group as $\mathrm{SU}(F)$.

So $G = \mathrm{SU}(F) \simeq \mathrm{SU}(2, 2)$ and the isomorphism is given by the change of basis of the diagonalization.

Based on the isomorphism that we have just computed, the following proposition establishes an alternative description of K^θ.

Proposition 2.11.3. *Let us consider the set of null planes of T with respect to the hermitian form (2.72)*

$$\mathcal{N} = \{\pi \in G(2,4) \mid (u,v) = 0 \ \forall u, v \in \pi\}.$$

Then, $K^\theta = \mathcal{N}$.

Proof. It is clear that $\pi_0 = \text{span}\{e_1, e_2\}$ belongs to $\mathcal{N}$. It is easy to check that, for any $U \in G$, the plane $\pi = U\pi_0$ is also a null plane. The action of G on K^θ is transitive because the action of $\mathrm{SO}(Q_\mathbb{R})$ is so (see Proposition 2.11.1). Then $K^\theta \subset \mathcal{N}$.

On the other direction, let π be a null plane and let $\langle\,\cdot\,,\,\cdot\,\rangle$ be the standard hermitian form in $\mathbb{C}^4$, so

$$(u,v) = \langle u, Fv\rangle\,.$$

Let $\{f_1, f_2\}$ be an orthonormal basis in π so

$$\langle f_i, f_j\rangle = \delta_{ij}\,, \qquad i,j = 1,2\,.$$

The plane $F\pi$ is a plane that is orthogonal to π with the standard hermitian form. As it is easy to check, an orthonormal basis of $F\pi$ is given by the vectors

$$f_3 = -\mathrm{i}Ff_1\,, \qquad f_4 = -\mathrm{i}Ff_2\,.$$

The set $\{f_1, f_2, f_3, f_4\}$ is an orthonormal basis of T

$$\langle f_i, f_j\rangle = \delta_{ij}\,, \qquad i,j = 1,2,3,4\,,$$

and one can check that

$$(e_i, e_j) = (f_i, f_j)\,, \qquad i,j = 1,2,3,4\,.$$

Then, there is a transformation U that preserves the hermitian form $(\,\cdot\,,\,\cdot\,)$ and such that

$$U(e_i) = f_i\,, \qquad i = 1,2,3,4\,.$$

In particular

$$U(\pi_0) = \pi\,.$$

We can multiply U by a unitary complex number, $U' = cU$, $|c| = 1$ so $\det U' = 1$ and U' satisfies the same property

$$U'(\pi_0) = \pi\,.$$

So we have proven that any null plane is of the form $\pi = U'(\pi_0)$ with $U' \in G$ and then, $\mathcal{N} \subset K^\theta$.

We have our result $K^\theta = \mathcal{N}$. ∎

We now turn and examine how the real Minkowski space fits into this picture.

Recall that in Section 2.6, we have identified the complex Minkowski space as the big cell U_{12} inside the Grassmannian manifold. These are the finite points of the complex Klein quadric K in $\mathbb{P}(E) \cong \mathbb{P}^5(\mathbb{C})$.

In Proposition 2.11.1 we have characterized the real Klein quadric K^θ as the set of points $[y]$ in $\mathbb{P}(E)$ with a representative $y \in E_\mathbb{R}$ satisfying the condition $Q_\mathbb{R}(y) = 0$ (see (2.64)). It is then natural to expect that the real Minkowski space is identified with the finite points of $K_\mathbb{R}$. In the following proposition, we prove this and find the description of the real Poincaré group (2.13) inside the real, conformal group $\mathrm{SL}(T)^\Theta \cong \mathrm{SU}(2,2)$.

Proposition 2.11.4.

(1) The real Minkowski space $\mathbf{M}$ *is a dense open set in* K^θ.

(2) The group $G = \mathrm{SL}(T)^\Theta = \mathrm{SU}(F) \cong \mathrm{SU}(2,2)$ *acts conformally and transitively on* K^θ, *and the subgroup that leaves invariant the real Minkowski space is*

$$F_0' = \left\{ \begin{pmatrix} L & 0 \\ NL & {L^\dagger}^{-1} \end{pmatrix}, \ N \text{ hermitian and } \det(L) \in \mathbb{R}^\times \right\} \subset G\,.$$

Its action on $\mathbf{M}$ *is*

$$A \to N + {L^\dagger}^{-1} A L^{-1}\,.$$

Then

$$F_0' \cong (\mathrm{SL}_2(\mathbb{C})_\mathbb{R} \times \mathbb{R}^\times) \ltimes \mathbb{R}^4\,.$$

Proof. Let us take

$$U = \begin{pmatrix} L & M \\ NL & R \end{pmatrix} \in \mathrm{SU}(F)\,,$$

where F is the matrix (4.64) defining the Hermitian form (2.72). Reasoning as in the proof of Proposition 2.7.1, we consider the action of U on an element of $\mathbf{M}$.

$$\begin{pmatrix} L & M \\ NL & R \end{pmatrix} \begin{pmatrix} I \\ A \end{pmatrix} = \begin{pmatrix} L + MA \\ NL + RA \end{pmatrix}.$$

The matrix $L + MA$ must be invertible for all hermitian matrices A, hence, we must take $M = 0$. Then the condition $U \in \mathrm{SU}(F)$ i.e. $U^\dagger F U = F$ forces $R = (L^\dagger)^{-1}$ and $N^\dagger = N$. The form taken by the action $A \mapsto g \cdot A$ is an immediate calculation.

The conformal metric on K (see Section 2.8) is given, on $\mathbf{M}$, by the quadratic form (2.61),

$$(\mathrm{d}x^0)^2 - (\mathrm{d}x^1)^2 - (\mathrm{d}x^2)^2 - (\mathrm{d}x^3)^2 ,$$

which is real and of signature $(1, 3)$. Since $\det A$ goes over to $\det A(\det L)^{-2}$ under the action of F_0', it follows that we have a real conformal metric which is invariant under G.

The transitivity was proven in Proposition 2.11.1. The group $\mathrm{SU}(F)$ thus acts conformally and transitively on K^θ.

Finally, we have that $L \in \mathrm{GL}_2(\mathbb{C}) \cong \mathrm{SL}_2(\mathbb{C}) \times \mathbb{R}^\times$ and

$$N \in \{2 \times 2 \text{ Hermitian matrices}\} \cong \mathbb{R}^4 ,$$

so

$$F_0' \cong (\mathrm{SL}_2(\mathbb{C})_{\mathbb{R}} \times \mathbb{R}^\times) \ltimes \mathbb{R}^4 . \tag{2.73}$$

■

Remark 2.11.5. The CCI (connected component of the identity) of F_0' is $F_0 = (\mathrm{SL}_2(\mathbb{C}) \times \mathbb{R}^+)_{\mathbb{R}} \ltimes \mathbb{R}^4$. We will call F_0 the *real Poincaré group times dilations*. The dilations are the factor $\mathbb{R}^+$ in (2.73) and the Lorentz part is $\mathrm{SL}_2(\mathbb{C})_{\mathbb{R}}$, which is in fact the spin group of $\mathrm{SO}(1,3)^0$. The *real Poincaré group* is then defined as

$$P_0 = \mathrm{SL}_2(\mathbb{C}) \ltimes \mathbb{R}^4 .$$

■

This completes the interpretation of the finite points of K^θ as the real Minkowski space.

2.12 References

The derivation of the conformal algebra and the conformal transformations is standard and can be found in any book on conformal field theory. We have used [67]. Our main references for the treatment of the Minkowski and conformal spaces are [147] and [108]. The functor of points approach is found in [108] but also, with more details in [66], [26] and [27]. The treatment of the real form has been taken from [147].

Chapter 3

Supersymmetry in Physics

This chapter is devoted to understanding the meaning and the role played by supersymmetry in physics. From a mathematical point of view, supersymmetry can be presented just as a generalization of the ordinary Lie groups that are spacetime symmetries of physical theories, the Poincaré and the conformal groups. This generalization implies the substitution of the category of commutative algebras by the larger category of commutative superalgebras. Commutative superalgebras contain even and odd generators: the even ones commute among themselves while the odd ones, which are nilpotent generators, anticommute among themselves. In spite of the non commutativity of the odd generators, this generalization has nothing to do with quantum physics: it is a *classical* generalization.

The passage from classical to quantum physics is often summarized as a substitution of the commutative algebra of classical observables by a non commutative one. Under proper assumptions, the non commutative algebra of quantum physics can be seen as a deformation of the original algebra, and there is a parameter (identified with Planck's constant) which measures the non commutativity. The introduction of odd generators does not have this property: they are not obtained from even ones by a deformation (at least not one of physical significance); in fact, superalgebras can also be deformed from commutative to non commutative ones, hence they can be also quantized as we shall see in Chapter 5.

From the physics point of view the situation is different. It is impossible to understand the role played by supersymmetry in physics without making an excursion in the quantum world. Some of the properties of the supergroups used in physics may seem arbitrary to mathematicians, but they arise as a consequence of stringent physical requirements. It is the case, for example, of the spin-statistics connection, that cannot be fully appreciated without at least a glimpse to Quantum Field Theory (QFT).

It is in QFT that supersymmetry has played an important role: in the first place, as a way to circumvent the Coleman-Mandula theorem mentioned in the Introduction, but later also because of the good behaviour of supersymmetric theories as quantum theories. Let us just mention relatively recent investigations which suggest that supergravity may indeed be free of the infinities that prevented the quantization of Einstein's gravity as a standard field theory [17]. In any case, supersymmetric field theories are generally better behaved than ordinary ones. Finally, we have to mention string theory, proposed as a theory that unifies all known interactions and that needs supersymmetry for its internal consistency.

This chapter is organized as follows.

In Section 3.1 we will give a very brief introduction to quantum physics. It is by no means a comprehensive account of it: we try to emphasize the physical and mathematical aspects that we will need for our discussion.

In Section 3.2 we try to answer what is the spin in the context of special relativity: this will introduce us to the study of the unitary representations of the Poincaré group. This is perhaps the most technical section of this chapter.

QFT arises as the merging of relativity and quantum physics. Again, it would be impossible to give here a thorough account of it. What we try instead in Section 3.3 is to introduce some of the fundamental ideas of QFT with a toy model: we will see what is the precise meaning of 'particle' in this context and we will see the necessity of superalgebras to describe the two types of particles known, classified according to their statistical behaviour: bosons and fermions. These are the basic building blocks of supersymmetry and supergeometry. Even if supersymmetry has not been found in the laboratory, the fact that nature shows both types of behaviour very strongly suggests that there must be some kind of supersymmetry at some level. Supergeometry is such a natural construction that the trained eye would inevitably ask why it is not happening, and one must come up with a convincing reason for such selection rule. Also in this section, we will be able to understand the statement of the *spin-statistics theorem*. Under the hypothesis that identical particles are indistinguishable in quantum physics, and using the full power of QFT, one can prove that bosons are always associated to integer spin representations of the Poincaré group while fermions are associated to half-integer spin representations. The theorem is not proven here (see the original references [52,120,134]), but we will see its relation with the *Pauli exclusion principle*, discovered when the atomic structure was being investigated, in the first days of the quantum theory.

Finally, in Section 3.4 we try to give a non technical overview that situates supersymmetry in the context of the theories that try to unify the fundamental interactions. The goodness of supersymmetric fields theories has been the main motivation behind the theoretical effort to understand it. Many theorists still share the attitude that there is no fundamental reason for supersymmetry to be realized in nature, other than making finite or well behaved, say, string theory or other physical theories. The importance of string theory in physics cannot be underestimated, but still, we have a different point of view: for us, it is a question of naturalness.

Nevertheless, we would not have a complete picture without paying attention to the actual supersymmetric theories that have been proposed and the physical phenomena that they intend to describe.

The extent up to which supersymmetry will help in the construction of a unified theory of all interactions is still unknown. Meanwhile, we can explore its potential.

3.1 Quantum physics

Every classical mechanical system has an associated symplectic manifold $\mathfrak{P}$ called its *phase space*. Points in the phase space are in one-to-one correspondence with the possible states of the system.

An *observable* of a classical system is a physically measurable quantity. The result of a measurement of an observable in an experiment is a real number, the value of the observable in the actual state of the system. Observables then are real valued functions on the phase space. Any observable has a value in every state of the system and states are determined by the values that observables have on them.

$$\left\{\begin{matrix}\text{States of a classical}\\ \text{mechanical system}\end{matrix}\right\} \equiv \left\{\begin{matrix}\text{Phase space}\\ \mathfrak{P}\end{matrix}\right\},$$

$$\left\{\begin{matrix}\text{Observables of a classical}\\ \text{mechanical system}\end{matrix}\right\} \equiv \left\{\begin{matrix}\text{Functions on phase space}\\ \mathcal{F}(\mathfrak{P})\end{matrix}\right\}.$$

The evolution of the system is governed by a system of ordinary differential equations of the form

$$\frac{dx(t)}{dt} = X_H(x(t)), \tag{3.1}$$

where $x(t)$ is a curve on phase space parametrized by the time t, H is an observable called the *hamiltonian* of the system and X_H is the hamiltonian vector field of H with respect to the symplectic form of $\mathfrak{P}$. In this way, if we know the state of the system at the time t_0, $x(t_0)$, then we can determine the state of the system at the time t, $x(t)$.

In quantum mechanics this simple picture fades. Measurements of a given observable, for two different systems prepared in the same way, may give different values. It is generally assumed that an observable has a probability distribution of values on each state of the system.

The mathematical treatment of a quantum mechanical system associates to it an infinite dimensional, complex, separable Hilbert space $\mathcal{H}$. The states or, more precisely, the *pure states*[1] of a quantum system are points in the projective space of $\mathcal{H}$, $\mathbb{P}(\mathcal{H})$.

The physical observables are represented by self-adjoint operators of the space $\mathcal{H}$. The probability that an observable A takes its value in a Borel set $B \subset \mathbb{R}$ when the system is in the state represented by the vector $\psi \in \mathcal{H}$ is given in terms of the resolution of the identity of A, that is, the projection operators $E_A(B)$ appearing in the spectral decomposition of A. One of the postulates of quantum mechanics states that this probability is

$$P_{A,\psi}(B) = \|E_A(B)\psi\|^2 = \langle E_A(B)\psi, \psi\rangle ,$$

where $\langle\, , \,\rangle$ is the inner product in $\mathcal{H}$.

The physical evolution of the system is governed by a particular self-adjoint operator, the hamiltonian H, and the Schrödinger equation

$$i\hbar \frac{d\psi_t}{dt} = H\,\psi_t\,. \tag{3.2}$$

We notice here the appearance of Planck's constant

$$h = 6.62606896(33) \times 10^{-34}\, J \cdot s\,, \qquad \hbar = \frac{h}{2\pi}\,.$$

[1]Often, in quantum mechanical experiments, systems are prepared in such way that we only know the probability that the system is in a certain pure state. We say then that the system is in a *mixed state*. This probability distribution is not inherent to quantum mechanics. In fact, also in classic statistical mechanics 'mixed' states are considered, and points in phase space could then be called 'pure' states. What is new in quantum mechanics is the existence of probability distributions associated to observables and pure states. These are unavoidable.

This universal constant first appeared as the proportionality constant between the energy (E) and the frequency (ν) of electromagnetic radiation

$$E = h\nu\,. \tag{3.3}$$

This was the breakthrough hypothesis that Planck introduced back in 1900 to explain the black body radiation and inaugurated the new era of quantum physics.

The *expected value* or *mean value* of an observable A in a state ψ is defined as

$$\langle A\rangle_\psi = \int_{-\infty}^{+\infty} t\,\mathrm{d}P_{A,\psi}(t)\,.$$

The *variance* of an observable A in a state ψ is defined as the squared deviation from the expected value

$$\mathrm{var}(A)_\psi = \langle A^2\rangle_\psi - \big(\langle A\rangle_\psi\big)^2,$$

which gives an idea on how spread the probability distribution is. In physics, one calls the square root of the variance the *uncertainty* of the observable in the given state,

$$\Delta_\psi A = \sqrt{\mathrm{var}(A)_\psi}.$$

It is a consequence of these definitions that for any two observables A and B

$$\Delta_\psi A \cdot \Delta_\Psi B \geq \frac{1}{2}|\langle [A,B]\rangle_\psi|\,, \tag{3.4}$$

where $[A,B]$ is the commutator of the two observables (which is not in general a self-adjoint operator). Equation (3.4) is known as the *Heisenberg uncertainty principle* of quantum mechanics. It tells us that, no matter how carefully we prepare a system, it is not possible to reduce arbitrarily and simultaneously the uncertainty, in a given state, of two observables that do not commute.

Think for example on a single particle system, whose Hilbert space is $\mathcal{H} = \mathcal{L}^2(\mathbb{R})$ and the operators of position $\hat{x}$ and momentum $\hat{p}$ are defined as

$$\hat{x}\,\psi(x) = x\,\psi(x), \qquad \hat{p}\psi(x) = -\mathrm{i}\hbar\,\frac{\mathrm{d}}{\mathrm{d}x}\psi(x),$$

with commutator

$$[\hat{x},\hat{p}] = \mathrm{i}\hbar\,\mathrm{id}\,. \tag{3.5}$$

Equation (3.4) implies that

$$\Delta_\psi \hat{x} \cdot \Delta_\Psi \hat{p} \geq \frac{\hbar}{2}. \tag{3.6}$$

So one cannot make the probability distribution of the position operator sharply concentrated in one value without spreading at the same time the probability distribution of the momentum operator.

We have summarized here the mathematical frameworks of classical and quantum mechanics, which have since long time been well established. A physicist, nevertheless, will not be satisfied with these two different (though self consistent) points of view.

There is a generic principle in physics known as the *correspondence principle* that states that a new physical theory, introduced to explain new phenomena, should explain also all of the phenomena for which the old theory was valid and shall determine the circumstances under which the old theory is a valid approximation. The new theory must reduce, under the appropriate circumstances, to the old theory. For example, special relativity reduces to galilean relativity when the velocity v of the bodies involved in the experiment is much smaller than the velocity of light c,

$$\frac{v}{c} << 1. \tag{3.7}$$

(refer also to Chapter 2, Section 2.1 for our brief introduction to Einstein's relativity).

A quantum mechanical system must have a limit in which the system acquires a classical behavior. But how to relate the two radically different pictures shown here by some kind of limit process?

Around 1920, in the days of the old quantum theory, Bohr stated the correspondence principle in quantum mechanics in this way:

> *When the quantum numbers of a quantum mechanical system are large, the quantum system behaves, in a good approximation, as a classical system.*

Let us analyze a quantum system that Bohr had in mind by then. We consider the model of the hydrogen-like atoms circulating in those days (Bohr's model in fact). Nowadays, the model hydrogen atom is a quantum system in the precise sense that we described above. Nevertheless, Bohr's naive model gives essentially correct results[2] and reproduces the facts of

[2]Several corrections of higher order in perturbation theory can be incorporated into the model.

the quantum theory that we want to illustrate with a few straightforward calculations.

In Bohr's model, the atom consisted of electrons (negatively charged particles of charge $-e$) orbiting around a positively charged nucleus (of charge $+Ze$, where Z is the atomic number) which is much heavier than the electron. The electron moves under the influence of the Coulomb potential of the nucleus

$$V = -\frac{1}{4\pi\epsilon_0}\frac{Ze}{r} \tag{3.8}$$

where r is the distance between the electron and the nucleus, the constant $\epsilon_0 = 8.854187817\cdots \times 10^{-12}\text{F/m}$ is the permittivity of vacuum and $e = 1.602176565(35) \times 10^{-19}\text{C}$ is the (absolute value of the) charge of the electron.

The only orbits allowed for the electrons are circular orbits such that the module of the angular momentum of the electron (a constant of the classical motion) is a (positive) integer multiple of $\hbar$,

$$J = m_e|\vec{r} \times \vec{v}| = m_e rv = n\hbar, \qquad n = 1, 2, 3, \ldots \tag{3.9}$$

where $m_e = 9.10938291(40) \times 10^{-31}$ kg is the mass of the electron. This is the *quantization rule* of the old quantum theory.

Newton's third law for the centripetal force

$$m_e\frac{v^2}{r} = \frac{Ze^2}{r^2},$$

together with (3.9) determine the allowed radius and velocity to be

$$r_n = \frac{4\pi\epsilon_0\hbar^2}{Ze^2 m_e}n^2, \qquad v = \frac{Ze}{4\pi\epsilon_0}\frac{1}{n}. \tag{3.10}$$

One could then compute the energy

$$E_n = \frac{m_e}{2}v^2 + V = -\frac{m_e}{2}\left(\frac{Ze^2}{4\pi\epsilon_0\hbar}\right)^2\frac{1}{n^2}, \qquad n = 1, 2, 3, \ldots. \tag{3.11}$$

In a more modern interpretation, the E_n would be eigenvalues of the Hamiltonian operator.

So the allowed orbits or energy levels are labeled by a positive integer n called the *principal quantum number*. Classically, an accelerated charge (as the electron moving in a non rectilinear orbit) would emit electromagnetic radiation. The charge then loses energy and this causes the radius of its orbit to decrease. Hence, the electron would end up collapsing over the nucleus: in such model the atom would not be stable.

It was then postulated that the electron was only emitting or absorbing electromagnetic radiation when jumping from one level to another. The frequency of the electromagnetic radiation was computed in terms of the difference between the energies of the two levels by Planck's formula (3.3),

$$h\nu = E_n - E_m = \frac{m_e}{2}\Big(\frac{Ze^2}{4\pi\epsilon_0\hbar}\Big)^2\Big(\frac{1}{m^2} - \frac{1}{n^2}\Big).$$

This formula is the *Rydberg formula*, well known from nineteenth century spectroscopy. The simple Bohr model for the atom was the first explanation of the mysterious atomical spectra.

Notice that there is an orbit with a minimum energy, the one with $n = 1$. This is the *fundamental state* of the atom. The electron cannot decay to a lower energy state by emitting electromagnetic radiation, so the atom is stable.

We now want to figure out what the correspondence principle means in this simple system. We will consider quantum numbers n, m that are very large. More precisely, if $\delta = n - m$, we will consider

$$\delta << n, m.$$

In this approximation, and substituting $n = m + \delta$ we obtain

$$h\nu \simeq m_e\Big(\frac{Ze^2}{4\pi\epsilon_0\hbar}\Big)^2 \frac{\delta}{m^3}, \qquad \delta = 1, 2, 3, \dots \tag{3.12}$$

Let us compare this result with the classical system. An electron moving in a circular orbit of radius r_m will emit electromagnetic radiation whose frequency (*cyclotron frequency*) is an integer multiple of the fundamental frequency associated to the circular motion[3]. The fundamental frequency is the inverse of the period

$$T = \frac{2\pi r_m}{v},$$

and this can be computed from (3.10). One obtains

$$\nu_0 = \frac{1}{T} = \frac{m_e}{4\pi\hbar^4}\Big(\frac{Ze^2}{4\pi\epsilon_0}\Big)^2 \frac{1}{m^3},$$

which can now be compared with the quantum frequency (3.12).

What can we learn from this example? If we go back to the quantization rule (3.9), we observe that a large quantum number n means an angular momentum, whose module is very large compared to $\hbar$. As in special relativity (3.7), we have a relation

$$\frac{\hbar}{J} << 1. \tag{3.13}$$

[3]To be precise, this circular motion is only meant to be instantaneous, since the actual orbit is a spiral. This is, in fact, an approximation.

Equations (3.7) and (3.13) are formally similar: we have a dimensionless variable that becomes very small. An expansion in this variable around zero shall bring the desired approximation. Often, both limits are written as

$$c \to \infty, \qquad \hbar \to 0.$$

The parameters c and $\hbar$ are dimensionfull quantities, so in principle, it makes no sense to say 'c is very big' or '$\hbar$ is very small', since their actual values depend on the units chosen. They are big or small compared to the magnitudes of velocities and angular momentum (respectively) appearing in the equations. Nevertheless, expansions in these parameters give the correct results provided that dimension counting in the final equations matches. The reason is that the limits ∞ or 0 are the same in any unit system.

While the classical limit procedure is straightforward for special relativity, in quantum mechanics we have to work a bit more to relate the Hilbert space and operator formalism with the phase space approach of classical mechanics. To see this, let us come back for a moment to the uncertainty principle in (3.6). This is the toy model that can give us the clue to solve this question. Position and momentum operators are basic observables appearing in quantum theory. For a single particle moving in a potential (and for spin 0) the other observables can be expressed in terms of the two basic ones $\hat{x}$ and $\hat{p}$. The uncertainty principle relies on the non commutativity of the observables, and the commutator of $\hat{x}$ and $\hat{p}$ (3.5) is proportional to $\hbar\,\mathrm{id}$. In the classical limit $\hbar \to 0$ this commutator goes to zero. In fact, this is the crucial difference between observables in classical and quantum mechanics. Classical observables commute, that is, we can measure them accurately and simultaneously, while quantum observables have a commutator proportional to the fundamental parameter $\hbar$.

To make a bit more precise the statement, let us consider arbitrary operators on the Hilbert space $\mathcal{O}(\mathcal{H})$ and not only the hermitian ones. These form an algebra that, in some way, should reduce to the complexified algebra of classical observables $\mathcal{F}(\mathfrak{P})^{\mathbb{C}}$ for $\hbar \to 0$. The self-adjointness condition is then imposed by introducing a $*$ operation on $\mathcal{O}(\mathcal{H})$ which in the limit will reduce to the reality condition on $\mathcal{F}(\mathfrak{P})$. The $*$ operation brings along the concept of $\mathbb{C}^\star$ algebra, which indeed appeared as a model of quantum observables, although nowadays one can refer to them abstractly, without explicit connection to Hilbert spaces or operators. This is the basic concept of non commutative geometry, a mathematical subject that is interesting by itself, for the possibility of deepening our knowledge of the quantum theory, and even throwing some light on quantum gravity (see for example [34]).

A less heavy machinery is used with the approach of deformation quantization [12,47,95]. At the expense of losing momentarily the Hilbert space picture, one tries to deform the algebra of classical observables into a non commutative algebra, the deformation depending on a parameter that one identifies with Planck's constant. The substitution $\hbar = 0$ gives us back the classical theory. This approach is easier but also has a drawback: one is forced in most cases to consider formal deformations, that is, deformations that are written as not necessarily convergent power series in the parameter (in other words, the product is not necessarily well defined on the subspace of convergent series). There are very important results on deformation theory that classify the space of possible formal deformations of an algebra, but it is hard in this approach to regain the Hilbert space picture, and then, precise numerical predictions.

We are interested in merging special relativity with the quantum theory, but the Schrödinger equation (3.2) is not relativistic, since it treats space and time differently. This problem is solved in the framework of Quantum Field Theory, that we will explore in Section 3.3.

But before going to that, we will address a more concrete problem. As we have seen, quantum observables are self-adjoint operators. The symmetries of a quantum theory are represented by unitary operators: if $A \in \mathcal{O}(\mathcal{H})$ is self-adjoint, and $U \in \mathcal{O}(\mathcal{H})$ is unitary, the transformed operator

$$A' = UAU^\dagger$$

is also self-adjoint.

If we have to make relativity compatible with the quantum theory we are then interested in studying the unitary representations of the Poincaré group[4].

We will approach the issue with some detail in the next section.

3.2 Particles and spin

The principle of relativity (see Chapter 2) forces the transformations between inertial systems to be Poincaré transformations. Such principle states

[4]Strictly, we need *projective, unitary representations* of the Poincaré group, but this detail is not relevant for us at this moment.

that the description of a physical system should be the same independently from the chosen inertial frame. In particular, the laws of electromagnetism and the fact that the velocity of light in vacuum has the value $c = 299792458$ m/s should hold in any inertial frame.

Another way of rephrasing the principle of relativity is by saying that the transformations of the Poincaré group are symmetries of the physical laws.

In order to realize the Poincaré group as the group of symmetries of a quantum theory we then need a unitary representation of it. We have defined the real Poincaré group (2.11.5) as the semidirect product of the semisimple, non compact Lie group $\mathrm{SL}_2(\mathbb{C})_\mathbb{R}$ times the abelian factor $\mathbb{R}^4$,

$$P_0 = \mathrm{SL}_2(\mathbb{C})_\mathbb{R} \ltimes \mathbb{R}^4 \,.$$

The classification of the irreducible, unitary representations of such a group is not straightforward. It was Wigner [161] who first carried out the analysis of the representations of the Poincaré group using the concept of *induced representation*. Later on, Mackey did the general study for locally compact groups [105].

We are going to explain here the theorem of Mackey in the case of semidirect products, which is the case most interesting to us. Instead of giving an exhaustive discussion, we will try to give an intuitive approach to the construction of these representations, though keeping intact the mathematical rigour. For a modern and accurate account of this theory, we refer the reader to Ref. [148], Chapter IX.

Let $P = G \ltimes A$ be a locally compact, topological group, with A an abelian group. There is an action of G on A^*, induced by the action of G on A. Let $\mathcal{O}$ be the orbit of G in the dual space A^* of the point $p_r \in A^*$ and let G_{p_r} be its stabilizer:

$$G_{p_r}\, p_r = p_r \,, \qquad \text{and} \qquad \mathcal{O} \approx G/G_{p_r} \,.$$

G_{p_r} is called the *little group* and it plays a key role in the development of this theory. If p_r' is another point of the same orbit, then there exists a $g \in G$ such that $p_r{}' = gp_r$ and then $G_{p_r'} = gG_{p_r}g^{-1} \cong G_{p_r}$.

We first construct the representation space. Let $\mathcal{H}$ be a Hilbert space with inner product

$$\langle\ ,\ \rangle_\mathcal{H} : \mathcal{H} \times \mathcal{H} \to \mathbb{C} \tag{3.14}$$

and suppose that there is a unitary, irreducible representation of G_{p_r} in $\mathcal{H}$

$$R : G_{p_r} \to \mathrm{End}(\mathcal{H}) \,.$$

Let $E = G \times \mathcal{H}$ be the trivial vector bundle with basis G and fiber $\mathcal{H}$. G acts trivially on E as follows:

$$\begin{aligned} G \times E &\longrightarrow E \\ (g, (g', v)) &\longrightarrow (gg', v) . \end{aligned} \tag{3.15}$$

On E we introduce the equivalence relation

$$(g, v) \sim (g', v') \Leftrightarrow g' = gr \quad \text{and} \quad v' = R(r^{-1})v ,$$

with $(g, v), (g', v') \in E$ and $r \in G_{p_r}$. The space E/G_{p_r} is a vector bundle with basis $G/G_{p_r} \cong \mathcal{O}$ and fiber $\mathcal{H}$. There is a well defined action of G on E/G_{p_r} induced from (3.15).

Let $\Gamma(E/G_{p_r})$ be the space of global sections of E/G_{p_r}. There is a representation of G on $\Gamma(E/G_{p_r})$,

$$U(g)\varphi(x) = g\varphi(g^{-1}x) , \qquad x \in \mathcal{O},\ \varphi \in \Gamma(E/G_{p_r}),\ g \in G .$$

To convert $\Gamma(E/G_{p_r})$ into a Hilbert space we need an inner product. Let μ be a measure in $\mathcal{O}$ and let $\mathfrak{H} = \mathcal{L}^2(E/G_{p_r}, \mu)$ be the space of square integrable functions on E/G_{p_r}. Then we can define an inner product in $\mathfrak{H}$ in terms of (3.14),

$$\langle \varphi, \psi \rangle_{\mathfrak{H}} = \int_{\mathcal{O}} \langle \varphi(p), \psi(p) \rangle_{\mathcal{H}} \, \mathrm{d}\mu, \qquad \varphi, \psi \in \mathfrak{H} .$$

The measure μ cannot be an arbitrary measure, though, if we want the inner product to be preserved by the action of the group. From the unique, invariant measure on G (the *Haar measure*) one can define on $\mathcal{O}$ a left invariant measure. We do not give here this construction: it is enough to know that such measure exists. Then, $\mathfrak{H}$ is a Hilbert space.

We want to remark that when $\mathcal{H}$ is finite dimensional the construction that we describe offers little difficulty and it is carried on adding a few more details to what we have outlined. However, when $\mathcal{H}$ is infinite dimensional, things become substantially harder and it is necessary to take care of some topological questions. We are not going to discuss such difficulties here, sending the reader to [147, 148] and the references therein. Besides, there are neither true nor hypothetical physical particles corresponding to such representations.

The group G is unitarily represented on $\mathfrak{H}$. In order to describe the representations of $P = G \ltimes A$, we now need to provide a representation of the abelian part A on $\mathfrak{H}$ in a way that fits with the unitary representations of G. This is done in the following way. Let us consider a point $p \in A^*$. Then, the map

$$\begin{aligned} A &\longrightarrow \mathbb{C}^\times \\ a &\longrightarrow \mathrm{e}^{\mathrm{i}p(a)} \end{aligned}$$

is an homomorphism, that is, a *character* of A. Since A is abelian, its irreducible representations are one dimensional, i.e. they are characters.

We are now ready to state Mackey's result for semidirect products.

Theorem 3.2.1. *(Mackey). Let $P = G \ltimes A$ be a locally compact, topological group, with A an abelian group. Let $\mathcal{O}$ be an orbit of G in the dual space A^* and let $p_r \in \mathcal{O}$ be a point in the orbit with G_{p_r} its stabilizer (its little group). Let $R : G_{p_r} \to End(\mathcal{H})$ be a unitary, irreducible representation on the Hilbert space $\mathcal{H}$ and let $E = G \times \mathcal{H}$ and E/G_{p_r} be defined as before. Let*

$$\mathfrak{H} = \mathcal{L}^2(E/G_{p_r}, \mu)$$

be the Hilbert space of square integrable sections of E/G_{p_r} with respect to the invariant measure μ on $\mathcal{O}$.

Then,

(1) The representation of P, $U^R_{\mathcal{O}} : P \to End(\mathfrak{H})$ defined by

$$U^R_{\mathcal{O}}(g,a)\varphi(p) = \mathrm{e}^{\mathrm{i}p(a)} g\,\varphi(g^{-1}p), \qquad (g,a) \in G \ltimes A\,, \quad p \in \mathcal{O}\,,\ \varphi \in \mathfrak{H}\,,$$

is a unitary, irreducible representation of G. This representation is said to be induced by the representation R of the little group G_{p_r}.

(2) Two such representations are equivalent if and only if the points p_r and p'_r belong to the same orbit and the representations R and R' are equivalent.

(3) If the orbit structure of G is smooth, then every irreducible representation of P is equivalent to one constructed in this way.

■

We now apply the method to the Poincaré group. The action of $\mathrm{SL}_2(\mathbb{C})_\mathbb{R}$ on the space of translations, $T \cong \mathbb{R}^4$ is, via the spin map, the fundamental representation of $\mathrm{SO}(1,3)^0$. By the way in which representations are induced, it is clear that the representations of both, $\mathcal{P}^0 = \mathrm{SO}(1,3)^0 \ltimes \mathbb{R}^4$, the component connected to the identity of $\mathcal{P}$, and its double cover $P_0 = \mathrm{SL}_2(\mathbb{C})_\mathbb{R} \ltimes \mathbb{R}^4$ are the same. We can then consider directly $\mathcal{P}^0$ instead of P_0. Moreover, since in this case it has physical significance, we will enlarge the group to include both connected components of SO(1, 3), that is, we will consider the group $\mathrm{SO}(1,3) \ltimes \mathbb{R}^4$.

The fundamental representation of SO(1, 3) is the same as the action of SO(1, 3) on the Minkowski space and the dual or contragradient representation on $T^* \cong \mathbb{R}^4$ is an equivalent representation. In fact, if $\Lambda \in \mathrm{SO}(1,3)$,

$$\Lambda^T \eta \Lambda = \eta\,, \qquad \text{which means that} \qquad \Lambda = \eta {\Lambda^T}^{-1} \eta\,,$$

so Λ and ${\Lambda^T}^{-1}$ are matrices of equivalent representations. Physicists call T^* the *momentum space*. Let us denote as (p_0, p_1, p_2, p_3) an arbitrary point in T^*. The orbits of this representation are classified as follows:

$$\mathcal{O}_m^+ = \{p \in T^* \mid {p_0}^2 - {p_1}^2 - {p_2}^2 - {p_3}^2 = m^2\,,\ m > 0\,,\ p_0 > 0\}\,,$$

$$\mathcal{O}_m^- = \{p \in T^* \mid {p_0}^2 - {p_1}^2 - {p_2}^2 - {p_3}^2 = m^2\,,\ m > 0\,,\ p_0 < 0\}\,,$$

$$\mathcal{O}_0^+ = \{p \in T^* \mid {p_0}^2 - {p_1}^2 - {p_2}^2 - {p_3}^2 = m^2\,,\ m = 0\,,\ p_0 > 0\}\,,$$

$$\mathcal{O}_0^- = \{p \in T^* \mid {p_0}^2 - {p_1}^2 - {p_2}^2 - {p_3}^2 = m^2\,,\ m = 0\,,\ p_0 < 0\}\,,$$

$$\mathcal{O}_0^0 = \{0\}\,,\ \text{which implies}\ \ m = 0\,,\ p_0 = 0\,,$$

$$\mathcal{O}_{\mathrm{im}} = \{p \in T^* \mid {p_0}^2 - {p_1}^2 - {p_2}^2 - {p_3}^2 = -m^2\,,\ m > 0\}\,.$$

Essentially, the orbits are classified by the value of the invariant ${p_0}^2 - {p_1}^2 - {p_2}^2 - {p_3}^2$ plus some signs that take into account different connected components.

We give a list of the little groups of these orbits in Table 3.1.

Table 3.1 Little groups of the orbits of $\mathrm{SL}_2(\mathbb{C})_\mathbb{R}$ on $T^* = \mathbb{R}^4$.

Orbit	**point**	**Little group**
$\mathcal{O}_m^+$	$(m, 0, 0, 0)$	SU(2)
$\mathcal{O}_m^-$	$(-m, 0, 0, 0)$	SU(2)
$\mathcal{O}_0^+$	$(1, 0, 0, 1)$	$\mathrm{E}_2 = \mathrm{U}(1) \ltimes \mathbb{R}^2$
$\mathcal{O}_0^-$	$(-1, 0, 0, -1)$	$\mathrm{E}_2 = \mathrm{U}(1) \ltimes \mathbb{R}^2$
$\mathcal{O}_0^0$	$(0, 0, 0, 0)$	$\mathrm{SL}_2(\mathbb{C})_\mathbb{R}$
$\mathcal{O}_{\mathrm{im}}$	$(0, m, 0, 0)$	$\mathrm{SL}_2(\mathbb{R})$

The derivation of this table is not difficult and, besides the original work of Wigner [161], one can find a detailed discussion in Ref. [148], Chapter IX.

Let us take, for example, the orbit $\mathcal{O}_m^+$. The little group is SU(2), whose finite dimensional representations are unitary and labelled by a half-integer number $s \geq 0$. If V_s is the representation space, then $\dim(V_s) = 2s+1$. There is then a class of representations of the Poincaré group labelled by the two positive numbers (m, s). In physics, the number m is the *mass* of the particle and s is called the *spin*. For $\mathcal{O}_m^-$ we have an identical classification. Let $(+, m, s)$ denote a representation in the first class and let $(-, m, s)$ be the corresponding representation in the second class. Then, these representations are equivalent through an antiunitary operator[5]. They can be used indistinctly to describe the same particle. The antiunitary equivalence is very often enough to have physical equivalence.

Particles as electrons, positrons, quarks, antiquarks, neutrinos (all with $s = 1/2$), mesons ($s = 0$), $W^\pm$ and Z bosons ($s = 1$) and even the Higgs boson ($s = 0$) fall in this class. Also, in string theory, massive particles of arbitrary spin appear as excitations of the fundamental string.

For the orbits $\mathcal{O}_0^\pm$, we have a different situation. The little group is itself a semidirect product and the method of induced representations could be applied to it too. As a consequence, the representations of the little group could be already infinite dimensional. The representations induced by infinite dimensional representations of the little group do not have an interpretation in physics (at least, not yet). Sometimes in the literature they are referred to as particles with *continuous spin spectrum.*

There is one case in which the representation of the little group is finite dimensional, namely, for the orbit $(0, 0) \in \mathbb{R}^2$. The little group is then the whole U(1), whose irreducible representations are unitary and one-dimensional. They are of the form

$$L^n(\theta) = \mathrm{e}^{in\theta}, \qquad \mathrm{e}^{\mathrm{i}\theta} \in \mathrm{U}(1), \quad n \in \mathbb{Z}.$$

Notice that they correspond simply to trivially represent the translation part of E_2.

For these representations we do have a physical interpretation. They correspond to particles of zero mass. The number n is called the *helicity* of

[5] An antiunitary operator is an operator A satisfies

$$\langle A\psi, A\varphi\rangle = \overline{\langle \psi, \varphi\rangle}.$$

the particle. As before, the orbits $\mathcal{O}_0^+$ and $\mathcal{O}_0^-$ correspond to representations that are antiunitarily equivalent.

Usually, in physics, one considers the direct sum of two such representations, say, L^{+n} and L^{-n}. One then says that the representation has two states of *helicity*. Photons and gravitons belong to this category.

The little group of the orbits $\mathcal{O}_{im}$ is $\mathrm{SL}_2(\mathbb{R})$, which is semisimple but non compact. All its unitary representations, other than the trivial one, are infinite dimensional. If realized in the physical world they would represent particles having an imaginary mass. A fundamental relation of relativistic mechanics says that the kinetic energy of a particle is

$$T = \frac{m}{\sqrt{1-(v/c)^2}}\,.$$

One can see that if the mass m is substituted by the imaginary quantity im, then, in order to have a real kinetic energy, the velocity of the particle should be always greater than the velocity of light.

This is a weird behaviour, since we were convinced that a particle could not reach a velocity greater than the velocity of light. However, while it is not possible to accelerate an ordinary particle to reach such velocities, it is conceivable that there exist particles traveling always at velocities greater than the velocity of light. These hypothetic particles are called *tachyons* and have not been observed. Tachyons were introduced in physics by Feinberg [48]. In spite of being unobserved, they play a role in some theories, like the *bosonic string*, which has a tachyonic mode.

Finally, the representations corresponding to the orbit $\mathcal{O}_0^0$ are just unitary irreducible representations of the little group $\mathrm{SO}(1,3)$, which is the Lorentz group itself. The translation part is represented trivially. These representations are all infinite dimensional except for the trivial one. They are classified in Ref. [8]. The case of the trivial representations is thought to be the *vacuum*, i.e., the state with no particles. It is, in fact, invariant under the Poincaré group. Other representations would correspond to an hypothetical vacuum with infinite degeneracy, since the representation would be infinite dimensional.

In physics, the particle representations are often described in terms of *wave equations*. These are systems of linear, partial differential equations on spacetime whose solutions form a vector space that can be identified with the Hilbert space $\mathfrak{H}$. The wave equations were found by searching for equations that were invariant under the Poincaré group; in this way, the space of solutions would support a representation of it. The simplest of such equations is the *Klein-Gordon equation*, namely

$$(\Box + m^2)\phi = 0\,, \qquad \Box = \frac{\partial^2}{\partial {x^0}^2} - \frac{\partial^2}{\partial {x^1}^2} - \frac{\partial^2}{\partial {x^2}^2} - \frac{\partial^2}{\partial {x^3}^2}\,, \tag{3.16}$$

which describes a particle of mass m and spin $s = 0$. ϕ is a real valued function over spacetime. In general, the functions over spacetime appearing in the wave equations are called *fields* in physics.

For spin $s = 1/2$ we have the *Dirac equation*, which we will introduce after we discuss Clifford algebras, (see Remark 4.3.2). The general case, for arbitrary spin, was written by Bargmann and Wigner [9].

The equivalence of both formulations is shown by performing a Fourier transform from the momentum space T^*, used in the previous description, to the Minkowski space, which is identified with the translations T (see also Ref. [148], Chapter IX, for a thorough account of this equivalence).

Finally, we want to remark that all these representations describe free particles, that is, particles that are not subject to any interaction. Interactions introduce non linearities in the wave equations and it is in fact a matter of active research how to write interacting equations for particles of spin higher than two (see for example Ref. [149] for a review).

3.3 The spin-statistic connection

As we have seen in the previous section, the wave equations as (3.16) are linear, relativistic equations, whose space of solutions carries a unitary representation of the Poincaré group. The functions appearing in the wave equations are maps or 'fields' from the Minkowski space $\mathbf{M}$ to $\mathbb{R}$ or $\mathbb{C}$. The fact that these equations are linear implies, physically, that they describe a free system, not subject to interactions.

Classical Field Theory studies more general equations. For example, let ϕ be a real field; one can consider the non linear equation

$$\Box\phi(x) + m^2\phi(x) + V\big(\phi(x)\big) = 0\,, \tag{3.17}$$

where $V(\phi)$ is an arbitrary function $V : \mathbb{R} \to \mathbb{R}$ called the *potential*. The presence of the potential term breaks the linearity of the Klein-Gordon equation, although it does not break the Poincaré invariance. The Poincaré group still acts on the space of solutions of the equation (3.17), but such space is not anymore a linear space. In physics, one says that a non linear term in the field equation (or wave equation) is an *interaction term*. In this example, the Klein-Gordon field is interacting with itself, but there are many examples of theories where two or more different fields have mutual interaction.

In the free case, the transition to the quantum theory (or *quantization*) is done by first reducing the wave equation to a system of coupled oscillators and then quantizing the oscillators, which is a problem that we know how to solve. The interaction terms can be also introduced in this scheme, but for the general characteristics that we want to describe, it will be enough to consider the free theory.

Let us focus for a moment on a simple linear system, the Klein-Gordon equation (3.16) with $m = 0$ in one spacial and one temporal dimension

$$\frac{\partial^2\phi(x,t)}{\partial x^2} - \frac{1}{c^2}\frac{\partial^2\phi(x,t)}{\partial t^2} = 0 \,. \qquad (3.18)$$

This is the wave equation of a vibrating string. We want to see how quantization works in this toy model. For simplicity, we will denote

$$\dot{\phi} = \frac{\partial\phi}{\partial t}, \qquad \phi' = \frac{\partial\phi}{\partial x}.$$

The lagrangian from which (3.18) stems out is

$$L(\dot{\phi},\phi') = \frac{M}{2}\int_0^l \mathrm{d}x\,\left(\dot{\phi}^2 - c^2\phi'^2\right).$$

Here l is the length of the string and M an arbitrary normalization constant that we can choose so the lagrangian has units of energy. The theory is independent of such constant since the wave equation is found as the points were the functional action

$$S[\phi] = \int_{t_1}^{t_2} \mathrm{d}t\, L\left(\dot{\phi}(x,t), \phi'(x,t)\right)$$

is an extreme[6], $\delta S = 0$.

We take boundary conditions

$$\phi(0,t) = \phi(l,t) = 0\,.$$

Then, performing a Fourier expansion of ϕ we get

$$\phi(x,t) = \sum_{n=1}^{\infty} \phi_n(t)\sin\frac{\omega_n x}{c}, \qquad \omega_n = \frac{nc\pi}{l}.$$

[6] We denote with square brackets the functional dependence, as opposed to the curved brackets of ordinary dependence. δS is the differential of the functional $S[\phi]$ and the variational problem is formulated assuming fixed ends, $\delta\phi(x,t_1) = \delta\phi(x,t_2) = 0$.

The coefficients of the Fourier expansion $\phi_n(t)$ are the *normal modes* and they satisfy the equation of the harmonic oscillator

$$\frac{\mathrm{d}^2\phi_n(t)}{\mathrm{d}t^2} + \omega_n^2\phi_n(t) = 0, \qquad n = 1, 2, \dots \tag{3.19}$$

We have then obtained a system of classical, decoupled oscillators with frequencies ω_n which is equivalent to the wave equation (plus the boundary conditions). The solutions can be expressed as

$$\phi_n(t) = A_n \mathrm{e}^{\mathrm{i}\omega_n t} + \bar{A}_n \mathrm{e}^{-\mathrm{i}\omega_n t}\,. \tag{3.20}$$

The bar stands for complex conjugation, since ϕ is a real field.

One can compute the energy E of the system in terms of the hamiltonian

$$H(\dot{\phi}, \phi') = \frac{M}{2}\int_0^l \mathrm{d}x \left(\dot{\phi}^2 + c^2\phi'^2\right),$$

by substituting the general solution in H. One obtains

$$E = \sum_{n=1}^{\infty} 2M\omega_n^2 A_n \bar{A}_n\,, \tag{3.21}$$

which is a sum of contributions of each normal mode.

One can in fact check that each term in (3.21) is precisely the energy of one harmonic oscillator. The equation (3.19) for a single oscillator is the extreme of the functional action

$$S_n[\phi_n] = \int_{t_1}^{t_2} \mathrm{d}t\, L_n(\phi_n, \dot{\phi}_n), \qquad L_n(\phi_n, \dot{\phi}_n) = \frac{M}{2}\left[\dot{\phi}_n^2 - \omega_n\phi_n^2\right].$$

The canonical conjugate momentum to ϕ_n is

$$p_n = \frac{\partial L_n}{\partial \dot{\phi}_n} = M\dot{\phi}_n\,,$$

and the hamiltonian is the Legendre transform of the lagrangian

$$H(\phi_n, \dot{\phi}_n) = p_n\dot{\phi}_n - L_n = \frac{M}{2}(\dot{\phi}_n^2 + \omega_n^2\phi_n^2)\,.$$

As before, we substitute the general solution (3.20) and the energy is

$$E_n = 2M\omega_n^2 A_n \bar{A}_n\,,$$

which is just the contribution of the n mode to the total energy (3.21).

We now want to quantize these harmonic oscillators, where with the word 'quantize' we mean that we want to construct a quantum system out of a classical one. Essentially, we need to construct a Hilbert space

$\mathcal{H}$ whose vectors correspond to the quantum states of the system and to assign a self-adjoint operator on $\mathcal{H}$ for each observable.

There is neither a standard nor a unique way of quantizing a system. There has been a lot of effort in this direction along the years, going back to Weyl [160], Dirac [40], the *geometric quantization* of Kostant-Soriau [96,136] or the *deformation quantization* framework [12, 47, 95], but the problem is not yet satisfactorily solved with complete generality. Nevertheless, all we need is the quantization of the harmonic oscillator, which is not ambiguous, and then consider an infinite number of them.

We will briefly outline the procedure. The phase space $\mathfrak{P}$ is the space of classical states or the space of solutions of the wave equation. One can describe such space by associating to each solution its initial conditions,

$$\big(q_n(t), p_n(t)\big) \longrightarrow \big(q_n(t_0), p_n(t_0)\big)\,.$$

One denotes the coordinates in phase space simply as

$$\big(q_n(t_0), p_n(t_0)\big) \longrightarrow (q_n, p_n)\,.$$

Then, the solution (3.20) (together with $p_n(t) = M\dot{\phi}_n(t)$) is a curve on phase space.

So $\mathfrak{P}$ is the space of sequences of real numbers

$$(\phi_1, \phi_2, \ldots, \phi_n, \ldots\ , p_1, p_2, , \ldots, p_n, \ldots)$$

that have only a finite number of terms different from zero. This is a linear space of infinite dimension. There is a non degenerate Poisson bracket on $\mathfrak{P}$ defined as

$$\{f, g\} = \sum_{n=1}^{\infty}\left(\frac{\partial f}{\partial \phi_n}\frac{\partial g}{\partial p_n} - \frac{\partial f}{\partial p_n}\frac{\partial g}{\partial \phi_n}\right), \qquad f, g \in C^\infty(\mathfrak{P})\,.$$

The Dirac quantization [40] now goes as follows: we look for an injective, linear map from $C^\infty(\mathfrak{P})$ (often one will be satisfied with quantizing a subspace $C^\infty(\mathfrak{P})$) to the set of self-adjoint operators[7] in a certain Hilbert space $\mathcal{H}$

$$\begin{aligned} C^\infty(\mathfrak{P}) &\longrightarrow \mathcal{O}(\mathcal{H}) \\ f &\longrightarrow \hat{f}, \end{aligned} \tag{3.22}$$

called the *quantization map*, sending the constant function 1 to the identity and such that the basic commutators satisfy

$$\begin{aligned} [\hat{\phi}_n, \hat{p}_m] &= \mathrm{i}\hbar\,\widehat{\{\phi_n, p_m\}} = \mathrm{i}\hbar\,\delta_{nm}\,\mathrm{id}_{\mathcal{H}}, \\ [\hat{\phi}_n, \hat{\phi}_m] &= \mathrm{i}\hbar\,\widehat{\{\phi_n, \phi_m\}} = 0, \\ [\hat{p}_n, \hat{p}_m] &= \mathrm{i}\hbar\,\widehat{\{p_n, p_m\}} = 0\,. \end{aligned} \tag{3.23}$$

[7] In the physics literature self-adjoint operators are called hermitian operators.

The constant $\hbar$ is the reduced Planck's constant ($\hbar = h/2\pi$) and the quantization map depends on it.

The commutation relations (3.23) for one mode are the ones of the *Heisenberg Lie algebra*,

$$\begin{aligned} &[X,Y]=Z, \qquad [X,Z]=0, \qquad [Y,Z]=0, \qquad \text{with} \\ &\mathrm{i}\phi \to Y, \qquad \mathrm{i}p \to X, \qquad \mathrm{i}\hbar\,\mathrm{id}_{\mathcal{H}} \to Z\,. \end{aligned} \tag{3.24}$$

In this way X, Y, Z are anti self-adjoint if ϕ and p are self-adjoint. It is known that there are no finite dimensional irreducible representations of the Heisenberg Lie algebra in dimension greater than one. One has then to consider infinite dimensional representations, where the unitarity is not guaranteed.

Remark 3.3.1. The Dirac quantization *does not* require the property (3.23) relating commutators and Poisson brackets to hold for arbitrary observables: in fact, such requirement would be inconsistent, as it is proven, for example, in Theorem 5.4.9 of Ref. [1]. In real systems, the bracket $[\hat{f}, \hat{g}]$ is still related with the Poisson bracket $\{\hat{f}, \hat{g}\}$, but for general observables it has terms that depend on higher powers of $\hbar$. This property is better studied in the context of deformation quantization. ■

There are some extra conditions on the Hilbert space and the quantization map that restrict the possible choices (see Section 5.4 of Ref. [1] for a careful discussion of the problem). If we had a finite, fixed number of oscillators, then the phase space would be $\mathfrak{P} = \mathbb{R}^{2n}$ and the Hilbert space would be the set of square integrable functions $\mathcal{L}^2(\mathbb{R}^n)$, where $\mathbb{R}^n$ is the *configuration space*, with coordinates $(\phi_1, \ldots, \phi_n)$. Then one would have

$$\hat{\phi}_m \Psi = \phi_m \Psi, \qquad \hat{p}_m \Psi = -\mathrm{i}\hbar \frac{\partial}{\partial \phi_m} \Psi\,, \tag{3.25}$$

with $\Psi \in \mathcal{L}^2(\mathbb{R}^n)$ but sufficiently smooth. This certainly satisfies (3.23). This realization of the operators is called the *Schrödinger picture* and it is essentially the only irreducible, unitary representation of the Heisenberg Lie algebra in which the central element Z is represented by a non zero operator (up to equivalence). This is the content of the Stone-Von Neumann theorem [139, 152].

Even with (3.25), the quantization map is not determined. If f is a polynomial function on phase space, say $f = q_n^2 p_n$, we have many possible choices for $\hat{f}$,

$$\hat{q}_n^2 \hat{p}_n, \qquad \hat{q}_n \hat{p}_n \hat{q}_n, \qquad \hat{p}_n \hat{q}_n^2\,,$$

and also any convex linear combination of these. This is the well known *ordering problem* of quantum mechanics. We will be satisfied by finding one such ordering that gives a consistent quantum theory.

We need a point of view here different from the Schrödinger picture, though. We follow the method of Dirac [40] for the quantization of the harmonic oscillator. Let us perform a change of variables in phase space (through a complexification)

$$(\phi_n, p_n) \longrightarrow (A_n, \bar{A}_n),$$

for which we choose $t_0 = 0$ (although it is not necessary). Then the Poisson bracket is

$$\{A_n, \bar{A}_m\} = \frac{\mathrm{i}}{2M\omega_n}\delta_{mn}$$

and the condition (3.23) becomes

$$[\hat{A}_n, \hat{A}^\dagger_m] = -\frac{\hbar}{2m\omega_n}\delta_{nm}\,\mathrm{id}_{\mathcal{H}}\,.$$

As usual, '$\dagger$' means hermitian conjugate. A simple rescaling

$$a_n = \Big(\frac{2M\omega_n}{\hbar}\Big)^{1/2}A^\dagger_n, \qquad a^\dagger_n = \Big(\frac{2M\omega_n}{\hbar}\Big)^{-1/2}A_n\,,$$

brings the commutation relations to

$$\begin{aligned}&[a_n, a^\dagger_m] = \delta_{nm}\,\mathrm{id}_{\mathcal{H}},\\ &[a_n, a_m] = 0, \qquad [a^\dagger_n, a^\dagger_m] = 0\,.\end{aligned} \tag{3.26}$$

We have to be aware that a_n and $a^\dagger_n$ *are not* self-adjoint operators: they are complex linear combinations of $\hat{\phi}_n$ and $\hat{p}_n$ (we assume the latter to be self-adjoint). They are one the adjoint of the other. Therefore, they are not observables of the quantum theory. However, we will be able to construct observables with them. The operators a and $a^\dagger$ are called *ladder operators* and were introduced by Dirac [40]. For a single mode one can represent them in an infinite dimensional space $\mathcal{H}$ with basis $\psi_0, \psi_1, \ldots$ in the following way

$$a\psi_k = \sqrt{k}\,\psi_{k-1}, \qquad a^\dagger\psi_k = \sqrt{k+1}\,\psi_{k+1}\,.$$

It is easy to check that the commutation relations (3.26) are satisfied. Also, all vectors can be generated by $a^\dagger$ acting repeatedly on ψ_0,

$$\psi_k = \frac{1}{\sqrt{k!}}{a^\dagger}^k\psi_0\,.$$

One can define an inner product in this space

$$\langle \psi_k, \psi_l \rangle = \delta_{kl}$$

which is compatible with the commutation relations together with the condition that $a^\dagger$ is the adjoint of a.

For an arbitrary number of modes, the basis vectors are labelled by a multiindex $(k_1, k_2, \dots)$ and the action of the ladder operators reads as

$$a_n \psi_{k_1,\dots,k_n,\dots} = \sqrt{k_n}\, \psi_{k_1,\dots,k_n-1,\dots}, \tag{3.27}$$

$$a_n^\dagger \psi_{k_1,\dots,k_n,\dots} = \sqrt{k_n+1}\, \psi_{k_1,\dots,k_n+1,\dots}\,. \tag{3.28}$$

As before, we can generate all vectors of the basis by successive application of the operators $a_i^\dagger$:

$$\psi_{k_1,\dots,k_n,\dots} = (a_1^\dagger)^{k_1} \cdots (a_n^\dagger)^{k_n} \cdots \psi_0\,.$$

Due to commutation relations $[a_n^\dagger, a_m^\dagger] = \delta_{nm}$, we can interchange the order of the creation operators of different modes and the state is unchanged, so indeed the succession $(k_1, \dots, k_n, \dots)$, which has only a finite number of terms different from zero, identifies univocally the vector.

As before, the inner product is defined as

$$\langle \psi_{k_1,k_2,\dots,k_n,\dots}, \psi_{l_1,l_2,\dots,l_n,\dots} \rangle = \delta_{k_1 l_1} \cdots \delta_{k_n l_n} \cdots\,.$$

The completion, with the above norm, of this space is called the *Fock space* [68], and it is the Hilbert space of the quantum theory of our toy model. We will denote it as $\mathcal{F}^{\mathrm{s}}$; the meaning of the superindex will be clear in a moment.

To have a physical interpretation of the ladder operators, it is instructive to compute the Hamiltonian operator. We have to go back to (3.21) and we first encounter the ordering problem. One chooses the *normal ordering*, which is defined as the ordering having all of the operators a_n on the right of the operators $a_n^\dagger$. We will see that this choice gives a well defined quantum theory. The hamiltonian operator is then

$$\hat{H} := \sum_{n=1}^{\infty} \hbar\omega_n a_n^\dagger a_n\,,$$

which is self-adjoint (if ϕ_n and p_n are so). The Hamiltonian is diagonal in the basis (3.28), and

$$\hat{H}\psi_{k_1,k_2,\dots} = \left(\sum_{n=1}^{\infty} \hbar k_n \omega_n \right) \psi_{k_1,k_2,\dots}\,.$$

The eigenvalues of $\hat{H}$, which are the only values allowed for the physical energy, can be read as the sum of k_1 'quanta' of value $\hbar\omega_1$ plus k_2 'quanta' of value $\hbar\omega_2$ and successively so. We will denote the eigenvalues of the hamiltonian as

$$E_{k_1,k_2,\dots} = \sum_{n=1}^{\infty} k_n \hbar\omega_n \,.$$

So the action of a_n in one state of the basis decreases the energy of the state by $\hbar\omega_n$ and the action of $a_n^\dagger$ increases the energy by $\hbar\omega_n$. We interpret this by saying that the state $\psi_{k_1,k_2,\dots}$ is a state with k_n 'quanta' of energy $\hbar\omega_n$ and that a_n, $a_n^\dagger$ are operators that annihilate or create, respectively, one such quanta. In fact, computing the commutators

$$[H, a_n] = -\hbar\omega_n a_n, \qquad [H, a_n^\dagger] = \hbar\omega_n a_n^\dagger \,, \tag{3.29}$$

we have

$$\begin{aligned} H\, a_n\psi_{k_1,\dots,k_n,\dots} &= (E_{k_1,\dots,k_n,\dots} - \hbar\omega_n)\, a_n\psi_{k_1,\dots,k_n,\dots}, \\ H\, a_n^\dagger\psi_{k_1,\dots,k_n,\dots} &= (E_{k_1,\dots,k_n,\dots} + \hbar\omega_n)\, a_n^\dagger\psi_{k_1,\dots,k_n,\dots} \,. \end{aligned}$$

These quanta are the 'particles' of QFT. One can assign to them the mass and spin represented by the original wave equation. We will not treat here the issue of the Poincaré invariance of the quantum theory: it is a result that can be read in any elementary book on QFT and, doubtlessly, it is a fundamental result for physics. There other issues that we leave untouched here, as considering more general boundary conditions, the positivity of the hamiltonian and the existence of antiparticles. We refer, for example, to the excellent book by Weinberg [156].

Yet, we may reinterpret the Fock space in a different way. Let us consider the subspace of $\mathcal{F}^{\mathrm{s}}$ generated by the states that only have one quantum of energy:

$$\mathcal{H} := \mathrm{span}\{\varphi_l = a_l^\dagger\psi_0, \quad l = 1, 2, 3, \dots\}, \qquad H\,\varphi_l = \hbar\omega_l\,\varphi_l \,.$$

$\mathcal{H}$, with the restricted inner product and completed with respect to it, is a Hilbert space. This subset of states represents all states of definite energy in which a single particle can be. In more complicated theories, the quantum state of a particle may need more *quantum numbers* (that is, eigenvalues of other operators), to specify the state of the system, but the system that we consider here is very simple. We are going to reconstruct the Fock space from $\mathcal{H}$. We consider first the symmetrized tensor products

$$\mathcal{H}^{\odot n} = \mathrm{Sym}(\mathcal{H}^{\otimes n})$$

for arbitrary n. Then we can inject $H^{\odot n}$ in $\mathcal{F}^{\text{s}}$ by sending

$$\varphi_{i_1} \otimes \varphi_{i_2} \otimes \dots \varphi_{i_n} \longrightarrow a_{i_1}^{\dagger} a_{i_2}^{\dagger} \cdots a_{i_n}^{\dagger} \psi_0 \,.$$

It is now not difficult to see the isomorphism

$$\mathcal{F}^{\text{s}} \cong \mathbb{C} \oplus \mathcal{H} \oplus \text{Sym}(\mathcal{H} \otimes \mathcal{H}) \oplus \cdots \oplus \text{Sym}(\mathcal{H}^{\otimes n}) \oplus \cdots = \bigoplus_{n=0}^{\infty} \text{Sym}(\mathcal{H}^{\otimes n}) \,.$$

A couple of observations. The term $\mathbb{C}$ is generated by the vector ψ_0, the state with no particles that physicists call the vacuum. It is the state with lowest energy. Also, notice that taking the symmetrization of the tensor products $\text{Sym}(\mathcal{H}^{\otimes n})$ is essential to establish the isomorphism.

The symmetrization has a deeper implication. We can interpret it in the following way: particles of the same 'species', that is, originated in the same field, are identical, and therefore indistinguishable for a quantum observer. In classical physics, two particles may be identical, but one has no theoretical impediment to distinguish them: one is in principle allowed to 'follow' them in their trajectories and keep them separated. In quantum physics, the indistinguishability of particles has a profound meaning. All possible information must be deduced from the vector in the Hilbert space representing the physical state, so the state itself should be the same if we interchange the states of two particles. For example, among all linear combinations of

$$\varphi_1 \otimes \varphi_2, \qquad \varphi_2 \otimes \varphi_1 \,,$$

only the symmetric combination

$$\frac{1}{\sqrt{2}}(\varphi_1 \otimes \varphi_2 + \varphi_2 \otimes \varphi_1)$$

has physical meaning. It is now clear that the superindex in $\mathcal{F}^{\text{s}}$ stands for 'symmetric'.

There is a question, though, that can arise at this point. We know systems that behave very differently from the one that we have described here. The most obvious example is the model of the atom, and for this matter, even Bohr's simple model (see Section 3.1) serves to explain the point. Bohr's model for the atom is a non relativistic model. The electron evolves under the influence of the electrostatic interaction with the nucleus. As a result of the quantization, only the energies that are eigenvalues of the Hamiltonian operator are allowed for the electron. In fact, with successive refinements of the model, one finds that not only the Hamiltonian, but other observables (as the angular momentum and the spin) define the state of the system. In any case, one can extrapolate, at least qualitatively, the model to multielectron atoms. Then we end up in a situation similar to the one with the harmonic oscillators. Electrons are identical particles and should be indistinguishable from a quantum point of view. Differently from the oscillator model, it happens that no two electrons in a atom are found in the same quantum state. Electrons 'fill' the energy levels starting from the lowest one and increasing it progressively. This is known as Pauli's exclusion principle, formulated in 1925. This model explains the chemical properties of the different atoms, so we have to admit the existence of different types of particles whose statistical behaviors are very different.

Particles as the photons, that can be in arbitrary numbers in the same quantum state, are called *bosons*, and their statistical behavior is governed by the Bose-Einstein probability distribution. On the other hand we have also particles as electrons, that cannot be in the same quantum state. These are called *fermions* and their statistical behavior is governed by the Fermi-Dirac probability distribution. We will not obtain here these probability distributions: it is enough to realize that they are arise from radically different postulates.

The question is now whether the toy model that we have described previously can be modified in such way that it would describe also fermions. We will encounter later the relativistic wave equation for the electron, the Dirac equation, in Remark 4.3.2 after we study Clifford algebras, but there is a crucial observation that we can make here. The interpretation of the normal modes as quanta or particles was based on the commutation relations (3.26) and (3.29), that lead us to the construction of the Fock space $\mathcal{F}^{\mathrm{s}}$ with states that are symmetric under the exchange of identical particles.

We would like to obtain a similar construction, but with a Fock space that is antisymemtric under the exchange of identical particles, so for a state with n identical particles we will consider

$$\wedge^n \mathcal{H} = \mathrm{Ant}(\mathcal{H} \otimes \mathcal{H} \otimes \cdots \otimes \mathcal{H}) = \mathrm{Ant}(\mathcal{H}^{\otimes n})$$

where $\mathrm{Ant}(\mathcal{H} \otimes \mathcal{H} \otimes \cdots \otimes \mathcal{H})$ denotes the antisymmetric tensors. This would assure that a state of two identical particles in the same quantum state (with the same energy in our case) is identically zero.

There is a modification of the commutation relations (3.26) that can do the job. Let us postulate operators b and $b^\dagger$ with *anticommutation relations*[8]

$$[b_n, b_m^\dagger]_+ = b_n b_m^\dagger + b_m^\dagger b_n = \delta_{nm}\mathrm{id},$$
$$[b_n, b_n]_+ = 0, \qquad [b_n^\dagger, b_n^\dagger]_+ = 0\,. \tag{3.30}$$

Notice that the last two relations imply that

$$b_n b_n = 0, \qquad b_n^\dagger b_n^\dagger = 0\,.$$

The anticommutation relations (3.30) define a Lie superalgebra with only odd dimensions. It is the fermionic counterpart of the Heisenberg Lie algebra (3.24) or, equivalently, (3.26). For a single mode, one can represent the algebra

$$[b, b^\dagger]_+ = \mathrm{id}$$

in a two dimensional space with ordered basis $\{\eta_0, \eta_1\}$

$$b^\dagger = \begin{pmatrix} 0 & 0 \\ 1 & 0 \end{pmatrix}, \qquad b = \begin{pmatrix} 0 & 1 \\ 0 & 0 \end{pmatrix}.$$

They are one the adjoint of the other. Contrary to the bosonic case, the roles of b and $b^\dagger$ can be interchanged, since the bracket is symmetric. If η_0 represents the vacuum, then $b^\dagger$ is the creation operator of one particle and b the annihilation operator. Notice that we cannot 'create' more than one particle (in the same mode), since

$$b^\dagger b^\dagger\,\eta_0 = b^\dagger\,\eta_1 = 0\,.$$

We can now construct the antisymmetric Fock space $\mathcal{F}^{\mathrm{a}}$ adding infinite modes. Let $\mathcal{H}^{\mathrm{f}}$ denote the Hilbert space of one particle — fermion — states, then

$$\mathcal{F}^{\mathrm{a}} = \mathbb{C} \oplus \mathcal{H}^{\mathrm{f}} \oplus \wedge^2\mathcal{H}^{\mathrm{f}} \oplus \cdots \oplus \wedge^n\mathcal{H}^{\mathrm{f}} \oplus \cdots = \bigoplus_{n=0} \wedge^n\mathcal{H}^{\mathrm{f}}\,.$$

Finally, let us suppose that the hamiltonian operator has the form

$$\hat{H}^{\mathrm{f}} = \sum_{n=1}^{\infty} \hbar\omega_n b_n^\dagger b_n\,,$$

then

$$[\hat{H}^{\mathrm{f}}, b_n]_- = -\hbar\omega_n b_n, \qquad [\hat{H}^{\mathrm{f}}, b_n^\dagger]_- = \hbar\omega_n b_n^\dagger\,.$$

Notice that here we are computing *commutation relations* with the hamiltonian. With this we have completed the interpretation of $b_n^\dagger$, b_n as creation-annihilation operators.

[8]Here it is crucial to distinguish commutators from anticommutators, so we will denote them with a subindex $\pm$.

In the bosonic case, the classical limit of the algebra of operators on the Hilbert space is a set of functions on phase space, therefore, a commutative algebra, endowed with a Poisson bracket. In the fermionic case the anticommutation relations (3.30) define indeed a Clifford algebra (see Definition 4.2.1), which is a non commutative superalgebra, as we will see in Section 4.2. Then, the classical limit gives necessarily a commutative superalgebra endowed with a super Poisson bracket.

We have then understood the concepts of spin and particles in QFT and the difference between bosons and fermions. We are ready now to state the spin-statistics theorem.

Spin-Statistics Theorem.

- *The quantum state of a system of identical integer spin (or helicity) particles is symmetric under the exchange of two such particles. So particles with integer spin (or helicity) follow the Bose-Einstein statistics.*
- *The quantum state of a system of identical half-integer spin (or helicity) particles is antisymmetric under the exchange of two such particles. So particles with half-integer spin (or helicity) follow the Fermi-Dirac statistics.*

■

We do not intend to prove the spin-statistics theorem. One has to go to the full QFT to prove it, but the relation between spin and statistics was known much before physicists had the mathematical tools to prove it. In fact, physicists are so used to it that very often they speak about 'bosons' and 'fermions' only referring to the spin of the particles.

We will take it for granted and as a motivation for the study of supersymmetry. As we will see in the next sections, it highly constrains the superalgebras that can be symmetries of a QFT.

3.4 The role of supersymmetry in particle physics

We have seen that there are two types of particles, bosons and fermions, that behave very differently. The mere existence of electrons, or other fermions, implies that superalgebras have to be used in the description of ordinary quantum field theories. This strongly suggests the possibility of a more fundamental formulation, were supergeometry plays a crucial role.

Whenever we have a space with even and odd coordinates, the possibility of performing coordinate transformations mixing both types of coordinates is perfectly allowed. This may result in complicated versions of the same theory, but there is nothing wrong with it.

As we mentioned in the introduction, we can compare this situation to the one when a non inertial frame is used: this can bring along complicated descriptions including fictitious 'forces' and other similar devices, however, there is nothing wrong with such descriptions. In fact, sometimes the complications may be unavoidable, as for example in the case when we have the observer attached to the Earth and the Earth's rotation cannot be neglected.

Taking the analogy further on, one may ask what would happen if some of those supergeometric transformations had a fundamental role in a physical theory, i.e., if we assumed that a transformation that mixes the bosonic and fermionic coordinates is a symmetry of a physical theory. All symmetries of the classical (non super) theories have the bosonic and fermionic subspaces as invariant subspaces. One could conceive a symmetry transformation that would mix those subspaces. Certainly, the consequences of such assumption would not be minor.

Let us think for the moment on the infinitesimal symmetries of a physical theory. They form a Lie algebra. They do not mix the bosonic and fermionic subspaces. If we assign even parity to the bosonic subspace and odd parity to the fermionic one, the infintesimal transformations do not change the parity, or, in other words, they are even transformations. Clearly, an hypothetic symmetry generator linking the bosonic and the fermionic subspaces must be an odd generator. In the language of Lie algebras, this translates into the fact that odd generators satisfy anticommutation relations. We arrive then to the picture that the algebra of symmetries is, indeed, a Lie superalgebra.

In the spirit of the spin-statistics connection, the bosonic subspace is the space of integer spin particles, so the fields are in an ordinary (non spinorial) representation of the Lorentz group (or algebra). The fermionic subspace, consisting on half-integer spin particles, are represented by fields that are in a spinorial representation of the Lorentz group. The Lorentz algebra will always be inside the even part of the symmetry superalgebra. Then, from elementary representation theory, it follows that the odd generators of symmetry must be in a spinorial representation of the Lorentz algebra, while the even generators must be in a non spinorial representation. This is a physical, not a mathematical requirement, and it significantly restricts the superalgebras that we can consider.

We will then be interested in superalgebras that extend the known symmetry algebras and whose odd part transforms as a spinorial representation of the Lorentz algebra (see Section 4.1). Supersymmetric field theories are theories where the Poincaré or conformal algebras are extended to suitable superalgebras.

Natural as this scheme may seem, supersymmetry appeared in physics in a different way. Nowadays it is strongly linked to string theory, a theory of fundamental interactions that needs supersymmetry to be consistent. We want to devote the rest of this section to give an overview of historical developments in particle physics to situate supersymmetry in the context of the models of fundamental interactions.

The model of *electroweak interactions*, for which A. Salam, S. Glashow and S. Weinberg received the Nobel Prize in physics in 1972 and G. t'Hooft and M. Veltman in 1999, has had an amazing success in unifying electromagnetism and the *weak nuclear force*, while simultaneously incorporating the principles of the quantum theory. The weak force is responsible for the radioactive decay of nucleus and the nuclear fusion (the reaction that provides fuel for the Sun). Other than the photon γ, which is the massless, helicity one particle carrying the electromagnetic interaction, the model predicted other particles to carry the electroweak interaction, the *intermediate vector bosons* $W^{\pm}$ *and* Z. They are particles of spin 1 and 'intermediate' mass (about 80-90 GeV/c^2). They were discovered at CERN in the early 80s and C. Rubbia and S. van der Meer were awarded the Nobel Prize in physics in 1984 for their discovery.

The fact that, unlike the photon, the $W^{\pm}$ and Z bosons have non zero mass was a non trivial matter. The electroweak model is a *gauge theory*, which means that it possesses a very large group of symmetries, one in which the parameters are functions on spacetime. The easiest example of a gauge theory is electromagnetism itself, with gauge group the (abelian) group of maps

$$\mathbf{M} \xrightarrow{\ \Lambda\ } \mathrm{U}(1)\cdot$$

This is the well known invariance of the Maxwell equations under the following transformation of the electromagnetic potential

$$A_\mu(x) \longrightarrow A_\mu(x) + \partial_\mu \Lambda(x)\,.$$

The fact that the parameter of the transformation depends on x is expressed in physicists's language by saying that the gauge symmetry is a local symmetry, somehow as when in Section 2.2 we spoke about dilation versus conformal symmetry.

In the electroweak theory, the abelian group $U(1)$ is replaced by the non abelian group $\mathrm{SU}(2) \times \mathrm{U}(1)$. There are as many particles carrying the interaction as the dimension of this group, so we have four of them, $W^{\pm}$, Z and the photon γ. $W^{\pm}$ have electric charge ($\pm$ the charge of the electron): this is possible because the group is non abelian. Z and γ are electrically neutral.

In this picture, though, the masses of the carriers are zero, just as for the photon, which does not correspond with what is seen experimentally. In the 60s, the mechanism (later to be known as the *Higgs mechanism*) that would explain the masses of all particles, and in particular, the one of the intermediate bosons, was proposed. The idea is that the gauge symmetry is *spontaneously broken*. This means that although the full theory retains the symmetry, the state of lowest energy (the vacuum) is not invariant under it. This generates shifts in all fields which produce mass terms in the lagrangian. In this mechanism, an extra particle is predicted, the Higgs boson, that would then play a fundamental and singular role in the generation of masses. It eluded experimentalists for more than 40 years, until it was finally caught by the Large Hadron Collider at CERN in 2012. P. Higgs and F. Englert were awarded the Nobel Prize in physics in 2013, while R. Brout, collaborator of Englert, could not receive it because he died in 2011.

In spite of this history of success, there are many open issues in the physics at small scale, mainly the incorporation to this scheme of the other nuclear force, the *strong nuclear force*. This force is responsible for holding protons and neutrons together in the nucleus, in spite of the positive electric charges of the protons. It is described by Quantum Chromodynamics (QCD), also a gauge theory, in this case with the group SU(3). The strong force is carried by the 8 ($\dim\big(\mathrm{SU}(3)\big) = 8$) *gluons*, which are massless particles of helicity one, since QCD does not show spontaneous breaking of the gauge symmetry. In QCD, the protons and neutrons (and other hadrons) have more fundamental constituents, the *quarks*, that interact strongly inside the nucleus.

QCD has very peculiar properties. The first one is *confinement*, meaning that the force between quarks increases as quarks separate and, as a consequence, quarks cannot be found free but only bound together inside the hadrons. The other property is the *asymptotic freedom*, which is somehow the opposite effect: at very high energies quarks and gluons interact very weakly. This property of the strong force has been amply confirmed by experiments. D. Politzer, F. Wilczek and D. Gross found asymptotic freedom back in the 70s, and they were awarded the Nobel Prize in physics in 2004. A consequence of the asymptotic freedom of QCD is that at very high energy, quarks and gluons should form a kind of 'soup' which is indeed a new state of matter called the *quark-gluon plasma.* In existing particle accelerators, LHC at CERN and RHIC[9] heavy ions are smashed together in high energy collisions and then the quark-gluon plasma is created.

The combination of QCD with the electroweak theory is known as the Standard Model of Particle Physics. The group of the standard model is $SU(3) \times SU(2) \times U(1)$, although one cannot consider this a unified theory in the sense that the weak and electromagnetic force were unified in the $SU(2) \times U(1)$ theory. Very roughly, the reason is as follows. Each of the simple factors in $SU(3) \times SU(2) \times U(1)$ has a different *coupling constant* (a dimensionless constant that determines the strength of the interaction). While electromagnetism and the weak force 'share' the coupling constants of $SU(2)$ and $U(1)$ (electromagnetism does not correspond to the factor $U(1)$, but to a combination between this factor and a $U(1)$ inside $SU(2)$), the strong force has associated the coupling constant of the factor $SU(3)$, and it does not mix with the others.

A truly unified theory would have only one coupling constant, and the difference of strength in the observed, effective theories, would correspond to be in different phases of the *Grand Unified Theory* (GUT). There are many proposed GUTs, although none is universally accepted. Perhaps the simplest (and the first) proposed GUT [73] is a gauge theory with a $SU(5)$ gauge symmetry, which is the smallest simple group containing $SU(3) \times SU(2) \times U(1)$. Notice that $SU(5)$ has dimension 24, so one would have to account for 12 extra force carriers. But then, the $SU(5)$ gauge symmetry should be (partially) spontaneously broken. Some of these particles would have a very large mass and experiments that could show them are beyond the reach of our particle accelerators. This and other GUT's present problems: for example, they predict proton's decay. In the super-Kamiokande experiment (Japan), the world largest facility that investigates proton decay, not a single event has been registered and the half-life of the proton has been estimated to be larger than 8.2×10^{33} years.

There is a fourth fundamental force whose properties and mathematical description are radically different from the one proposed by the Standard Model. It is the gravitational interaction. The gravitational force is totally

[9]Relativistic Heavy Ion Collider at Brookhaven National Laboratory (USA).

negligible in the subatomic domain. One can compare it with the electric force, for example, with the following elementary calculation. Let us consider the hydrogen atom, composed by a proton and an electron whose masses and electric charges are (we give only approximate values)

$$m_e \simeq 9.1 \times 10^{-31}\text{kg}\,, \qquad q_e \simeq -1.6 \times 10^{-19}\text{C}\,,$$
$$m_p \simeq 1.7 \times 10^{-27}\text{kg}\,, \qquad q_p \simeq 1.6 \times 10^{-19}\text{C}\,, \qquad r \simeq 5.3 \times 10^{-11}\text{m}\,.$$

r is the *Bohr radius*, the most probable distance between electron and proton in the hydrogen atom. Using Coulomb's and Newton's law for the electrostatic and gravitational force between the proton and the electron, we can compute the magnitude of these forces,

$$F_C = \left|k\frac{q_e q_p}{r^2}\right| \simeq 8.2 \times 10^{-8}\text{N}\,, \qquad F_N = \left|G\frac{m_e m_p}{r^2}\right| \simeq 3.7 \times 10^{-47}\text{N}\,.$$

$k \simeq 9 \times 10^9\text{Nm}^2/\text{C}^2$ is Coulomb's constant and $G \simeq 6.7 \times 10^{-11}\text{Nm}^2/\text{kg}^2$ is Newton's constant. We see that the gravitational force is 39 orders of magnitude smaller than the electrostatic force, so it has no relevance to the description of the hydrogen atom. This is also true in the subnuclear domain, the weak force being even weaker than the electromagnetic force and the strong force stronger, but still both much stronger than gravitation. This statement could be justified by comparing their coupling constants, but we do not want to enter in such detailed calculations. We can safely assume that the effects of the gravitational force are negligible in particle physics.

Gravity shares with electromagnetism the property of being long range, capable of reaching astronomic distances. This is in contrast with the nuclear forces, that are very short ranged: they do not show significantly beyond the range of the nucleus. However, there are electric charges of both signs, which results in cancelation of the total charge of most macroscopic bodies, while there are no particles of negative mass and the gravitational force is always accumulative. The behaviour of matter at large (macroscopic or astronomical) distances is mainly governed by the gravitational interaction which, as we have already seen, has no visible effects in experiments with elementary particles.

There is an even more meaningful difference of gravity with the other interactions. Its mathematical description, Einstein's theory of General Relativity (GR), invoques only the geometry of spacetime, which is considered flat and invariable in the Standard Model. By virtue of the equivalence principle, matter is the source of the gravitational field, so the presence of matter determines the metric of spacetime. This is what physicists mean when they say that matter 'curves' spacetime. GR has resisted all attempts to be reconciled with the quantum theory. When one treats GR as a quantum field theory, inconsistencies appear everywhere in the form of observables attaining unphysical infinite values. It is this incongruence between the theory that governes physics at large scales and the quantum theory, relevant in the very small, that urges theorists to search for different frameworks where both conceptions, radically different, could be seen as limits of a more fundamental theory. It is perhaps needed a revolutionary change of paradigm.

String theory has been inspiring in the search for this new framework. It was originally proposed in the 60s as a theory of the strong nuclear force. One may think on this force as a rubber band (or string) that stretches between two quarks, a picture that would describe some characteristics of the strong force. However, the model had several unwanted features, one of them being the presence of a massless particle of helicity two in the spectrum. Theorists did not know what to do with it: helicity one, massless particles were all right (gluons are like that), but there was no use for a helicity two particle in this context. Then, a better theory, QCD, came along and the string theory of the strong interaction was forgotten.

Then, in the 70s, it was realized that the helicity two particle could be a *graviton*. If the attempts to quantize GR had been successful, then the graviton, that is, the particle carrying the gravitational force, would be a massless, helicity two particle. This opened the exciting possibility of using string theory as a theory of quantum gravity, a hope that string theorists still hold nowadays.

A few more things about string theory:

- Different elementary particles are seen as strings in different vibrational modes. This implies that strings are at the basis of every fundamental interaction and string theory must be a theory that unifies all interactions.
- Strings would be incredibly small objects, since they have not been directly observed. At the energies involved in our current experiments, we are unable to resolve their size, and strings look just like particles. String theory, then, becomes an 'effective' quantum field theory that should reproduce the Standard Model of particle physics. In fact, we cannot expect to resolve the size of the strings unless we are at energies where gravity would show signs of unification with the other forces.

This presumably will occur at energies of the order of the Planck mass,

$$m_P = \sqrt{\frac{\hbar c}{G}} \cong 1.2209 \times 10^{19} \mathrm{GeV}/c^2 \,,$$

which is well beyond the energies that we may eventually reach in any conceivable terrestrial laboratory. Notice that the LHC at CERN is expected to operate 'only' at at 10^4 GeV.

- To include fermions, the standard (bosonic) string is not enough and a supersymmetric generalization is called for: this is superstring theory. Since the Standard Model is not supersymmetric, then one has to assume that supersymmetry is spontaneously broken, just as the gauge symmetry was in the electroweak model. Supersymmetry also cures another important drawback of the bosonic string: the appearence in the spectrum of a tachyon (a particle of imaginary mass, see Section 3.2) that indicates an instability of the string. The tachyon disappears in the supersymmetric version.
- For internal consistency of string theory, one has to assume that spacetime has a dimension larger than four: it is 26 for the bosonic string and 10 for the superstring. Weird as it may sound, this presents no contradiction as long as one assumes that the extra dimensions are curled into a compact submanifold of size much smaller that what we can observe with our means. This is not a conceptual problem but an important practical one: the effective, four dimensional quantum field theory that we can expect to test in our labs at low energies strongly depends on the properties of the compact manifold. This results in an huge set of possibilities that makes incredibly difficult to come up with a realistic scenario that displays exactly the Standard Model, or, more precisely, the Minimal Supersymmetric Standard Model (MSSM), a supersymmetric generalization of the Standard Model.
- Contrary to the varied landscape of string theory at low energies, at high energies there is a kind of uniqueness: in the 90s, the different string models were shown to be related by certain maps called *dualities* and it was conjectured that they were all different limits of a unique theory, the so called *M-theory*, living in a spacetime of dimension 11 [162].

In recent times, string theory, powered by the gauge/gravity correspondence [107] (see the comments at the end of Section 2.2), has found applications in other fields such as condensed matter and then has expanded its area of influence. The correspondence establishes a map among gauge theories and gravity, so gravity methods could be now used to treat gauge theories such as, say, QCD. In fact, since string theory is a quantum theory of gravity, it is a nice turn of events that now strings can point to its original target: a quantum theory of the strong interaction. After all, it seems that the coincidence found between the transition amplitudes of the strongly interacting particles and the ones of strings were not casual. We needed only to look at them from the correct point of view.

3.5 References

Our main reference for unitary representations of the Poincaré group is the book by Varadarajan [148]. For the brief (incomplete) introduction to QFT see the book by Weinberg [156]. The spin-statistics theorem was first formulated in 1939 by M. Fierz, and later reworked out by W. Pauli. Later on, both J. Schwinger and R. Feyman have given their contribution and we refer the reader to [51] or [156] for a complete treatment.

Chapter 4

The Minkowski and Conformal Superspaces

In this chapter we want to describe a generalization of spacetime that appears in supersymmetric field theories. It is the *superspace*[1] or *super spacetime* in two versions: the Minkowski and the conformal superspaces.

This is the most beautiful presentation of supersymmetry. Spacetime is enlarged with *odd* or *fermionic* coordinates, which somehow encode the properties of fermions. The original coordinates of spacetime are then called *even* or *bosonic coordinates*. There are many versions of superspace, mainly depending on the number of these extra, odd variables. When this number is minimal, physicists call it $N = 1$ Minkowski or conformal superspace, and they denote with $N > 1$ superspaces with a larger number of odd coordinates. We just remark that the minimal number of odd coordinates happens to be four.

Although one can construct superspaces for arbitrary N, it is a bit disappointing that for theories with $N > 2$ no superspace formulation is available. Usually the problem is that when enlarging superspace the number of fields starts increasing and one cannot find physical meaning for all of them. Instead, other formulations result in more economical theories. Generally one tries to impose some consistent constraints on the fields, but finding them can be a very intrincated problem, or even one without solution.

We will not address this problem here. In spite of the difficulties, superspace appears to us as a natural object. We will deal with $N = 1$ superspace, the smallest version of it, which has a wide use in physics, even in phenomenological theories like the Minimal Supersymmetric Standard

[1] In Chapter 1 we gave a very general definition of superspace. In physics, though, the word is used for some concrete generalizations of the Minkowski and conformal spacetimes.

Model.

The enlargement of spacetime corresponds to an enlargement of the transformations groups: superspace is an homogeneous space for the Poincaré or the conformal supergroup. The first task that we will take up will be the description of such supergroups. In Sections 4.4 and 4.6 their infinitesimal counterpart, the Lie superalgebras, will be discussed. We will try an approach that is different from the way in which they are usually presented in physics. By considering the mathematical properties of the (non super) groups and some physical input (as, for example, the spin-statistics connection), we will try to perform an extension of them as supergroups in such way that the extensions have similar properties and are, somehow, minimal. As we have said, this is not the way in which supersymmetry is usually introduced, nor does it correspond to the historical development of the subject. However, this kind of argument has the advantage of proving that, under reasonable demands, there is no other choice.

Section 4.5 deals with technical issues concerning Borel and parabolic subalgebras of the conformal group and supergroup: these are pivotal concepts in our approach.

The Poincaré superalgebra appeared first in the literature with the work of Gol'fand and Likhtman (1971) [77]. Later on, the conformal superalgebra appeared in the work of Wess and Zumino (1974) [158]. The rigorous construction of the corresponding Lie supergroups, together with the actions on their homogeneous spaces came much later (see [108] and references therein).

The description of the supergroups and their homogeneous spaces i.e., the non linear counterpart of the theory, requires an extensive use of the machinery of the functor of points developed in Chapter 1 (see also the examples, in the ordinary setting, in Chapter 2). This approach allows us to think on the Poincaré and conformal supergroups as groups of matrices and to express their actions as left multiplication of matrices on a suitable space, thus recovering the intuitive picture, but leaving intact all the mathematical rigor.

We are obliged to remark that the rigorous approach of the functor of points is not totally alien to physics. In their heuristic and non rigorous way, physicists think of supergroups in terms of matrices with entries in an undetermined superalgebra. Although very imprecise, this idea is essentially the same that is behind the functor of points approach.

We start in Sections 4.1 and 4.3 with a treatment of the spin representations and Clifford algebras. These are technical issues that are heavily used in any formulation of supersymmetry.

4.1 Spinors and spin representations

In Chapter 2 we mentioned that the Lorentz algebra $\mathfrak{so}(1,3) \cong \mathfrak{sl}_2(\mathbb{C})_{\mathbb{R}}$ admits representations that integrate to representations of the spin group $\mathrm{SL}_2(\mathbb{C})_{\mathbb{R}}$ but that do not descend to give representations of the special orthogonal group $\mathrm{SO}(1,3) \cong \mathrm{SL}_2(\mathbb{C})_{\mathbb{R}}/\mathbb{Z}_2$. These are irreducible, finite dimensional, non unitary representations, which play a crucial role both, in mathematics and physics. The basic ones, with the lowest dimension, are called *spin representations*. The elements in the representation space are called *spinors*. By tensoring the spin representations one obtains irreducible subrepresentations that may, or may not, integrate to representations of the orthogonal group. In the latter case we call them *spinorial representations*.

As irreducible representations of semisimple algebras, the spin representations can be characterized in terms of *weights* on the *Dynkin diagram*. It is the purpose of this section to give a brief descriptive account of them. For a complete and thorough treatment of the representations of semisimple Lie groups we send the reader to Ref. [146]. The reader acquainted with the terminology and the theory may very well jump ahead to the next section, were we turn to a more explicit construction with the help of Clifford algebras.

Let us consider the complex case. The Lie algebras $\mathfrak{so}(n)$ belong to two of the classical series of semisimple Lie algebras. For $n = 2l+1$ $(l \geq 1)$ we have that $\mathfrak{so}(2l+1) = B_l$, whose Dynkin diagram is

(4.1)

The Lie algebra B_l has rank l, which means that it has l *roots*, that correspond to the l nodes of the Dynkin diagram. As Cartan showed, the finite dimensional irreducible representations of such Lie algebra are characterized, up to equivalence, by the l-tuples of non negative integers $\lambda = (\lambda_1, \dots, \lambda_l)$, each of which can be thought as corresponding to one of the nodes of the Dynkin diagram. Such l-tuples are called *weights*. We denote by π_i the representation having weight such that $\lambda_i = 1$ and $\lambda_j = 0$ for $j \neq i$ (such a weight is called a *fundamental weight*). All the irreducible representations are associated with a weight λ and can be realized as subrepresentations of a tensor product

$$\pi_1^{\otimes\lambda_1} \otimes \cdots \otimes \pi_l^{\otimes\lambda_l}.$$

Moreover, the representations π_i with $i = 1, \dots l-1$, are obtained as irreducible subrepresentations of tensor products of the defining, (or *vector*) representation in $\mathbb{C}^{2l+1}$. This leaves out π_l, which cannot be obtained in this way and is the *spin representation* of B_l. It has dimension 2^l. As for any Lie algebra representation, it integrates to the unique complex, simply connected Lie group, in this case the spin group $\mathrm{Spin}(n)$. However, this representation, contrary to what happens for the other ones corresponding to the π_i, $i \neq l$, does not descend to the special orthogonal group. We denote with S its representation space. An element of S is then a spinor.

For $n = 2l$ $(l \geq 3)$ we have that $\mathfrak{so}(2l) = D_l$ and the Dynkin diagram is:

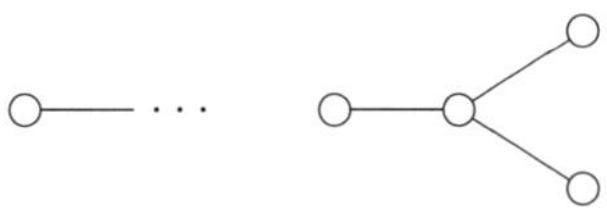

This case is very much similar to B_l, except that here we have two special roots, α_{l-1} and α_l, corresponding to the last two nodes on the Dynkin diagram. This gives rise to two inequivalent spinor representations, π_{l-1} and π_l. They have both dimension 2^{l-1}. The representation spaces are (indistinctively) denoted as $S^{\pm}$. The elements of $S^{\pm}$ sometimes are called *half spinors*, although we prefer to call them also spinors.

In the next two sections we will give a more practical description of the spin representations using the Clifford algebra associated to an orthogonal group. We will also treat the real case.

4.2 The complex Clifford algebra and the spin group

In this section we review some well known facts on Clifford algebras in general, focusing our attention on the complex case. The details on this classical material can be found, for example, in Refs. [72, 147].

Our ground field k will be $\mathbb{C}$ or $\mathbb{R}$, although most of these constructions hold in a more general setting. We start with the definition of Clifford algebra.

Definition 4.2.1. *Clifford algebra.* Let V be a real or complex vector space of finite dimension n, equipped with a non degenerate quadratic form

Q. We define the Clifford algebra associated to the vector space V with quadratic form Q as the associative algebra generated by all the $v \in V$, subject to the relation

$$v^2 = Q(v)\,\mathrm{id}\,.$$

It will be denoted as $C(Q)$.

If $\{e_i\}_{i=1}^n$ is a basis of V and B is the symmetric bilinear form associated to Q, we have that $C(Q)$ is generated, as an algebra, by the e_i's subject to the relations

$$e_i e_j + e_j e_i = 2B(e_i, e_j)\,\mathrm{id}\,, \qquad i, j = 1, \ldots, n\,.$$

■

We have that $V \subset C(Q)$ and that the elements

$$1, \quad e_{i_1} \cdots e_{i_r}, \qquad 1 \leq i_1 < \cdots < i_r \leq n \tag{4.2}$$

form a basis for $C(Q)$, so $C(Q)$ has dimension 2^n.

The Clifford algebra can also be realized as a quotient of the full tensor algebra over V:

$$C(Q) \;=\; T(V) \,/\, \big(x \otimes y + y \otimes x - 2B(x, y)\mathrm{id}\big)\,. \tag{4.3}$$

$C(Q)$ is also a superalgebra in the following way. The linear map $v \mapsto -v$ on V preserves $Q(v)$ hence it gives an involutive automorphism ι of $C(Q)$. Since ι, as any involution, has eigenvalues ± 1, $C(Q)$ splits into the direct sum of the corresponding eigenspaces

$$C(Q) = C^+(Q) \oplus C^-(Q)\,. \tag{4.4}$$

This defines, respectively, the even and odd subspaces of $C(Q)$.

Notice that the exterior algebra $\wedge V$, which is realized as the quotient

$$\wedge V \;=\; T(V) \,/\, (x \otimes y + y \otimes x)\,,$$

has the same dimension as $C(Q)$, in fact the elements in the form (4.2) describe also a basis of $\wedge V$. Moreover, $\wedge V$ is a superalgebra and a $\mathbb{Z}$-graded algebra, the two gradings being compatible.

Over the complex field there always exists an orthonormal basis $\{e_i\}$ of $V \cong \mathbb{C}^n$, i.e. a basis such that

$$B(e_i, e_j) = \delta_{ij}\,.$$

The quadratic form in this basis becomes

$$Q(x) = x_1^2 + \cdots + x_n^2 .$$

Then, there is only one Clifford algebra (up to isomorphism) for every dimension. We denote it as C_n.

Over the reals the story is different. In fact, if Q has signature (p, q) we can always choose a basis for $V \cong \mathbb{R}^{p+q}$ such that

$$B(e_i, e_j) = \eta_{ij} , \qquad \eta = \begin{pmatrix} \mathrm{id}_p & 0 \\ 0 & -\mathrm{id}_q \end{pmatrix} , \tag{4.5}$$

so that we can write

$$Q(x) = x_1^2 + \cdots + x_p^2 - x_{p+1}^2 - \cdots - x_{p+q}^2 .$$

The vector space $\mathbb{R}^{p+q}$ together with the quadratic form Q specified above is denoted by $\mathbb{R}^{p,q}$ and we write $C_{p,q}$ for the Clifford algebra of $\mathbb{R}^{p,q}$. These algebras are non isomorphic. Any real Clifford algebra is isomorphic to one of the $C_{p,q}$.

We focus now on the complex case. The following theorem describes completely the structure for Clifford algebras over an algebraically closed field. For more details and the proof, we refer the reader to Ref. [147], Section 5.3 or to Ref. [72], Chapter 20.

Theorem 4.2.2. *Let $C(Q)$ be the Clifford algebra of (V, Q), where V is a vector space over an algebraically closed field and Q a non degenerate quadratic form.*

- *Let $\dim(V) = 2m$. Then $V = W \oplus W'$, where W and W' are m-dimensional isotropic subspaces, i.e. $Q|_W = Q|_{W'} = 0$, and $C(Q)$ is isomorphic to a full matrix superalgebra.*

$$C(Q) \cong \mathrm{End}(\wedge W) .$$

- *Let $\dim(V) = 2m + 1$. Then $V = W \oplus W' \oplus U$, where W and W' are m-dimensional isotropic subspaces and U is a one dimensional space orthogonal to them. Furthermore:*

$$C(Q) \cong \mathrm{End}(\wedge W) \oplus \mathrm{End}(\wedge W') .$$

■

From this theorem we can conclude that there are algebra isomorphisms

$$C_{2m} \cong \mathrm{End}(\mathbb{C}^{2^m}) \cong \mathrm{M}_{2^m}(\mathbb{C}) , \tag{4.6}$$

$$C_{2m+1} \cong \mathrm{End}(\mathbb{C}^{2^m}) \oplus \mathrm{End}(\mathbb{C}^{2^m}) \cong \mathrm{M}_{2^m}(\mathbb{C}) \oplus \mathrm{M}_{2^m}(\mathbb{C}) , \tag{4.7}$$

where $\mathrm{M}_n(\mathbb{C})$ is the set of $n \times n$ matrices over $\mathbb{C}$.

Moreover, since for any n the only irreducible, non trivial representation (up to equivalence) of $\mathrm{M}_n(\mathbb{C})$ is the defining representation, we can conclude from (4.6) that C_{2m} has only one class of irreducible representations and that it is of dimension 2^m. Instead, from (4.7) we conclude that C_{2m+1} has two classes of irreducible representations, each corresponding to one of the surjective homomorphisms $C_{2m+1} \to \mathrm{M}_{2^m}(\mathbb{C})$, one for each of the factors in (4.7).

We now want to introduce the *Clifford group*, which is a subset of the Clifford algebra. The Clifford group contains, as a subgroup, the spin group $\mathrm{Spin}(Q)$, that is, the double cover of the special orthogonal group $\mathrm{SO}(Q)$ associated with the quadratic form Q. The representations of the Clifford algebra restrict to representations of the Clifford group and of the spin group, and they coincide with the spin representations that we described in Section 4.1.

We first need some definitions. We denote by $C(Q)^{\times}$ the set of invertible elements in $C(Q)$, and similarly, we write $C(Q)^{\pm\times}$.

Definition 4.2.3. As before, let V be a vector space over an algebraically closed field and Q a non degenerate quadratic form on it. We define the *even Clifford group* as the following subgroup of $C(Q)^{+\times}$:

$$\Gamma^{+} = \{x \in C(Q)^{+\times} \quad | \quad xvx^{-1} \in V, \quad \forall v \in V\}\,.$$

■

The elements of Γ^+ act on V as

$$\begin{aligned} \Gamma^{+} \times V &\longrightarrow \quad V \\ (x, v) &\longrightarrow xvx^{-1}\,. \end{aligned}$$

This action preserves the quadratic form, since

$$Q(xvx^{-1})\mathrm{id} = xvx^{-1} \cdot xvx^{-1} = xv^{2}x^{-1} = xQ(v)\mathrm{id}x^{-1} = Q(v)\mathrm{id}\,.$$

Then, there is an homomorphism h:

$$\Gamma^{+} \xrightarrow{\quad h \quad} \mathrm{O}(Q)$$

of the even Clifford group into the orthogonal group $\mathrm{O}(Q)$ associated with the quadratic form Q.

In fact we have more (see Ref. [147]): there exists an exact sequence

$$1 \longrightarrow \mathbb{C}^{\times}\,\mathrm{id} \longrightarrow \Gamma^{+} \longrightarrow \mathrm{SO}(Q) \longrightarrow 1\,.$$

We are now going to identify a subgroup of Γ^+ as the spin group. We first need to define an involution on $C(Q)$ that will play an essential role. Let us consider the space of tensors over V, $T(V)$. We define the following involution

$$\begin{array}{ccc} T(V) & \xrightarrow{\beta} & T(V) \\ v_1 \cdots v_r & \longrightarrow & v_r \cdots v_1 \, , \end{array} \qquad v_i \in V \, .$$

Since the generators of the ideal in (4.3) are fixed by β, then β descends to an involution on $C(Q)$. Clearly $\beta^2 = \mathrm{id}$ and $\beta(x \cdot y) = \beta(y) \cdot \beta(x)$, so it is an antiautomorphism of algebras.

Moreover, let $x = v_1 \cdots v_r$. We have that

$$x\beta(x) = v_1 \cdots v_r \cdot v_r \cdots v_1 = Q(v_1) \cdots Q(v_r)\mathrm{id} \, .$$

This can be extended to an arbitrary x. In fact, we have seen that the transformation $h(x)$ is an orthogonal transformation on V. The Cartan-Dieudonné theorem states that any orthogonal transformation can be decomposed as a product of $r \leq \dim V$ reflections. If the transformation is in $\mathrm{SO}(Q)$ then r is even. This implies that there exist unit vectors $v_1, \ldots, v_r \in V$ such that

$$h(x) = h(v_1) \cdots h(v_r)$$

and then the exact sequence above implies that

$$x = c\, v_1 \cdots v_r \, , \qquad c \in \mathbb{C}.$$

So we can conclude that there is an homomorphism

$$\begin{array}{ccc} C(Q)^+ & \xrightarrow{\beta} & \mathbb{C}\,\mathrm{id} \\ x & \longrightarrow & x\beta(x) \, . \end{array}$$

Finally, we summarize our discussion, together with further results, with the following proposition on the complex spin group.

Proposition 4.2.4. *Let V be a complex vector space with a quadratic form Q and let $\dim(V) \geq 2$. Consider the following set inside $C^+(Q)$*

$$G(Q) = \{x \in C^+(Q) \mid Q(x) = 1, \quad x \cdot V \cdot x^{-1} \subset V\} \, .$$

Then $G(Q)$ is a connected Lie group and h is a double covering morphism:

$$\begin{array}{ccc} G & \longrightarrow & \mathrm{SO}(Q) \\ x & \longrightarrow & h(x) \end{array}$$

with $h(x)v = x \cdot v \cdot x^{-1}$. Hence $G \cong \mathrm{Spin}(Q)$. ■

In the next proposition, in analogy with Theorem 4.2.2, we establish the structure of $C^+(Q)$. This will allow us to construct the spin representations, in terms of the representations of $C^+(Q)$.

Proposition 4.2.5. *Let $C(Q)$ be the Clifford algebra of (V, Q), where V is a vector space over an algebraically closed field and Q a non degenerate quadratic form.*

- *Let $\dim(V) = 2m$. Then $V = W \oplus W'$, where W and W' are m-dimensional isotropic subspaces and there is an algebra homomorphism*

$$C^+(Q) \cong \mathrm{End}(\wedge^{\mathrm{even}} W) \oplus \mathrm{End}(\wedge^{\mathrm{odd}} W) \cong \mathrm{M}_{2^{m-1}}(\mathbb{C}) \oplus \mathrm{M}_{2^{m-1}}(\mathbb{C})\,.$$

- *Let $\dim(V) = 2m+1$. Then $V = W \oplus W' \oplus U$, where W and W' are m-dimensional isotropic subspaces and U is a one dimensional space orthogonal to them. There is then an algebra homomorphism*

$$C^+(Q) \cong \mathrm{End}(\wedge W) \cong \mathrm{M}_{2^m}(\mathbb{C})\,.$$

■

Incidentally, notice that if $n = \dim(V)$, then $C_n^+ \cong C_{n-1}$.

We conclude the treatment of the complex case by identifying the spin representations of $\mathrm{Spin}(Q)$ in terms of representations of $C^+(Q)$

Proposition 4.2.6. *Let $C(Q)$ be the Clifford algebra of (V, Q), where V is a vector space over an algebraically closed field and Q a non degenerate quadratic form. Let the notation be as in Proposition 4.2.5*

- *Let $\dim(V) = 2m$. Then the vector spaces*

$$S^+ := \wedge^{\mathrm{even}} W, \qquad S^- := \wedge^{\mathrm{odd}} W\,,$$

support the only (up to equivalence) irreducible representations of $C^+(Q)$. Their restrictions to $G \cong \mathrm{Spin}(Q)$ correspond to the spin (or half spin) representations $S^{\pm}$.
- *Let $\dim(V) = 2m+1$. Then the vector space*

$$S := \wedge W\,,$$

supports the only (up to equivalence) irreducible representation of $C^+(Q)$. Its restrictions to $G \cong \mathrm{Spin}(Q)$ corresponds to the spin representation S.

In the even case, the direct sum $S^+ \oplus S^-$ extends to the full irreducible representation of the Clifford algebra. In the odd case, S itself extends to one of the non faithful, irreducible representations of the Clifford algebra. In physics, the elements of these complex representations that extend to the Clifford algebra are called *Dirac spinors*. The elements of $S^{\pm}$ and S (even or odd case respectively) are called *Weyl spinors*, so in the odd case Weyl spinors are also Dirac spinors. Weyl spinors of S^+ are also called *right handed* spinors and Weyl spinors of S^- are called *left handed* spinors.

4.3 The real Clifford algebra and the spin group

The situation we described in the previous section becomes more complicated over the reals. We shall give here an account of it but skipping the proofs. We will focus our attention in the low dimensional cases, which are the ones most relevant for us.

We invite the reader to consult Refs. [3, 21, 29, 30, 35] for the complete story.

The real Clifford algebras $C_{p,q}$ are classified by the signature $\rho = p - q$ modulo 8. We list the results in Table 4.1, where we have used the following notation:

$$\mathbf{2}E = E \oplus E, \quad D = \dim V = p + q\,.$$

Notice that D has the same parity as ρ.

$\mathrm{M}_p(\mathbb{H})$ denotes the $p \times p$ complex matrices satisfying the quaternionic condition

$$M^* = -\Omega M \Omega\,, \qquad \Omega \text{ antisymmetric and } \Omega^2 = -\mathrm{id}\,. \tag{4.8}$$

Using the appropriate matrix Ω one can check that this condition is equivalent to have a $p/2 \times p/2$ matrix M whose entries are quaternions (in the standard, two dimensional, complex representation of quaternions).

Table 4.1 Real Clifford algebras.

ρ, D even, $D = 2m$				
ρ	0	2	4	6
$C_{p,q}$	$\mathrm{M}_{2^m}(\mathbb{R})$	$\mathrm{M}_{2^m}(\mathbb{R})$	$\mathrm{M}_{2^m}(\mathbb{H})$	$\mathrm{M}_{2^m}(\mathbb{H})$

ρ, D odd, $D = 2m+1$				
ρ	1	3	5	7
$C_{p,q}$	$\mathbf{2}\mathrm{M}_{2^m}(\mathbb{R})$	$\mathrm{M}_{2^m}(\mathbb{C})$	$\mathbf{2}\mathrm{M}_{2^m}(\mathbb{H})$	$\mathrm{M}_{2^m}(\mathbb{C})$

We now want to understand better the table. By looking at it, we observe that there are basically three mechanisms to impose a reality condition on the complex Clifford algebra, which we know (from Theorem 4.2.2) it is either a matrix algebra or the sum of two matrix algebras.

(1) Let us consider the matrix algebra $\mathrm{M}_n(\mathbb{C})$ and $E = \mathbb{C}^n$. Let $\sigma : E \to E$ be a conjugation (that is, an antilinear map with $\sigma^2 = \mathrm{id}$) of E. Then, the set of elements $M \in \mathrm{M}_n(\mathbb{C})$ that commute with the conjugation,

$$M\sigma(e) = \sigma(M(e)) \qquad \forall e \in E\,,$$

is the same as the set of matrices satisfying

$$\sigma \circ M \circ \sigma = M\,. \tag{4.9}$$

The antilinear map

$$\begin{array}{ccc} \mathrm{M}_n(\mathbb{C}) & \stackrel{\hat{\sigma}}{\longrightarrow} & \mathrm{M}_n(\mathbb{C}) \\ M & \longrightarrow & \hat{\sigma}(M) = \sigma \circ M \circ \sigma \end{array}$$

is a conjugation on $\mathrm{M}_n(\mathbb{C})$ and the set of its fixed points (4.9) is a real form of the algebra $\mathrm{M}_n(\mathbb{C})$. The real algebra has a real representation on the set of fixed points

$$E^\sigma = \{e \in E \mid \sigma(e) = e\}\,,$$

which is a real vector space. If $\sigma(e) = e^*$, that is, if σ is the standard complex conjugation, then we obtain in this way the real form $\mathrm{M}_n(\mathbb{R})$.

(2) Let us consider again $\mathrm{M}_n(\mathbb{C})$. Suppose now that σ is a pseudoconjugation, that is, it is an antilinear map such that $\sigma^2 = -1$. Then the map

$$\sigma \circ M \circ \sigma = M$$

is still a conjugation and defines a real form of $\mathrm{M}_n(\mathbb{C})$. But the set of fixed points E^σ contains only the zero, since the condition

$$\sigma(e) = e \quad \text{implies} \quad e = -e\,.$$

The real algebra then does not have a real representation on a real form of E. In order to find a real representation we need to double the dimension: we consider $E \oplus E$. On such complex vector space, we impose the reality condition

$$e_1 = \sigma(e_2)\,, \qquad (e_1, e_2) \in E \oplus E.$$

If $\sigma(e) = \Omega e^*$, with Ω an antisymmetric matrix such that $\Omega^2 = -1$, we obtain the real form $\mathrm{M}_n(\mathbb{H})$.

(3) Let us now view the complex algebra $\mathrm{M}_n(\mathbb{C})$ as a real algebra $\mathrm{M}_n(\mathbb{C})_{\mathbb{R}}$. Its complexification is $\mathbf{2}\mathrm{M}_n(\mathbb{C}) = \mathrm{M}_n(\mathbb{C}) \oplus \mathrm{M}_n(\mathbb{C})$. We can obtain a real representation on $E \oplus E$ via the map

$$M = A + \mathrm{i}B \longrightarrow \begin{pmatrix} A & B \\ -B & A \end{pmatrix} \qquad A, B \in \mathrm{M}_n(\mathbb{R})\,.$$

We now wish to describe in some key examples how we can realize in practice these real representations. We first introduce some physicists' terminology.

Assume that we have a set of matrices $(\gamma_i)_{i=1}^n$ satisfying the defining relations of the Clifford algebra:

$$\gamma_i\gamma_j + \gamma_j\gamma_i = 2\eta_{ij} \qquad i, j = 1, \dots, n$$

where η_{ij} is the bilinear form (see (4.5)). Then we have immediately a representation of the full Clifford algebra. Such sets are called in the physics literature sets of *gamma matrices* for the corresponding Clifford algebra.

We now are ready to describe some real representations of the real Clifford algebra via the language of gamma matrices.

Example 4.3.1. *Real Clifford algebras in dimension 2.* We want to study the case of $V = \mathbb{R}^2$. Let us consider the Pauli matrices

$$\sigma_0 = \begin{pmatrix} 1 & 0 \\ 0 & 1 \end{pmatrix} = \mathrm{id}, \quad \sigma_1 = \begin{pmatrix} 0 & 1 \\ 1 & 0 \end{pmatrix}, \quad \sigma_2 = \begin{pmatrix} 0 & -\mathrm{i} \\ \mathrm{i} & 0 \end{pmatrix}, \quad \sigma_3 = \begin{pmatrix} 1 & 0 \\ 0 & -1 \end{pmatrix}.$$

They satisfy the relations

$$\sigma_i\sigma_j = \delta_{ij}\mathrm{id} + \mathrm{i}\epsilon_{ijk}\sigma_k\,, \qquad i, j, k = 1, 2, 3\,,$$

where, as usual, ϵ_{ijk} is the antisymmetric tensor and it assumes the value 1 if the permutation $(1, 2, 3) \to (i, j, k)$ is even, and the value -1 if the permutation is odd.

Starting from the Pauli matrices we now construct three sets of gamma matrices, one for each signature, that yield the representations of the real Clifford algebras in dimension 2.

- $C_{2,0} \cong \mathrm{M}_2(\mathbb{R})$. A set of gamma matrices is

$$\gamma_1 = \sigma_1\,, \quad \gamma_2 = \sigma_2\,. \tag{4.10}$$

A basis of the full Clifford algebra is then

$$\sigma_0 = \mathrm{id}, \quad \sigma_1, \quad \sigma_2, \quad \sigma_1\sigma_2 = -\mathrm{i}\sigma_3\,. \tag{4.11}$$

The isomorphism of $C_{2,0}$ with $\mathrm{M}_2(\mathbb{R})$ is then obtained by sending the ordered basis (4.11) to

$$\mathrm{id} = \begin{pmatrix} 1 & 0 \\ 0 & 1 \end{pmatrix}, \quad \sigma_1 = \begin{pmatrix} 0 & 1 \\ 1 & 0 \end{pmatrix}, \quad -\sigma_3 = \begin{pmatrix} -1 & 0 \\ 0 & 1 \end{pmatrix}, \quad \mathrm{i}\sigma_2 = \begin{pmatrix} 0 & 1 \\ -1 & 0 \end{pmatrix}.$$

We can use now (4.10) to construct sets of gamma matrices in the other signatures. We just have to multiply conveniently by i some of them.

- $C_{1,1} \cong \mathrm{M}_2(\mathbb{R})$. A set of gamma matrices is

$$\gamma_1 = \sigma_1 \,, \quad \gamma_2 = \mathrm{i}\sigma_2 \,.$$

The basis of $C_{1,1}$ is

$$\sigma_0 = \mathrm{id}, \quad \sigma_1, \quad \mathrm{i}\sigma_2, \quad \sigma_3 \,,$$

which is real, so the isomorphism with $\mathrm{M}_2(\mathbb{R})$ is trivial.

- $C_{2,0} \cong \mathrm{M}_2(\mathbb{H}) = \mathbb{H}$. A set of gamma matrices is

$$\gamma_1 = \mathrm{i}\sigma_1 \,, \quad \gamma_2 = \mathrm{i}\sigma_2 \,.$$

The basis of $C_{2,0}$ is

$$\sigma_0 = \mathrm{id}, \quad \mathrm{i}\sigma_1, \quad \mathrm{i}\sigma_2, \quad \mathrm{i}\sigma_3 \,,$$

which satisfies the identities defining the quaternion algebra.

■

If we interchange p and q, the orthogonal groups (and spin groups too) corresponding to $C_{p,q}$ and $C_{q,p}$ are isomorphic, but this isomorphism will not extend to the whole Clifford algebras.

For example, in the case of the Minkowskian metric, that is $(p,q) = (1,3)$, we have $\rho = -2$, which modulo 8 is $\rho \sim 6$, so the real Clifford algebra is $C_{1,3} \cong \mathrm{M}_4(\mathbb{H})$.

Interchanging p and q we get $\rho = 2$, which gives $C_{3,1} \cong \mathrm{M}_4(\mathbb{R})$. We can then use two different and non isomorphic Clifford algebras to describe the spin representations of $\mathrm{Spin}(1,3)$: in one case the matrices of the representation are quaternionic, and in the other case they are real. The spin groups are isomorphic, $\mathrm{Spin}(1,3) \cong \mathrm{Spin}(3,1)$, and the spin representations are equivalent.

We now want to give some sets of gamma matrices for $C_{3,1}$ [70], in order to relate our treatment with the terminology which is used in the physics literature.

The first set of gamma matrices that we examine is constructed with the Pauli matrices.

$$\gamma_0 = \mathrm{i}\sigma_1 \otimes \mathrm{id}_2\,, \quad \gamma_1 = \sigma_2 \otimes \mathrm{id}_2\,, \quad \gamma_2 = \sigma_3 \otimes \sigma_1\,, \quad \gamma_3 = \sigma_3 \otimes \sigma_2\,.$$

The gamma matrices in this signature have indices $0, 1, 2, 3$, the zero subscript corresponding always to the time direction, $\gamma_0^2 = -\mathrm{id}$.

Notice that since $C_{3,1} \cong \mathrm{M}_4(\mathbb{R})$, there should be a set of gamma matrices for this Clifford algebra such that all the matrices are real. One can check that this is the case for the set

$$\gamma_0 = \mathrm{i}\sigma_2 \otimes \mathrm{id}_2\,, \quad \gamma_1 = \mathrm{i}\sigma_3 \otimes \mathrm{id}_2\,, \quad \gamma_2 = \mathrm{i}\sigma_1 \otimes \sigma_1\,, \quad \gamma_3 = \mathrm{i}\sigma_1 \otimes \sigma_3\,. \quad (4.12)$$

There is still another set of gamma matrices that is used in physics:

$$\gamma_0 = \begin{pmatrix} 0 & \mathrm{id}_2 \\ \mathrm{id}_2 & 0 \end{pmatrix}, \ \gamma_1 = \begin{pmatrix} 0 & -\sigma_1 \\ \sigma_1 & 0 \end{pmatrix}, \ \gamma_2 = \begin{pmatrix} 0 & -\sigma_2 \\ \sigma_2 & 0 \end{pmatrix}, \gamma_3 = \begin{pmatrix} 0 & -\sigma_3 \\ \sigma_3 & 0 \end{pmatrix}. \quad (4.13)$$

We now determine the spinor subspaces $S^{\pm}$ introduced in the previous section, in these representations. It is hard to say, only by inspection, what are the subspaces $S^{\pm}$ in the decomposition $S = S^+ \oplus S^-$ for each of the representations of the Clifford algebra detailed above. There is a way, though, of constructing the projectors of the spaces $S^{\pm}$. Let us consider the following element of the algebra

$$\gamma_5 =_{\mathrm{def}} \gamma_0\gamma_1\gamma_2\gamma_3\,.$$

It is easy to check that

$$\gamma_5\gamma_\mu + \gamma_\mu\gamma_5 = 0, \qquad \gamma_5^2 = -\mathrm{id}\,.$$

In particular, γ_5 commutes with all $C_{3,1}^+$, which leaves invariant $S^{\pm}$. We can then construct the projectors

$$P_+ = \frac{1}{2}(\mathrm{id} + \mathrm{i}\gamma_5)\,, \qquad P_- = \frac{1}{2}(\mathrm{id} - \mathrm{i}\gamma_5)\,,$$

satisfying

$$P_+^2 = P_+\,, \qquad P_-^2 = P_-\,, \qquad P_+P_- = 0\,.$$

These properties are enough to conclude that$S^+ = P_+S$ and $S^- = P_-S$.

In particular, in the real representation (4.12), the projectors are not real, which means that the Weyl spinors are not real representations. Then, as a real representation, $S = S^+ \oplus S^-$ is irreducible.

In the representation (4.13) the splitting is the easiest since

$$\gamma_5 = \mathrm{i}\begin{pmatrix} \mathrm{id} & 0 \\ 0 & -\mathrm{id} \end{pmatrix} .$$

We now want to look at the spin representations from another point of view. Since the Weyl representations are complex, we can, for many purposes, look at the complex group, where things are easier. We recall that $\mathrm{Spin}(1,3) \cong \mathrm{SL}_2(\mathbb{C})_{\mathbb{R}}$, so the complexification will just be

$$\mathrm{Spin}(1,3)^c = \mathrm{Spin}(4) \cong (\mathrm{SL}_2(\mathbb{C})_{\mathbb{R}})^c = \mathrm{SL}_2(\mathbb{C}) \times \mathrm{SL}_2(\mathbb{C}) . \tag{4.14}$$

The irreducible representations of $\mathrm{SL}_2(\mathbb{C})$ can be parametrized by a non negative half-integer m and the dimension of the irreducible representation corresponding to the half integer m is $2m+1$. The representations of $\mathrm{Spin}(1,3)^c$ are then the tensor product of representations of each of the two factors. They are parametrized by a couple of non negative half-integers and denoted as $D(m,n)$.

We also know that $\mathrm{Spin}(4) \subset C_4^+ \cong C_3 \cong \mathrm{M}_2(\mathbb{C}) \oplus \mathrm{M}_2(\mathbb{C})$. One can prove that each of the factors $\mathrm{SL}_2(\mathbb{C})$ in (4.14) sits inside one of the factors $\mathrm{M}_2(\mathbb{C})$. Then, one can identify $S^+ \cong D(1/2,0)$ and $S^- \cong D(0,1/2)$.

Remark 4.3.2. At the end of Section 3.2 we mentioned that the unitary irreducible representations of the Poincaré group could be also described, by applying a Fourier transform to the representation space, in terms of wave equations over spacetime that are invariant under the Poincaré group. We wrote explicitly the Klein-Gordon equation for a particle of mass m and spin $s=0$, which is the simplest of these,

$$(\Box + m^2)\phi = 0 , \qquad \Box = \frac{\partial^2}{\partial x^{0^2}} - \frac{\partial^2}{\partial x^{1^2}} - \frac{\partial^2}{\partial x^{2^2}} - \frac{\partial^2}{\partial x^{3^2}} .$$

We want now to write the *Dirac equation*, which is the wave equation for a particle of mass m and spin $s = 1/2$. The physical *field* that one has to consider in this case is a function from spacetime to the space of Dirac spinors, that is, to the direct sum $S^+ \oplus S^-$ of the spin representations of $\mathrm{Spin}(3,1)$, so

$$\mathbf{M} \xrightarrow{\ \psi\ } S^+ \oplus S^- .$$

Let γ_μ be a set of gamma matrices. Then the Dirac equation is

$$\hbar c \gamma^\mu \frac{\partial \psi}{\partial x^\mu} + mc^2 \psi = 0 , \tag{4.15}$$

where c is the velocity of light in vacuum and $\hbar = h/2\pi$, with h being Planck's constant.

We want to recall that $S^+ \oplus S^-$ admits a reality condition compatible with the action of Spin(3, 1). In particular, one can chose the representation where the gamma matrices are all real (4.12) and then the reality condition is just

$$\psi^* = \psi \, .$$

We are not proving here that the space of solutions of this wave equation carries the above mentioned representation of the Poincaré group (see Ref. [148] for a complete proof), but with the information that we have provided here, it is not difficult to check that (4.15) is invariant under the action of the Poincaré group. Moreover, as a general rule, particles with half-integer spin make use, in their wave equations, of fields that are valued in a spinorial representation of Spin(3, 1), while particles with integer spin use representations that are also representations of SO(3, 1).

■

The real properties of the spin representations that we have just described will have a dramatic impact on the construction of the *super Poincaré algebra*, which is a generalization of the real Poincaré algebra. The next sections will be devoted to the study of such generalization.

In the next section we shall start our construction of the Minkowski superspace by giving the complex conformal Lie superalgebra.

4.4 The Wess-Zumino superalgebra

In this section we want to describe the complex Wess-Zumino or conformal superalgebra and its real form. This superalgebra was originally introduced by Wess and Zumino [158] in the context of physical field theories. As we will see, the real Wess-Zumino superalgebra is the minimal Lie superalgebra whose even part is the conformal algebra $\mathfrak{su}(2,2)$ and whose odd part is a direct sum of spin representations of the Lorentz subalgebra $\mathfrak{sl}_2(\mathbb{C})_\mathbb{R} \subset \mathfrak{su}(2,2)$. Its uniqueness comes provided we require that the commutator of the odd part with itself has some non zero intersection with the even part, otherwise the generalization would be trivial.

This was not the first superalgebra introduced in physics: the first one was the generalization of the Poincaré algebra [77, 151] that we will describe in the next section.

Here we want to deduce the uniqueness of the superconformal algebra from some natural mathematical and physical requirements that we have outlined above. In this way, the construction appears to be more natural.

Let us now start with the definition of the *Wess-Zumino* or *conformal superalgebra*. As it often happens, it is easier to start with its complexification.

Definition 4.4.1. The complex *Wess-Zumino* or *conformal* superalgebra is the complex Lie superalgebra given by

$$\mathfrak{sl}(4|1) = \left\{ \begin{pmatrix} p & \alpha \\ \beta & c \end{pmatrix} \mid p \in \mathrm{M}_4(\mathbb{C}),\ \alpha, \beta^t \in \mathbb{C}^4,\ c \in \mathbb{C},\ \mathrm{tr}\, p = c \right\}, \qquad (4.16)$$

with

$$\mathfrak{sl}(4|1)_0 = \left\{ \begin{pmatrix} p & 0 \\ 0 & c \end{pmatrix} \right\} = \mathfrak{sl}_4(\mathbb{C}) \oplus \mathbb{C}\,, \qquad \mathfrak{sl}(4|1)_1 = \left\{ \begin{pmatrix} 0 & \alpha \\ \beta & 0 \end{pmatrix} \right\} = \mathbf{4}^{-1} \oplus \overline{\mathbf{4}}^{+1}\,.$$

Here $\mathbf{4}$ is the fundamental representation of $\mathfrak{sl}_4(\mathbb{C})$ and $\overline{\mathbf{4}}$ its contragradient. The superindices specify the representation of $\mathbb{C}$, namely

$$\left[\begin{pmatrix} 0 & 0 \\ 0 & c \end{pmatrix}, \begin{pmatrix} 0 & \alpha \\ 0 & 0 \end{pmatrix} \right] = -c \begin{pmatrix} 0 & \alpha \\ 0 & 0 \end{pmatrix}, \qquad \left[\begin{pmatrix} 0 & 0 \\ 0 & c \end{pmatrix}, \begin{pmatrix} 0 & 0 \\ \beta & 0 \end{pmatrix} \right] = c \begin{pmatrix} 0 & 0 \\ \beta & 0 \end{pmatrix}.$$

■

We observe that the even part $\mathfrak{sl}(4|1)$ contains, not only the conformal algebra, but also an extra factor $\mathbb{C}$. We will see in Proposition 4.4.5 that this factor is unavoidable.

We now want to define the real forms of this Lie superalgebra whose even part contains the real conformal algebra $\mathfrak{su}(2,2)$. Consider the 4×4 matrix

$$F = \mathrm{i} \begin{pmatrix} 0 & -\mathrm{id}_2 \\ \mathrm{id}_2 & 0 \end{pmatrix}.$$

Then, we define the following conjugations on $\mathfrak{sl}(4|1)$:

$$\begin{array}{ccc} \mathfrak{sl}(4|1) & \xrightarrow{\sigma_\pm} & \mathfrak{sl}(4|1) \\ \begin{pmatrix} p & \alpha \\ \beta & c \end{pmatrix} & \longrightarrow & \begin{pmatrix} -Fp^\dagger F & \pm \mathrm{i} F \beta^\dagger \\ \pm \mathrm{i} \alpha^\dagger F & -\bar{c} \end{pmatrix}. \end{array} \qquad (4.17)$$

We leave to the reader the straightforward verification that $\sigma_\pm$ are antilinear maps satisfying $\sigma_\pm^2 = \mathrm{id}$, so they are conjugations and they preserve the super bracket on $\mathfrak{g}$. The sets of fixed points of these conjugations are

$$\mathfrak{su}(2,2|1)_\pm = \left\{ \begin{pmatrix} p & \alpha \\ \beta & \mathrm{i}z \end{pmatrix} \mid p \in \mathrm{M}_4(\mathbb{C}),\ \alpha, \beta^t \in \mathbb{C}^4,\ z \in \mathbb{R}; \right.$$
$$\left. Fp + p^\dagger F = 0,\ \mathrm{tr}\, p = \mathrm{i}z,\ \alpha = \pm \mathrm{i} F \beta^\dagger \right\}. \tag{4.18}$$

This defines in fact two different, non isomorphic, real forms. In Ref. [90] they are called *dual* real forms, while in Ref. [147] they are called *isomers*. These are real Lie superalgebras whose complexification is $\mathfrak{sl}(4|1)$.

In the Example (1.12.7) F is written in diagonal form, which clearly gives isomorphic algebras,

$$\mathfrak{su}_+(2,2|1) \cong \mathfrak{su}(2,2|0,1), \qquad \mathfrak{su}_+(2,2|1) \cong \mathfrak{su}(2,2|1,0)\,.$$

We have that

$$\mathfrak{su}(2,2|1)_{\pm,0} = \left\{ \begin{pmatrix} p & 0 \\ 0 & \mathrm{i}z \end{pmatrix} \right\} = \mathfrak{su}(2,2) \oplus \mathbb{R}\,,$$
$$\mathfrak{su}(2,2|1)_{\pm,1} = \left\{ \begin{pmatrix} 0 & \alpha \\ \beta & 0 \end{pmatrix} \mid \alpha = \pm \mathrm{i} F \beta^\dagger \right\} = (\mathbf{4}^{-1} \oplus \overline{\mathbf{4}}^{+1})_r\,.$$

The odd part of the Lie algebra is in the complex representation $\mathbf{4}^{-1} \oplus \overline{\mathbf{4}}^{+1}$ together with a reality condition denoted with the subindex r that commutes with the action of $\mathfrak{su}(2,2) \oplus \mathbb{R}$.

The two isomers $\mathfrak{su}(2,2|1)_\pm$ are the only two real forms (up to isomorphism) of the complex Wess-Zumino superalgebra $\mathfrak{sl}(4|1)$ whose even part contains the conformal algebra $\mathfrak{su}(2,2)$. One can easily check this statement by looking at the table of real forms of complex Lie superalgebras in Ref. [90]. We therefore call both of them *real conformal Lie superalgebras*, *real superconformal algebras* or *real Wess-Zumino superalgebras*.

The two isomers are related by the complex map

$$\begin{array}{ccc} \mathfrak{su}(2,2|1)_+ & \longrightarrow & \mathfrak{su}(2,2|1)_- \\ & & \\ \begin{pmatrix} p & \alpha \\ \beta & \mathrm{i}z \end{pmatrix} & \longrightarrow & \begin{pmatrix} p & \mathrm{i}\alpha \\ \mathrm{i}\beta & \mathrm{i}z \end{pmatrix}. \end{array} \tag{4.19}$$

Remark 4.4.2. Most physicists seem not to be aware that there are two non isomorphic, real, superconformal algebras. Looking at the complex map (4.19) one sees that one can switch from one algebra to another by multiplying the odd part of the algebra by the imaginary unit i. It turns out that, in a physical theory, it is irrelevant which generator is associated with the physical fields, either u or iu are possible choices and then both algebras describe the same physics. ■

We now want to see how it is possible to obtain the Wess-Zumino superalgebra from general principles, imposing some natural conditions on a generic complex Lie superalgebra $\mathfrak{h} = \mathfrak{h}_0 + \mathfrak{h}_1$, with $\mathfrak{sl}_4(\mathbb{C}) \subset \mathfrak{h}_0$. We will work over the complex field. We start with an observation on the odd part of $\mathfrak{h}$.

Observation 4.4.3. Let $\mathfrak{h} = \mathfrak{h}_0 + \mathfrak{h}_1$ be a superalgebra such that $\mathfrak{h}_0$ contains $\mathfrak{sl}_4(\mathbb{C})$ (the complexification of $\mathfrak{su}(2,2)$). Then, $\mathfrak{h}_1$ carries a representation of $\mathfrak{sl}_4(\mathbb{C})$ and in turn, of $\mathfrak{sl}_2(\mathbb{C})\times\mathfrak{sl}_2(\mathbb{C}) \subset \mathfrak{sl}_4(\mathbb{C})$, where $\mathfrak{sl}_2(\mathbb{C})\times\mathfrak{sl}_2(\mathbb{C})$ is the complexification of the real Lorentz algebra $\mathfrak{so}(1,3)$ and it is embedded into $\mathfrak{sl}_4(\mathbb{C})$ as the block-diagonal matrices.

According to the spin-statistics theorem (see Section 3.3) and the discussion in Section 3.4, we must require that the representation of $\mathfrak{sl}_2(\mathbb{C})\times\mathfrak{sl}_2(\mathbb{C})$ on $\mathfrak{h}_1$ is a spinorial representation. We choose the simplest possibility, with $\mathfrak{h}_1$ being the direct sum of a number of representations $\mathbf{4}$ and $\overline{\mathbf{4}}$.

Let us focus now on the commutators of the odd part. We impose that $[\mathfrak{h}_1, \mathfrak{h}_1] \cap \mathfrak{h}_0 \neq \emptyset$ in order to have a non trivial extension. In fact the symmetric linear map

$$\begin{array}{ccc} \mathfrak{h}_1 \otimes \mathfrak{h}_1 & \xrightarrow{F} & \mathfrak{h}_0 \\ Q \otimes Q' & \longrightarrow & [Q, Q'] \end{array} \tag{4.20}$$

is, by the Jacobi identity, an *equivariant map* under the action of $\mathfrak{h}_0$. This means that for any $x \in \mathfrak{h}_0$

$$\big[x, \big(F(Q \otimes Q')\big)\big] = F\big([x, Q] \otimes Q'\big) + F\big(Q \otimes [x, Q']\big)\,.$$

Since $\mathfrak{sl}_4(\mathbb{C})$ is simple, the adjoint action is irreducible and then $[\mathfrak{h}_1, \mathfrak{h}_1] \cap \mathfrak{sl}_4(\mathbb{C}) = \mathfrak{sl}_4(\mathbb{C})$. So the commutator $[\mathfrak{h}_1, \mathfrak{h}_1]$ must contain the adjoint representation of $\mathfrak{sl}_4(\mathbb{C})$.

Let us denote as $\mathbf{n}$ the representation of $\mathfrak{sl}_4(\mathbb{C})$ of dimension n sitting in some tensor power of the fundamental representation and by $\overline{\mathbf{n}}$ its contragredient representation. We have the decompositions:

$$\begin{aligned} \mathbf{4} \otimes \mathbf{4} &= \mathbf{6} \oplus \mathbf{10}\,, \\ \bar{\mathbf{4}} \otimes \bar{\mathbf{4}} &= \mathbf{6} \oplus \overline{\mathbf{10}}\,, \\ \mathbf{4} \otimes \bar{\mathbf{4}} &= \mathbf{1} \oplus \mathbf{15}\,. \end{aligned}$$

The only possibility of having the adjoint representation $\mathbf{15}$ in $[\mathfrak{h}_1, \mathfrak{h}_1]$ is that $\mathfrak{h}_1$ contains both, $\mathbf{4}$ and $\overline{\mathbf{4}}$. ■

In the next proposition, we shall prove the, perhaps surprising, fact that a Lie superalgebra whose even part is the conformal algebra and whose odd part corresponds to spin representations of the Lorentz subalgebra, does not exist.

Proposition 4.4.4. *There is no Lie superalgebra $\mathfrak{h}$ with the following properties:*

(1) $\mathfrak{h}_0 = \mathfrak{sl}_4(\mathbb{C})$,
(2) $\mathfrak{h}_1$ *consists of a sum of representations* $\mathbf{4}$ *and* $\overline{\mathbf{4}}$,
(3) $[\mathfrak{h}_1, \mathfrak{h}_1]$ *has a non zero intersection with* $\mathfrak{h}_0$.

Proof. Let us assume that such superalgebra exists. We first notice that the condition (3) implies that the commutator $[\mathfrak{h}_1, \mathfrak{h}_1] = \mathfrak{h}_0$ (see also Observation 4.4.3).

Because of Observation 4.4.3, we know that $\mathfrak{h}_1$ contains $\mathbf{4} \oplus \overline{\mathbf{4}}$ and that $[\mathfrak{h}_1, \mathfrak{h}_1] = \mathfrak{h}_0$. So, in particular, the subspace $\mathfrak{h}' = \mathfrak{sl}_4(\mathbb{C}) \oplus \mathbf{4} \oplus \overline{\mathbf{4}} \subset \mathfrak{h}$ would be a Lie superalgebra with the same properties as $\mathfrak{h}$. Since the even part is exactly $\mathfrak{sl}_4(\mathbb{C})$ we must have a symmetric, $\mathfrak{sl}_4(\mathbb{C})$-equivariant map as (4.20). The Jacobi identity implies the cubic condition (see Chapter 1) for Lie superalgebras, that is,

$$[Q, [Q, Q]] = 0\,, \qquad \forall Q \in \mathfrak{h}_1\,. \tag{4.21}$$

$\mathfrak{sl}_4(\mathbb{C})$-equivariant maps are unique up to multiplication by a non zero factor. In particular, the Lie superalgebra $\mathfrak{sl}(4|1)$ also contains such map in the odd commutator (i.e. the commutator of its odd part), so we can obtain it by projecting $[\mathfrak{h}_1, \mathfrak{h}_1]$ on the $\mathfrak{sl}_4(\mathbb{C})$ factor.

With the notation of (4.16),

$$\left[\begin{pmatrix} 0 & \alpha \\ \beta & 0 \end{pmatrix}, \begin{pmatrix} 0 & \alpha \\ \beta & 0 \end{pmatrix}\right] = \begin{pmatrix} 2\alpha\beta & 0 \\ 0 & 2\beta\alpha \end{pmatrix}$$

$$= \begin{pmatrix} 2\alpha\beta - \frac{1}{2}(\beta\alpha)\mathrm{id} & 0 \\ 0 & 0 \end{pmatrix} + \begin{pmatrix} \frac{1}{2}(\beta\alpha)\mathrm{id} & 0 \\ 0 & 2\beta\alpha \end{pmatrix}.$$

In the right hand side of the above equation, the first term belongs to the $\mathfrak{sl}_4(\mathbb{C})$ subalgebra and the second term to the central factor $\mathbb{C}$. Let us denote as (u, v) an arbitrary element of $\mathfrak{h}'_1$. Then

$$[(\alpha, \beta), (\alpha, \beta)] = 2\alpha\beta - \frac{1}{2}(\beta\alpha)\mathrm{id}\,.$$

The right hand side is a matrix of $\mathfrak{sl}_4(\mathbb{C})$.

We recall now that $\mathfrak{h}'_1 = \mathbf{4} \oplus \overline{\mathbf{4}}$, so

$$\big[(\alpha, \beta), \big[(\alpha, \beta), (\alpha, \beta)\big]\big] = \big(\frac{3}{2}(\beta\alpha)\alpha\,, -\frac{3}{2}(\beta\alpha)\beta\big)\,.$$

This is non zero, which implies that, under the required conditions, there is no bracket that can convert $\mathfrak{h}'$ into a Lie superalgebra.

So there is no Lie superalgebra with the given properties. ■

This result forces us to consider a larger Lie superalgebra. We then introduce a central factor in the even part of the Lie superalgebra; this will give us immediately the Lie superalgebra $\mathfrak{sl}(4|1)$. In the next proposition we show that $\mathfrak{sl}(4|1)$ is indeed the unique minimal Lie superalgebra satisfying all the requirements.

Proposition 4.4.5. $\mathfrak{sl}(4|1)$ *is, up to isomorphism, the unique Lie superalgebra* $\mathfrak{g}$ *of minimal dimension such that:*

- $\mathfrak{sl}_4(\mathbb{C}) \subset \mathfrak{g}_0 = [\mathfrak{g}_1, \mathfrak{g}_1]$;
- $\mathfrak{g}_1$ *consists of a sum of representations* $\mathbf{4}$ *and* $\overline{\mathbf{4}}$.

Proof. By the discussion in Proposition 4.4.4 and Observation 4.4.3, the dimension of any Lie superalgebra with the properties above must be at least $16|8$; if it is exactly $16|8$, then the even part is the direct sum of $\mathfrak{sl}_4(\mathbb{C})$ and a one-dimensional center. We shall now show that, up to isomorphism, $\mathfrak{g} = \mathfrak{sl}(4|1)$ is the only Lie superalgebra in dimension $16|8$ satisfying these conditions.

Suppose that there exists another super bracket $[\ ,\ ]'$ on $\mathfrak{g}$ satisfying these conditions. We will prove that $(\mathfrak{g}, [\ ,\ ]')$ and $(\mathfrak{g}, [\ ,\ ])$ are isomorphic as Lie superalgebras.

We have that

$$[p, (\alpha, \beta)]' = [p, (\alpha, \beta)] \qquad \forall p \in \mathfrak{sl}_4(\mathbb{C}), \quad (\alpha, \beta) \in \mathfrak{g}_1 = \mathbf{4} \oplus \bar{\mathbf{4}} .$$

Let c denote now a generator of the central factor in $\mathfrak{g}_0 = \mathfrak{sl}_4(\mathbb{C}) \oplus \mathbb{C}$. It acts as a scalar on each summand in $\mathfrak{g}_1$,

$$[c, (\alpha, 0)] = a(\alpha, 0) , \qquad [c, (0, \beta)] = b(0, \beta) , \qquad a, b \in \mathbb{C} .$$

Let now $x \in \mathfrak{g}_0$. Then

$$x = x_{\mathfrak{sl}} + x_c \qquad x_{\mathfrak{sl}} \in \mathfrak{sl}_4(\mathbb{C}) \quad x_c \in \mathbb{C}c .$$

We use this decomposition in the odd commutator and, in view of the projective uniqueness of the equivariant maps, we have that

$$[(\alpha_1, \beta_1), (\alpha_2, \beta_2)]' = f[(\alpha_1, \beta_1), (\alpha_2, \beta_2)]_{\mathfrak{sl}} + g[(\alpha_1, \beta_1), (\alpha_2, \beta_2)]_d .$$

We apply now the cubic condition (4.21). It gives the relations

$$a = b, \qquad f = ag .$$

Thus there are non zero a, g such that

$$[c, (\alpha, \beta)]' = a[c, (\alpha, \beta)]$$
$$[(\alpha_1, \beta_1), (\alpha_2, \beta_2)]' = ag[(\alpha_1, \beta_1), (\alpha_2, \beta_2)]_{\mathfrak{sl}} + g[(\alpha_1, \beta_1), (\alpha_2, \beta_2)]_c .$$

The linear map

$$\begin{array}{ccc} (\mathfrak{g}, [\ ,\ ]') & \longrightarrow & (\mathfrak{g}, [\ ,\ ]) \\ p, c, (\alpha, \beta) & \longrightarrow & p, ac, (ag)^{\frac{1}{2}}(\alpha, \beta) \end{array}$$

is an automorphism of superalgebras. ■

Remark 4.4.6. We want to remark that the superalgebra $\mathfrak{sl}(4|1)$ is a *simple superalgebra*, in fact, one of the *classical superalgebras* described, for example, in Ref. [90] and in our Appendix C. This was not an assumption. The assumptions were minimality and the spin-statistics connection. Then the simplicity comes as a consequence. ■

4.5 Borel and parabolic subalgebras

The Poincaré algebra plus dilations is a *parabolic subalgebra* of the conformal algebra $\mathfrak{sl}_4(\mathbb{C})$. Parabolic subalgebras of a semisimple algebra are the Lie algebras of the isotropy groups of *flag varieties*, an example of which is the complex conformal space, the Grassmannian $G(2,4)$.

The goal is to see if there is one or more parabolic subalgebras of $\mathfrak{sl}(4|1)$ that could be suitable candidates for the Poincaré superalgebra plus dilations, so the construction of the non super case can, in some way, be transported to the super setting. Obviously, the even part of it must be the Poincaré algebra plus dilations, but there are other conditions that one may require.

In the next section we will see that these reasonable requirements determine a unique parabolic sub superalgebra: it coincides with the well known Poincaré superalgebra plus dilations from the physical literature.

This section is devoted to the study of the *Borel* and the *parabolic subalgebras* of $\mathfrak{sl}_4(\mathbb{C})$ and $\mathfrak{sl}(4|1)$. The non super and super cases have enough differences to deserve some explanation. We will not prove our assertions, since they are well known results that can be found in the classical literature on the subject, but the treatment is detailed enough so the reader can reproduce the calculations with little effort. The treatment of the parabolic subalgebras in the non super case can be found in a number of books on semisimple group theory. For the super case, the reader can consult Refs. [88, 109, 123, 153].

We will be working in the complex case, so $k = \mathbb{C}$. We start with some definitions.

Let $\mathfrak{g}$ be a semisimple Lie algebra over $\mathbb{C}$ and let $\mathfrak{h} \subset \mathfrak{g}$ be a Cartan subalgebra. Let Δ be the *root system* associated to $\mathfrak{h}$ and let $\Delta^+ \subset \Delta$ be a choice of positive roots. For a root $\alpha \in \Delta$, we denote as $\mathfrak{g}_\alpha$ the corresponding *root space*. Then, the vector space $\mathfrak{g}$ has the following *root decomposition*

$$\mathfrak{g} = \mathfrak{h} \oplus \left(\bigoplus_{\alpha \in \Delta} \mathfrak{g}_\alpha \right). \tag{4.22}$$

Definition 4.5.1. Let $\mathfrak{g}$ be a semisimple Lie algebra over $\mathbb{C}$, $\mathfrak{h}$ a Cartan subalgebra, Δ its root system and $\Delta^\pm$ the subsets of positive or negative roots respectively. We have the root decomposition (4.22). The *Borel subalgebras* $\mathfrak{b}^{\pm}$ associated to our choice of positive system are the subalgebras consisting of the direct sum of the Cartan subalgebra and the root spaces corresponding to $\Delta^{\pm}$ respectively,

$$\mathfrak{b}^{+} = \mathfrak{h} \oplus \left(\bigoplus_{\alpha \in \Delta^{+}} \mathfrak{g}_{\alpha} \right), \qquad \mathfrak{b}^{-} = \mathfrak{h} \oplus \left(\bigoplus_{\alpha \in \Delta^{-}} \mathfrak{g}_{\alpha} \right).$$

■

Borel subalgebras are equivalently defined as maximal, solvable subalgebras of $\mathfrak{g}$. If G is a connected Lie group[2] with Lie algebra $\mathfrak{g}$, all the Borel subalgebras of $\mathfrak{g}$ are conjugated by an element $g \in G$

$$\mathfrak{b}' = g\, \mathfrak{b}\, g^{-1} \qquad \text{for some } g \in G\,.$$

In particular, the Borel subalgebras $\mathfrak{b}_{\pm}$ are one conjugated to the other.

Definition 4.5.2. Let $\mathfrak{g}$ be a semisimple Lie algebra over $\mathbb{C}$. A *parabolic subalgebra* of $\mathfrak{g}$ is any subalgebra $\mathfrak{f}$ that contains a Borel subalgebra $\mathfrak{b}$ and such that

$$\mathfrak{b} \subseteq \mathfrak{f} \subsetneq \mathfrak{g}\,.$$

If $\mathfrak{g}$ is a real Lie algebra we say that $\mathfrak{f}$ is a parabolic subalgebra if $\mathfrak{f} \otimes \mathbb{C}$ is a parabolic subalgebra of $\mathfrak{g} \otimes \mathbb{C}$. ■

Let Π be a set of simple roots and let Φ be a proper subset of Π. We consider the subset $\widehat{\Phi}$ of Δ^{+} generated by Φ, that is, the set of all positive roots, whose decomposition in terms of simple roots in Φ consists only of roots in $\widehat{\Phi}$:

$$\widehat{\Phi} = \left\{ \alpha \in \Delta^{+} \,|\, \alpha = m_1\alpha_1 + \cdots + m_n\alpha_n, \quad \alpha_i \in \Phi, \quad m_i \in \mathbb{Z}_{\geq 0} \right\}.$$

The subalgebras

$$\mathfrak{b}^{+} \oplus \left(\bigoplus_{\alpha \in \widehat{\Phi}} \mathfrak{g}_{-\alpha} \right), \qquad \mathfrak{b}^{-} \oplus \left(\bigoplus_{\alpha \in \widehat{\Phi}} \mathfrak{g}_{\alpha} \right), \tag{4.23}$$

are parabolic subalgebras. Moreover, every parabolic subalgebra containing $\mathfrak{b}^{+}$ (resp. $\mathfrak{b}^{-}$) is of that form.

[2]It is in fact enough to consider the *adjoint group* of $\mathfrak{g}$, or group of inner automorphisms of $\mathfrak{g}$.

We want to focus our attention on $\mathfrak{g} = \mathfrak{sl}_n(\mathbb{C})$. A Cartan subalgebra of $\mathfrak{g}$ is the set of traceless diagonal matrices

$$\begin{pmatrix} a_1 & 0 & \dots & 0 \\ \vdots & \ddots & & \vdots \\ 0 & \dots & \ddots & 0 \\ 0 & 0 & \dots & a_n \end{pmatrix}, \qquad a_1 + a_2 + \cdots + a_n = 0 \,.$$

This algebra has a root system

$$\Delta = \{\alpha_{ij} = \epsilon_i - \epsilon_j \; i, j = 1, \dots, n \,, \; i \neq j\} \,,$$

$$\Delta^+ = \{\alpha_{ij} = \epsilon_i - \epsilon_j \; i, j = 1, \dots, n \,, \; i < j\} \,,$$

$$\Delta^- = \{\alpha_{ij} = \epsilon_i - \epsilon_j \; i, j = 1, \dots, n \,, \; i > j\} \,,$$

$$\Pi = \{\alpha_{i\,i+1} = \epsilon_i - \epsilon_{i+1} \; i = 1, \dots, n-1\} \,,$$

where ϵ_i is the linear function $\epsilon_i : \mathrm{diag}(a_1 \dots a_n) \mapsto a_i$. The root space for α_{ij} is

$$\mathfrak{g}_{\alpha_{ij}} = \mathrm{span}_{\mathbb{C}} \{E_{ij}\}$$

where, as usual, E_{ij} denotes the elementary matrix with 1 in the position (i, j) and zero everywhere else. The Dynkin diagram for this system of roots is

With respect to this simple system, $\mathfrak{b}^-$ is the algebra of lower triangular matrices. All the other Borel subalgebras of $\mathfrak{sl}_n(\mathbb{C})$ are conjugated to this one.

A simple system of $\mathfrak{sl}_4(\mathbb{C})$ is

$$\Pi_{\mathfrak{sl}_4} = \{\alpha_{12}, \alpha_{23}, \alpha_{34}\} \,.$$

The Borel subalgebra $\mathfrak{b}^-$ consists of matrices of the type

$$\begin{pmatrix} * & 0 & 0 & 0 \\ * & * & 0 & 0 \\ * & * & * & 0 \\ * & * & * & * \end{pmatrix} . \tag{4.24}$$

The parabolic subalgebras, instead, are conjugated to lower, block triangular matrices. An example of parabolic subalgebra of $\mathfrak{sl}_4(\mathbb{C})$ is the algebra of matrices of the type

$$\begin{pmatrix} * & * & 0 & 0 \\ * & * & 0 & 0 \\ * & * & * & * \\ * & * & * & * \end{pmatrix} = \begin{pmatrix} A & 0 \\ B & C \end{pmatrix}, \qquad A, B, C \in \mathrm{M}_2(\mathbb{C}). \tag{4.25}$$

This corresponds to the subset $\Phi_0 \subset \Pi_{\mathfrak{sl}_4}$

$$\Phi_0 = \{\alpha_{12}, \alpha_{34}\}. \tag{4.26}$$

The algebra (4.25) is the Lie algebra of the complex Poincaré group plus dilations defined in Chapter 2, Definition 2.7.2. In Table 4.2 the reader can consult the complete list of parabolic subalgebras of $\mathfrak{sl}_4(\mathbb{C})$ containing $\mathfrak{b}^-$.

Remark 4.5.3. It is clear that the list in Table 4.2 depends on the lower Borel subalgebra $\mathfrak{b}^-$, which is linked to the choice of a positive system. If we choose $\mathfrak{b}^+$ instead of $\mathfrak{b}^-$, the parabolic subalgebras of $\mathfrak{sl}_4(\mathbb{C})$ containing $\mathfrak{b}^+$ are upper triangular (block-triangular) matrices. The subalgebras thus obtained are isomorphic (in fact, conjugated) to the ones we have described. ■

Parabolic subgroups of a semisimple Lie group G are subgroups whose Lie algebra is a parabolic subalgebra of $\mathfrak{g} = \mathrm{Lie}(G)$, and the same for Borel subgroups. So the Poincaré group times dilations of Definition 2.7.2 is a parabolic subgroup of the conformal group. We have the following result that we give without proof (see [18]).

Proposition 4.5.4. *Let G be a closed subgroup of the algebraic semisimple group G. Then F is a parabolic subgroup if and only if G/F is a projective algebraic variety.* ■

For example, the group I_0 defined in (2.45) is the group of upper block-triangular matrices

$$\begin{pmatrix} L & M \\ 0 & R \end{pmatrix}, \qquad L, M, R \in \mathrm{M}_2(\mathbb{C}), \quad \det L \det R = 1,$$

and it is the isotropy group of the homogeneous space

$$G(2,4) \cong \mathrm{SL}_4(\mathbb{C})/I_0, \quad \det L \det R = 1.$$

The group I_0 is related by conjugation to the group of lower block triangular matrices. For example, one has the map

$$g = \begin{pmatrix} 0 & \mathrm{id} \\ -\mathrm{id} & 0 \end{pmatrix}, \qquad g \begin{pmatrix} L & M \\ 0 & R \end{pmatrix} g^{-1} = \begin{pmatrix} R & 0 \\ -M & L \end{pmatrix}. \tag{4.27}$$

So I_0 is a parabolic subgroup and Proposition 4.5.4 assures us that $G(2,4)$ is a projective variety.

The Poincaré group times dilations was defined as the subgroup of $\mathrm{SL}_4(\mathbb{C})$ preserving the big cell of $G(2,4)$. If the isotropy group is the group of the upper, block triangular matrices, it is natural that the Poincaré group times dilations arises in our construction as the group of lower, block triangular matrices. Both groups are parabolic subgroups and isomorphic, though.

More generally, for $\mathrm{SL}_n(\mathbb{C})$ the parabolic subgroups are isotropy groups of *flag varieties*:

Definition 4.5.5. Let V be a finite dimensional vector space of dimension n over $\mathbb{C}$. A *flag* in V is a sequence of subspaces $V^0, V^1, \dots, V^n$ of V such that

$$\{0\} = V^0 \subsetneq V^1 \subsetneq V^2 \subsetneq \cdots \subsetneq V^k = V\,.$$

If $\dim V^i = d_i$, then the set of such flags is denoted with $\mathcal{F}\ell(d_1, d_2, \dots, d_k)$, with $d_k = n$.

For the case in which $d_i = i$, $i = 0, \dots, n$ and V is identified with $\mathbb{C}^n$, then $\mathcal{F}\ell(1, 2, 3, \dots, n)$ is called the *complete flag*. ■

If $F_{d_1 \dots d_k}$ is the subgroup of $\mathrm{SL}_4(\mathbb{C})$ that stabilizes the vector spaces in Definition 4.5.5

$$F_{d_1 \dots d_k} = \{X \in \mathrm{SL}_4(\mathbb{C}) \mid X(V^i) \subset V^i,\ i = 1, \dots, k\}\,, \tag{4.28}$$

then $F_{d_1 \dots d_k}$ is a parabolic subgroup and

$$\mathcal{F}\ell(d_1 \dots d_k) = \mathrm{SL}_4(\mathbb{C})/F_{d_1 \dots d_k}\,.$$

The parabolic subgroup $F_{d_1 \dots d_k}$ corresponds to the choice $\mathfrak{b}^+$, in other words, $F_{d_1 \dots d_k}$ consists of upper, block triangular matrices. To switch to the description in terms of lower triangular matrices, it is enough to substitute in (4.28) X by the transpose matrix X^t. Also, for the complete flag, F_{1234} is the upper Borel subgroup, that is $\mathfrak{b}^+ = \mathrm{Lie}(F_{1234})$.

In Table 4.2 one can also read the flag associated to each parabolic subalgebra. The flags with only one subspace V^d are called Grassmannians and are denoted as $G(d, n)$.

We now turn to examine the super case. We could proceed to define the Borel and parabolic sub superalgebras similarly to what we did for the

Table 4.2 Parabolic subalgebras of $\mathfrak{sl}_4(\mathbb{C})$ and flags.

Parabolic subalgebra	$\mathbf{\Phi} \subset \mathbf{\Pi}_{\mathfrak{sl}_4}$	**Flag**
$\begin{pmatrix} * & 0 & 0 & 0 \\ * & * & 0 & 0 \\ * & * & * & 0 \\ * & * & * & * \end{pmatrix}$	$\emptyset$	$\mathcal{F}\ell(1,2,3,4)$
$\begin{pmatrix} * & * & 0 & 0 \\ * & * & 0 & 0 \\ * & * & * & 0 \\ * & * & * & * \end{pmatrix}$	$\{\alpha_{12}\}$	$\mathcal{F}\ell(2,3,4)$
$\begin{pmatrix} * & 0 & 0 & 0 \\ * & * & * & 0 \\ * & * & * & 0 \\ * & * & * & * \end{pmatrix}$	$\{\alpha_{23}\}$	$\mathcal{F}\ell(1,3,4)$
$\begin{pmatrix} * & 0 & 0 & 0 \\ * & * & 0 & 0 \\ * & * & * & * \\ * & * & * & * \end{pmatrix}$	$\{\alpha_{34}\}$	$\mathcal{F}\ell(1,2,4)$
$\begin{pmatrix} * & * & * & 0 \\ * & * & * & 0 \\ * & * & * & 0 \\ * & * & * & * \end{pmatrix}$	$\{\alpha_{12}, \alpha_{23}\}$	$G(3,4) := \mathcal{F}\ell(3,4)$
$\begin{pmatrix} * & 0 & 0 & 0 \\ * & * & * & * \\ * & * & * & * \\ * & * & * & * \end{pmatrix}$	$\{\alpha_{23}, \alpha_{34}\}$	$G(1,4) := \mathcal{F}\ell(1,4)$
$\begin{pmatrix} * & * & 0 & 0 \\ * & * & 0 & 0 \\ * & * & * & * \\ * & * & * & * \end{pmatrix}$	$\{\alpha_{12}, \alpha_{34}\}$	$G(2,4) := \mathcal{F}\ell(2,4)$

ordinary setting. But the issue is a bit more involved (see Refs. [88, 109, 123, 153] for details).

Since we are especially interested in the superalgebra $\mathfrak{sl}(4|1)$ we prefer to consider only the Lie superalgebra $\mathfrak{sl}(m|n)$ with $m \neq n$ [3] and $m+n \geq 3$.

[3] $\mathfrak{sl}(n|n)$ is not a simple Lie algebra. It has a center given by matrices the $\lambda \mathrm{id}_{2n}$, $\lambda \in \mathbb{C}$. One has to quotient by this center in order to get a simple Lie superalgebra (refer to Appendix C).

The parabolic subalgebras of $\mathfrak{sl}(m|n)$ are also related with the *flag supervarieties* of the supervector space $\mathbb{C}^{m|n}$. In Sections 4.7 and 4.11 we construct explicitly two examples of superflags, the super Grassmannian $G(2|0,4|1)$ and the superflag $\mathcal{F}\ell(2|0,2|1,4|1)$ as supervarieties and as homogeneous spaces by means of their functors of points. Here we want to give a definition of the superflag in purely linear algebraic terms, mimicking Definition 4.5.5. This will be enough to establish the relation with the parabolic subalgebras.

Definition 4.5.6. Let V be a finite dimensional super vector space of dimension $m|n$ over $\mathbb{C}$

$$V = V_0 \oplus V_1 .$$

A *superflag* in V is a sequence of super vector subspaces $V^0, V^1, \dots, V^k$ of V such that

$$V^i = (V^i)_0 \oplus (V^i)_1 , \qquad (V^i)_0 = V^i \cap V_0 , \quad (V^i)_1 = V^i \cap V_1 ,$$

and

$$\{0\} = V^0 \subsetneq V^1 \subsetneq V^2 \subsetneq \cdots \subsetneq V^n = V .$$

If we denote $\dim V_i = p_i|q_i$ and we identify $V \cong \mathbb{C}^{m|n}$, then the flag defined by the sequence above is denoted as $\mathcal{F}\ell(p_1|q_1, p_2|q_2, \dots, m|n)$. If

$$p_{i+1}|q_{i+1} = (p_i + 1)|q_i \quad \text{or} \quad p_{i+1}|q_{i+1} = p_i|q_i + 1$$

and $k = m+n$, then $\mathcal{F}\ell(p_1|q_1, p_2|q_2, \dots, m|n)$ is called a *complete superflag*. ■

Notice that, differently to the non super case, there are more than one complete superflag.

Remark 4.5.7. It is immediate to realize that $\mathcal{F}\ell(p_1|q_1, p_2|q_2, \dots, m|n)$ can be identified with the product of classical flags $\mathcal{F}\ell(p_1, p_2, \dots, m) \times \mathcal{F}\ell(q_1, q_2, \dots, n)$ of V_0 and V_1 respectively. Each of the classical flags has a manifold structure, hence, in particular, it is a topological space. It is on this topological space that we shall build in Section 4.11 (in the special case of $V_0 = \mathbb{C}^4$ and $V_1 = \mathbb{C}$) a supermanifold and a supervariety structure. Our procedure can easily be generalized to arbitrary superflags. ■

We can associate to a superflag $\mathcal{F}\ell(p_1|q_1, p_2|q_2, \dots, m|n)$ a subalgebra $\mathfrak{f}$ of $\mathfrak{sl}(m|n)$ consisting of all the elements stabilizing the subspaces in $\mathcal{F}\ell(p_1|q_1, p_2|q_2, \dots, m|n)$

$$\mathfrak{f} = \{X \in \mathfrak{sl}(m|n) \mid X(V^i) \subset V^i,\ (V_1, \dots V_k) \in \mathcal{F}\ell(p_1|q_1, p_2|q_2, \dots, m|n)\} .$$

As in the even case, we shall call such subalgebras $\mathfrak{f}$ the *parabolic subalgebras* [88].

We want to study the parabolic subsuperalgebras of $\mathfrak{sl}(m|n)$. We first fix a convenient basis and then describe the root system of $\mathfrak{sl}(m|n)$. Let

$$\{e_1, \dots, e_m, \epsilon_1, \dots, \epsilon_n\} \tag{4.29}$$

be a basis of $\mathbb{C}^{m|n}$ compatible with the $\mathbb{Z}_2$ grading, so $\{e_r, r = 1, \dots, m\}$ are even vectors and $\{\epsilon_p, p = 1, \dots, n\}$ are odd vectors.

A Cartan subalgebra of $\mathfrak{sl}(m|n)$ is the set of traceless diagonal matrices

$$\begin{pmatrix} a_1 & \cdots & 0 & \cdots & 0 & 0 \\ \vdots & \ddots & 0 & \cdots & 0 & \vdots \\ 0 & 0 & a_m & 0 & \cdots & 0 \\ 0 & \cdots & 0 & b_{m+1} & \cdots & 0 \\ \vdots & \cdots & 0 & \vdots & \ddots & \vdots \\ 0 & \cdots & 0 & 0 & 0 & b_{m+n} \end{pmatrix} \qquad a_1 + \cdots + a_m = b_{m+1} + \cdots + b_{m+n} \,. \tag{4.30}$$

We also assume that each subspace V^i is spanned by a subset of the above basis and then say that the basis is *adapted* to the superflag. It is clear that each superflag has a basis adapted to it and compatible with the $\mathbb{Z}_2$-grading. Also, the Cartan subalgebra (4.30) preserves any superflag for which (4.29) is an adapted basis.

The root system of $\mathfrak{sl}(m|n)$ is $\Delta = \Delta_0 \cup \Delta_1$, where Δ_0 is the root system of the reductive Lie algebra $\mathfrak{sl}(m|n)_0 = \mathfrak{sl}(m) \oplus \mathfrak{sl}(n) \oplus \mathbb{C}$ and Δ_1 is the system of weights of the representation of $\mathfrak{sl}(m|n)_0$ on $\mathfrak{sl}(m|n)_1$.

$$\Delta_0 = \{\alpha_{rs} = \epsilon_r - \epsilon_s,\ r, s = 1, \dots, m,\ r \neq s\}$$

$$\cup \{\beta_{pq} = \delta_p - \delta_q,\ \ p, q = m+1, \dots, m+n,\ p \neq q\}\,,$$

$$\Delta_1 = \{\gamma_{rp} = \epsilon_r - \delta_p,\ \gamma_{pr} = \delta_p - \epsilon_r,\ r = 1, \dots, m,\ p = m+1, \dots, m+n\}\,.$$

where $\epsilon_i : \mathrm{diag}(a_1, \dots, a_m, b_1, \dots, b_n) \mapsto a_i$ and $\delta_j : \mathrm{diag}(a_1, \dots, a_m, b_1, \dots, b_n) \mapsto b_j$.

The root spaces are

$$g_{\alpha_{pq}} = \mathrm{span}\{E_{pq}\}\,, \qquad\qquad g_{\beta_{rs}} = \mathrm{span}\{E_{rs}\}\,,$$

$$g_{\gamma_{rp}} = \mathrm{span}\{E_{rp}\}\,, \qquad\qquad g_{\gamma_{pr}} = \mathrm{span}\{E_{pr}\}\,.$$

The Weyl group of a semisimple Lie superalgebra is the Weyl group of its reductive, even part. Due to the presence of odd roots, the Weyl group does not act transitively on the simple systems. Each simple system generates an associated positive system. Extending the Definition 4.5.1 to superalgebras, each conjugacy class of simple systems leads to a different Borel subalgebra. This is ultimately the reason why we have several Dynkin diagrams associated with the same simple Lie superalgebra of classical type.

For $\mathfrak{sl}(m|n)$ there are $\binom{m+n}{m}$ conjugacy classes of simple systems. The associated Borel subalgebras correspond to different complete flags. We are going to see this relation explicitly in the example of interest for us, $\mathfrak{sl}(4|1)$.

The *standard positive system* (see Ref. [69] for its definition) of roots of $\mathfrak{sl}(m|n)$ is the following:

$$\Delta_0^+ = \{\alpha_{rs} = \epsilon_r - \epsilon_s,\ r, s = 1, \ldots, m,\ r < s\}$$

$$\cup \{\beta_{pq} = \delta_p - \delta_q,\ p, q = m+1, \ldots, m+n,\ p < q\},$$

$$\Delta_1^+ = \{\gamma_{rp} = +(\epsilon_r - \delta_p),\ r = 1, \ldots, m,\ p = m+1, \ldots, m+n\},$$

for which a simple system is

$$\Pi = \{\alpha_{12}, \alpha_{23}, \ldots, \gamma_{m\,m+1}, \beta_{m+1\,m+2}, \ldots, \beta_{m+n-1\,m+n}\}.$$

The standard Dynkin diagram is

(4.31)

where the node $\otimes$ stands for the odd root $\gamma_{m\,m+1}$.

Let us work out the case of $\mathfrak{sl}(4|1)$. There are five conjugacy classes of simple systems and of positive systems. We list them in Table 4.3. The even part of the positive system is always the same,

$$\Delta_0^+ = \{\alpha_{12}, \alpha_{13}, \alpha_{14}, \alpha_{23}, \alpha_{24}, \alpha_{34}\},$$

so we will list only Δ_1^+ for each system. In the Dynkin diagram, the nodes are in the same order than the roots are listed in the positive system $\Pi^{(a)}$.

$\Pi^{(1)}$ is the standard simple system.

Each of the positive system generates a Borel subalgebra, corresponding to a complete superflag. We list them in Table 4.4.

Table 4.3 Simple and positive systems of $\mathfrak{sl}(4|1)$.

$\mathbf{\Pi}^{(\mathbf{a})}$	**Simple system**	**Odd pos. system**	**Dynkin diagram**
$\mathbf{\Pi}^{(1)}$	$\alpha_{12}, \alpha_{23}, \alpha_{34}, \gamma_{45}$	$\gamma_{15}, \gamma_{25}, \gamma_{35}, \gamma_{45}$	○—○—○—⊗
$\mathbf{\Pi}^{(2)}$	$\alpha_{12}, \alpha_{23}, \gamma_{35}, \gamma_{54}$	$\gamma_{15}, \gamma_{25}, \gamma_{35}, \gamma_{54}$	○—○—⊗—⊗
$\mathbf{\Pi}^{(3)}$	$\alpha_{12}, \gamma_{25}, \gamma_{53}, \alpha_{34}$	$\gamma_{15}, \gamma_{25}, \gamma_{53}, \gamma_{54}$	○—⊗—⊗—○
$\mathbf{\Pi}^{(4)}$	$\gamma_{15}, \gamma_{52}, \alpha_{23}, \alpha_{34}$	$\gamma_{15}, \gamma_{52}, \gamma_{53}, \gamma_{54}$	⊗—⊗—○—○
$\mathbf{\Pi}^{(5)}$	$\gamma_{51}, \alpha_{12}, \alpha_{23}, \alpha_{34}$	$\gamma_{51}, \gamma_{52}, \gamma_{53}, \gamma_{54}$	⊗—○—○—○

Table 4.4 Borel subalgebras of $\mathfrak{sl}(4|1)$ and complete superflags.

Simple system	**Borel subalgebra**	**Complete superflag**
$\mathbf{\Pi}^{(1)}$	$\left(\begin{array}{cccc\|c} * & 0 & 0 & 0 & 0 \\ * & * & 0 & 0 & 0 \\ * & * & * & 0 & 0 \\ * & * & * & * & 0 \\ \hline * & * & * & * & * \end{array}\right)$	$\mathcal{F}\ell(1\|0, 2\|0, 3\|0, 4\|0, 4\|1)$
$\mathbf{\Pi}^{(2)}$	$\left(\begin{array}{cccc\|c} * & 0 & 0 & 0 & 0 \\ * & * & 0 & 0 & 0 \\ * & * & * & 0 & 0 \\ * & * & * & * & * \\ \hline * & * & * & 0 & * \end{array}\right)$	$\mathcal{F}\ell(1\|0, 2\|0, 3\|0, 3\|1, 4\|1)$
$\mathbf{\Pi}^{(3)}$	$\left(\begin{array}{cccc\|c} * & 0 & 0 & 0 & 0 \\ * & * & 0 & 0 & 0 \\ * & * & * & 0 & * \\ * & * & * & * & * \\ \hline * & * & 0 & 0 & * \end{array}\right)$	$\mathcal{F}\ell(1\|0, 2\|0, 2\|1, 3\|1, 4\|1)$
$\mathbf{\Pi}^{(4)}$	$\left(\begin{array}{cccc\|c} * & 0 & 0 & 0 & 0 \\ * & * & 0 & 0 & * \\ * & * & * & 0 & * \\ * & * & * & * & * \\ \hline * & 0 & 0 & 0 & * \end{array}\right)$	$\mathcal{F}\ell(1\|0, 1\|1, 2\|1, 3\|1, 4\|1)$
$\mathbf{\Pi}^{(5)}$	$\left(\begin{array}{cccc\|c} * & 0 & 0 & 0 & * \\ * & * & 0 & 0 & * \\ * & * & * & 0 & * \\ * & * & * & * & * \\ \hline 0 & 0 & 0 & 0 & * \end{array}\right)$	$\mathcal{F}\ell(0\|1, 1\|1, 2\|1, 3\|1, 4\|1)$

We now look for parabolic subalgebras whose even part is (4.25). They will be our candidates for the Poincaré superalgebra plus dilations. The construction of the parabolic subalgebras containing a Borel subalgebra is as in the non super case (4.23). They are in one-to-one correspondence with proper subsets of the simple system corresponding to the Borel subalgebra. In the super case we have five different simple systems that can give non isomorphic parabolic subalgebras.

Let $\Phi^{(a)} \subset \Pi^{(a)}$. We want the even part of the parabolic subalgebra to be (4.25), so it seems clear that the roots in $\Phi_0 = \{\alpha_{12}, \alpha_{34}\}$ (4.26) must be in $\hat{\Phi}^{(a)}$, the set of positive roots generated by $\Phi^{(a)}$. Also, no other even roots can appear, since this would enlarge the even part of the subalgebra.

We list all the cases that satisfy these constraints in Table 4.5. We obtain a total of five different parabolic superalgebras,

$$\mathfrak{f}_1 = \left\{ \begin{pmatrix} * & * & 0 & 0 & 0 \\ * & * & 0 & 0 & 0 \\ * & * & * & * & 0 \\ * & * & * & * & 0 \\ \hline * & * & * & * & * \end{pmatrix} \right\}, \qquad \mathfrak{f}_2 = \left\{ \begin{pmatrix} * & * & 0 & 0 & 0 \\ * & * & 0 & 0 & 0 \\ * & * & * & * & * \\ * & * & * & * & * \\ \hline * & * & * & * & * \end{pmatrix} \right\},$$

$$\mathfrak{f}_3 = \left\{ \begin{pmatrix} * & * & 0 & 0 & 0 \\ * & * & 0 & 0 & 0 \\ * & * & * & * & * \\ * & * & * & * & * \\ \hline * & * & 0 & 0 & * \end{pmatrix} \right\}, \qquad \mathfrak{f}_4 = \left\{ \begin{pmatrix} * & * & 0 & 0 & * \\ * & * & 0 & 0 & * \\ * & * & * & * & * \\ * & * & * & * & * \\ \hline * & * & 0 & 0 & * \end{pmatrix} \right\},$$

$$\mathfrak{f}_5 = \left\{ \begin{pmatrix} * & * & 0 & 0 & * \\ * & * & 0 & 0 & * \\ * & * & * & * & * \\ * & * & * & * & * \\ \hline 0 & 0 & 0 & 0 & * \end{pmatrix} \right\}. \tag{4.32}$$

In the next section we will impose other natural conditions so that $\mathfrak{f}_3$ is singled out from this set.

4.6 The Poincaré superalgebra

We want now to find a suitable definition of the Poincaré superalgebra as a subsuperalgebra of the Wess-Zumino superalgebra. This is, in some way, a different and somehow inverse path with respect to the one we have

Table 4.5 Parabolic subalgebras of $\mathfrak{sl}(4|1)$ whose even part is the Poincaré algebra plus dilations.

Simple system	Φ	**Parab. superalg.**	**Superflag**
$\Pi^{(1)}$	α_{12}, α_{34}	$\mathfrak{f}_1$	$\mathcal{F}\ell(2\|0, 4\|0, 4\|1)$
$\Pi^{(1)}$	$\alpha_{12}, \alpha_{34}, \gamma_{45}$	$\mathfrak{f}_2$	$\mathcal{F}\ell(2\|0, 4\|1)$
$\Pi^{(2)}$	$\alpha_{12}, \gamma_{35}, \gamma_{54}$	$\mathfrak{f}_2$	$\mathcal{F}\ell(2\|0, 4\|1)$
$\Pi^{(3)}$	α_{12}, α_{34}	$\mathfrak{f}_3$	$\mathcal{F}\ell(2\|0, 2\|1, 4\|1)$
$\Pi^{(3)}$	$\alpha_{12}, \alpha_{34}, \gamma_{25}$	$\mathfrak{f}_4$	$\mathcal{F}\ell(2\|1, 4\|1)$
$\Pi^{(4)}$	$\gamma_{15}, \gamma_{52}, \alpha_{34}$	$\mathfrak{f}_2$	$\mathcal{F}\ell(2\|1, 4\|1)$
$\Pi^{(2)}$	$\alpha_{12}, \gamma_{35}, \gamma_{54}$	$\mathfrak{f}_4$	$\mathcal{F}\ell(2\|1, 4\|1)$
$\Pi^{(5)}$	α_{12}, α_{34}	$\mathfrak{f}_5$	$\mathcal{F}\ell(0\|1, 2\|1, 4\|1)$
$\Pi^{(5)}$	$\alpha_{12}, \alpha_{34}, \gamma_{51}$	$\mathfrak{f}_4$	$\mathcal{F}\ell(2\|1, 4\|1)$

taken for the non super case. The Poincaré algebra (or the Poincaré group, see Definition 4.6.1) was found by complying with the physical requirements. It is the group of transformations between inertial reference frames. Then the conformal algebra (or group) was found by suitably enlarging the Poincaré algebra to allow for symmetries of some physical theories, like electromagnetism.

However, in the super setting, one is investigating possible symmetries of theories that do not have yet a direct and clear connection with the experiments, as we saw in Section 3.4. We found the Wess-Zumino superalgebra in Section 4.4 by imposing only the minimality of the extension and the restriction on the odd part coming from the spin-statistics theorem. The result is non trivial and was clearly stated in Proposition 4.4.5: there are no other superalgebras satisfying both conditions. Our strategy for the Poincaré superalgebra is similar. We will look for a subalgebra of the Wess-Zumino superalgebra that enjoys the mathematical properties that the Poincaré algebra has as a subalgebra of the conformal algebra. As we will see, these properties are quite natural and also lead to the uniqueness of the Poincaré superalgebra.

This is not the way in which the Poincaré superalgebra was found in physics, since its discovery preceded the one of the Wess-Zumino algebra, but we think that the results on uniqueness are sufficiently important to introduce a change in the way this story is usually told.

The first condition that we impose for the supersymmetric extension of the Poincaré algebra plus dilations is that it should be a parabolic subalgebra of the superconformal algebra $\mathfrak{sl}(4|1)$. As we already mentioned in

Section 4.5, the idea is to maintain the construction of the non super case, where the Poincaré algebra plus dilations is a parabolic subalgebra and the conformal space is a flag variety (a Grassmannian in fact). This is already a non trivial hypothesis that one may find justified for the sake of simplicity. It leaves us with five possible choices, the five parabolic subalgebras found in (4.32).

The next step is to impose the reality condition. We already determined the real form of the superconformal group in (4.34). Let us recall it here

$$\mathfrak{su}(2,2|1)_{\pm} = \left\{ \begin{pmatrix} p & \alpha \\ \beta & \mathrm{i}z \end{pmatrix} \mid p \in \mathrm{M}_4(\mathbb{C}),\ \alpha, \beta^t \in \mathbb{C}^4,\ z \in \mathbb{R}; \right. \tag{4.33}$$

$$\left. Fp + p^\dagger F = 0,\ \mathrm{tr}p = \mathrm{i}z,\ \alpha = \pm \mathrm{i}F\beta^\dagger \right\}. \tag{4.34}$$

The super Poincaré algebra plus dilations must have a real form that is embedded in $\mathfrak{su}(2,2|1)_{\pm}$. For the even part there is no problem: it corresponds to the Lie algebra of the real Poincaré group times dilations that we wrote in Proposition 2.11.4. For the odd part, we have the condition

$$\alpha = \pm \mathrm{i}F\beta^\dagger, \qquad F = \mathrm{i}\begin{pmatrix} 0 & -\mathrm{id} \\ \mathrm{id} & 0 \end{pmatrix},$$

which can only be satisfied by the parabolic subalgebra

$$\mathfrak{f}_3 = \left\{ \left(\begin{array}{cccc|c} * & * & 0 & 0 & 0 \\ * & * & 0 & 0 & 0 \\ * & * & * & * & * \\ * & * & * & * & * \\ \hline * & * & 0 & 0 & * \end{array} \right) \right\}$$

from the set of five given in (4.32). This singles out this parabolic subalgebra as the super Poincaré algebra plus dilations.

Definition 4.6.1. The complex *super Poincaré algebra plus dilations* is defined as the subalgebra of $\mathfrak{sl}(4|1)$

$$\mathfrak{f}^c = \left\{ \begin{pmatrix} l & 0 & 0 \\ n & r & \alpha \\ \beta & 0 & c \end{pmatrix},\ l, n, r \in \mathrm{M}_2(\mathbb{C}),\ \alpha, \beta^t \in \mathbb{C}^2,\ c \in \mathbb{C},\ \mathrm{tr}\, l + \mathrm{tr}\, r = c \right\}.$$

■

So we have

$$\mathfrak{f}_0^c = \left\{ \begin{pmatrix} l & 0 & 0 \\ n & r & 0 \\ 0 & 0 & c \end{pmatrix}, \quad l, n, r \in \mathrm{M}_2(\mathbb{C}), \quad c \in \mathbb{C}, \quad \mathrm{tr}\, l + \mathrm{tr}\, r = c \right\},$$

$$\mathfrak{f}_1^c = \left\{ \begin{pmatrix} 0 & 0 & 0 \\ 0 & 0 & \alpha \\ \beta & 0 & 0 \end{pmatrix}, \quad \alpha, \beta^t \in \mathbb{C}^2 \right\}.$$

Let us write the commutation relations explicitly. For the even part we just have

$$\left[\begin{pmatrix} l & 0 & 0 \\ 0 & r & 0 \\ 0 & 0 & c \end{pmatrix}, \begin{pmatrix} l' & 0 & 0 \\ 0 & r' & 0 \\ 0 & 0 & c' \end{pmatrix} \right] = \begin{pmatrix} [l, l'] & 0 & 0 \\ 0 & [r, r'] & 0 \\ 0 & 0 & 0 \end{pmatrix},$$

$$\left[\begin{pmatrix} l & 0 & 0 \\ 0 & r & 0 \\ 0 & 0 & c \end{pmatrix}, \begin{pmatrix} 0 & 0 & 0 \\ n' & 0 & 0 \\ 0 & 0 & 0 \end{pmatrix} \right] = \begin{pmatrix} 0 & 0 & 0 \\ -n'l + rn' & 0 & 0 \\ 0 & 0 & 0 \end{pmatrix},$$

and the rest zero. The commutation relations involving the odd part are

$$\left[\begin{pmatrix} l & 0 & 0 \\ n & r & 0 \\ 0 & 0 & c \end{pmatrix}, \begin{pmatrix} 0 & 0 & 0 \\ 0 & 0 & \alpha' \\ \beta' & 0 & 0 \end{pmatrix} \right] = \begin{pmatrix} 0 & 0 & 0 \\ 0 & 0 & r\alpha' - c\alpha' \\ c\beta' - \beta' l & 0 & 0 \end{pmatrix},$$

$$\left[\begin{pmatrix} 0 & 0 & 0 \\ 0 & 0 & \alpha \\ \beta & 0 & 0 \end{pmatrix}, \begin{pmatrix} 0 & 0 & 0 \\ 0 & 0 & \alpha' \\ \beta' & 0 & 0 \end{pmatrix} \right] = \begin{pmatrix} 0 & 0 & 0 \\ \alpha\beta' + \alpha'\beta & 0 & 0 \\ 0 & 0 & 0 \end{pmatrix}. \qquad (4.35)$$

The first thing that we can observe is that $\mathfrak{f}_1^c$ is in the representation

$$D(1/2, 0) \oplus D(0, 1/2) \cong S^+ \oplus S^-$$

of the Lorentz algebra $\mathfrak{sl}_2(\mathbb{C}) \oplus \mathfrak{sl}_2(\mathbb{C})$ (see (4.14) and the comments after). In fact, the second factor appears as the contragredient representation of $D(0, 1/2)$, but it is equivalent to $D(0, 1/2)$ itself. This is consistent with what we know about real spin representations. In order to impose a reality condition on a spin representation of the Lorentz algebra one needs the direct sum $S^+ \oplus S^-$ which, as a real representation is irreducible. Separately, $S^\pm$ are genuinely complex representations, so we cannot expect to obtain a real superalgebra with copies of only S^+ or S^- in the odd part. Both representations must always appear in pairs, and this is a condition that only $\mathfrak{f}_3$ in (4.32) satisfies.

Another thing that we can observe is that the *translation algebra*

$$\mathfrak{t}_0^c = \left\{ \begin{pmatrix} 0 & 0 & 0 \\ n & 0 & 0 \\ 0 & 0 & 0 \end{pmatrix}, \quad n \in \mathrm{M}_2(\mathbb{C}) \right\},$$

which is an abelian subalgebra, can be extended to a superalgebra, the *translation superalgebra*

$$\mathfrak{t}^c = \left\{ \begin{pmatrix} 0 & 0 & 0 \\ n & 0 & \alpha \\ \beta & 0 & 0 \end{pmatrix}, \quad n \in \mathrm{M}_2(\mathbb{C}), \quad \alpha, \beta^t \in \mathbb{C}^2 \right\}, \tag{4.36}$$

which is not abelian anymore. It is 2-step nilpotent, a property that somehow replaces abelianity in the super setting. The translation superalgebra plays a role similar to the one played by the translation algebra in the non super case.

Let us see the analogy. For the non super case, we have the Poincaré algebra plus dilations as a subalgebra of $\mathfrak{sl}_4(\mathbb{C})$

$$\mathfrak{f}_0^c = \left\{ \begin{pmatrix} l & 0 \\ n & r \end{pmatrix} \;\middle|\; l, n, r \in \mathrm{M}_2(\mathbb{C}), \quad \operatorname{tr} l + \operatorname{tr} r = 0 \right\},$$

$$\mathfrak{sl}_4(\mathbb{C}) = \left\{ \begin{pmatrix} l & m \\ n & r \end{pmatrix} \;\middle|\; l, m, n, r \in \mathrm{M}_2(\mathbb{C}), \quad \operatorname{tr} l + \operatorname{tr} r = 0 \right\}. \tag{4.37}$$

We can identify the quotient $\mathfrak{sl}_4(\mathbb{C})/\mathfrak{f}_0^c$ with

$$\mathfrak{sl}_4(\mathbb{C})/\mathfrak{f}_0^c \cong \left\{ \begin{pmatrix} 0 & m \\ 0 & 0 \end{pmatrix} \;\middle|\; m \in \mathrm{M}_2(\mathbb{C}) \right\},$$

which is again an abelian algebra, isomorphic to the translation algebra $\mathfrak{t}_0$. As we have explained in (4.27), the Poincaré group times dilations F_0^c is conjugated to the isotropy group I_0 of the conformal space $G(2,4)$, so one can write

$$G(2,4) \cong \mathrm{SL}_4(\mathbb{C})/F_0^c\,.$$

The translation algebra can then be identified with the tangent space at the identity coset of $G(2,4)$ and in turn, of the Minkowski space.

In the super case, we recall that

$$\mathfrak{sl}(4|1) = \left\{ \begin{pmatrix} l & m & \alpha_1 \\ n & r & \alpha_2 \\ \beta_1 & \beta_2 & c \end{pmatrix}, \quad l, m, n, r \in \mathrm{M}_2(\mathbb{C}), \quad \alpha_i, \beta_i^t \in \mathbb{C}^2, \quad c \in \mathbb{C}, \right.$$

$$\left. \operatorname{tr} l + \operatorname{tr} r = c \right\},$$

and

$$\mathfrak{sl}(4|1)/\mathfrak{f}^c \cong \left\{ \begin{pmatrix} 0 & m & \alpha_1 \\ 0 & 0 & 0 \\ 0 & \beta_2 & 0 \end{pmatrix}, \quad m \in \mathrm{M}_2(\mathbb{C}), \quad \alpha_1, \beta_2^t \in \mathbb{C}^2 \right\}. \tag{4.38}$$

The last one closes a subalgebra isomorphic to the translation superalgebra $\mathfrak{t}$.

According to Table 4.5, the superflag $\mathcal{F}\ell(2|0, 2|1, 4|1)$ corresponds to the conformal superspace. More precisely, we shall give to the topological space $\mathcal{F}\ell(2|0, 2|1, 4|1)$ a supermanifold structure (by an abuse of notation still denoted by $\mathcal{F}\ell(2|0, 2|1, 4|1)$), together with a natural action of the supergroup $\mathrm{SL}(4|1)$. In this way the supermanifold $\mathcal{F}\ell(2|0, 2|1, 4|1)$ becomes identified with a quotient of $\mathrm{SL}(4|1)$ by a parabolic subgroup and its tangent space at the identity element is $\mathfrak{sl}(4|1)/\mathfrak{f}^c$. We will give a thorough account of the construction of the superflag as a supermanifold in Section 4.11 and as an homogeneous space in Section 4.12, but we can already see that at the infinitesimal level, the quotient $\mathfrak{sl}(4|1)/\mathfrak{f}^c$ is the model of the conformal superspace (or the Minkowski superspace, which will also be defined as the big cell).

Let us now define the *real super Poincaré algebra plus dilations*. It is useful to write $j = \pm 1$ to deal with the alternative signs. The reality condition (4.34) for $\mathfrak{su}(2, 2|1)_\pm$ in the notation (4.37) becomes

$$\mathfrak{su}(2,2|1)_\pm = \left\{ \begin{pmatrix} l & m & j\beta_2^\dagger \\ n & -l^\dagger & -j\beta_1^\dagger \\ \beta_1 & \beta_2 & \mathrm{i}z \end{pmatrix}, \quad l, m, n, \in \mathrm{M}_2(\mathbb{C}), \quad \beta_i^t \in \mathbb{C}^2, \quad z \in \mathbb{R}, \right.$$
$$\left. 2\mathrm{Im}(\mathrm{tr}\, l) = z, \quad m^\dagger = m, \quad n^\dagger = n \right\}.$$

We have substituted $c = \mathrm{i}z$.

We have then the following definitions:

Definition 4.6.2.

- The *real Poincaré superalgebra plus dilations* $\mathfrak{f}$ is the Lie superalgebra given by the matrices

$$\begin{pmatrix} l & 0 & 0 \\ n & -l^\dagger & -j\beta^\dagger \\ \beta & 0 & \mathrm{i}z \end{pmatrix} \tag{4.39}$$

with $l, n \in \mathrm{M}_2(\mathbb{C})$ and $n = n^\dagger$, $z \in \mathbb{R}$ and $\beta^t \in \mathbb{C}^2$. It is not difficult to check that the two alternative signs that appear in (4.39) give isomorphic superalgebras, the isomorphism being realized by $n \longrightarrow -n$ and the rest unchanged.

- The *real translation superalgebra* $\mathfrak{t}$ is the Lie superalgebra given by matrices

$$\begin{pmatrix} 0 & 0 & 0 \\ n & 0 & -j\beta^\dagger \\ \beta & 0 & 0 \end{pmatrix}$$

with $n \in \mathrm{M}_2(\mathbb{C})$, $n = n^\dagger$ and $\beta^t \in \mathbb{C}^2$. Obviously, the comment on the two alternative signs also applies here. It is isomorphic to the superalgebra of matrices

$$\begin{pmatrix} 0 & m & j\delta^\dagger \\ 0 & 0 & 0 \\ 0 & \delta & 0 \end{pmatrix}$$

with $m \in \mathrm{M}_2(\mathbb{C})$, $m = m^\dagger$ and $\delta^t \in \mathbb{C}^2$.

■

We want to extend the infinitesimal results to obtain the Minkowski superspace time as the big cell inside a certain super flag manifold, realized as homogeneous space for the super conformal group.

4.7 The Grassmannian supermanifold $\mathrm{Gr}^{\mathrm{ch}}$

In this section we start the *global theory*, that is, we leave the Lie superalgebra environment, corresponding to the infinitesimal setting, and we concentrate our attention on the Poincaré and conformal supergroups and their homogeneous spaces: the *Minkowski and conformal superspaces*.

We first introduce the complex Grassmannian supermanifold of $2|0$ spaces in a $4|1$ dimensional complex vector space. This is the flag $G(2|0, 4|1) := \mathcal{F}\ell(2|0, 4|1)$ and it is associated to the parabolic subalgebra $\mathfrak{f}_2$ in (4.32). The reason why we start with this flag and not with $\mathcal{F}\ell(2|0, 4|0, 4|1)$ is twofold. On one hand, this is a simpler case with which we introduce the techniques. On the other hand, this is the prototype for the *chiral conformal superspace*, which has a considerable importance in physics.

The space $G(2|0,4|1)$ does not have a real form compatible with the action of the Poincaré group. To see this, we mimic what we did with $\mathcal{F}\ell(2|0,4|0,4|1)$ in Section 4.6: we will deduce from the superalgebra the infinitesimal structure of this space. We have that

$$\mathfrak{f}_2 = \left\{ \begin{pmatrix} l & 0 & 0 \\ n & r & \alpha \\ \beta_1 & \beta_2 & c \end{pmatrix}, \quad l, n, r \in \mathrm{M}_2(\mathbb{C}), \quad \alpha, \beta_i^t \in \mathbb{C}^2, \quad c \in \mathbb{C}, \quad \mathrm{tr}\, l + \mathrm{tr}\, r = c \right\},$$

so

$$\mathfrak{sl}(4|1)/\mathfrak{f}_2 \cong \left\{ \begin{pmatrix} 0 & m & \delta \\ 0 & 0 & 0 \\ 0 & 0 & 0 \end{pmatrix}, \quad m \in \mathrm{M}_2(\mathbb{C}), \quad \delta \in \mathbb{C}^2 \right\}. \tag{4.40}$$

This space is an abelian complex subalgebra of dimension $4|2$, but we see readily that the odd part contains only the spinorial representation $S^+ = D(1/2, 0)$ of the Lorentz subalgebra. Then, it cannot have a real form that is preserved by the Lorentz group.

In Section 4.16 we will explain how this space arises in physics. It always appears together with the *antichiral superspace*, which is related to $\mathfrak{f}_4$ and $\mathcal{F}\ell(2|1,4|1)$. As a consequence, neither $\mathfrak{f}_2$ nor $\mathfrak{f}_4$ have direct physical significance, since they do not represent symmetry transformations of any physical theory. Instead, their intersection

$$\mathfrak{f}_2 \cap \mathfrak{f}_4 = \mathfrak{f}_3$$

is, once again, the right superalgebra to consider.

Our strategy will be to define a supermanifold that locally is the complex analytic supermanifold $\mathbb{C}^{4|2}$; we will define it through the gluing conditions. This will be our candidate for $G(2|0,4|1)$ and we will denote it as $\mathrm{Gr}^{\mathrm{ch}}$. Next, we will define a functor that has encoded the geometrical meaning of Definition 4.5.6. We will then prove that it is, in fact, the functor of points of $\mathrm{Gr}^{\mathrm{ch}}$. Finally we will show that $\mathrm{Gr}^{\mathrm{ch}}$ can be seen as an homogeneous space for the superconformal group $\mathrm{SL}(4|1)$.

Let us start with $\mathrm{Gr}^{\mathrm{ch}}$. The underlying topological space is $G(2,4)$, which admits an open cover by the six open subsets U_{ij}, with $i < j$, $i, j = 1, \dots, 4$, described in Chapter 2, (2.46). On each of the subsets U_{ij} we define the supermanifold structure:

$$\mathcal{U}_{ij} = (U_{ij}, \mathcal{O}_{ij}), \qquad \mathcal{O}_{ij} = \mathcal{O}_{\mathbb{C}^{4|2}},$$

where $\mathcal{O}_{\mathbb{C}^{4|2}}$ is the structural sheaf of the complex analytic supermanifold $\mathbb{C}^{4|2}$. Hence $\mathcal{U}_{ij} \cong \mathbb{C}^{4|2}$ as supermanifolds. On the supermanifold $\mathcal{U}_{ij}$ we have six global coordinates that we denote as

$$u^{ij}_{k1}, \quad u^{ij}_{k2}, \quad \nu^{ij}_{51}, \quad \nu^{ij}_{52}, \qquad k \neq i,j, \quad 1 \leq k \leq 4\,.$$

They can be conveniently organized in a 2×5 matrix as in the in the following example:

$$\begin{pmatrix} u^{23}_{11} & u^{23}_{12} \\ 1 & 0 \\ 0 & 1 \\ u^{23}_{41} & u^{23}_{42} \\ \nu^{23}_{51} & \nu^{23}_{52} \end{pmatrix}. \tag{4.41}$$

In general, we organize the global coordinates as follows: the place of the rows i, j is occupied by the identity matrix and the coordinates are in the entries corresponding to the numbering of the subindices.

The structural sheaf $\mathcal{O}_{\mathrm{Gr}^{\mathrm{ch}}}$ of $\mathrm{Gr}^{\mathrm{ch}}$ is $\mathcal{O}_{\mathrm{Gr}^{\mathrm{ch}}}|_{U_{ij}} = \mathcal{O}_{ij}$ on each open set. We need now to give supermanifolds morphisms among the $\mathcal{U}_{ij}$ that will correspond to the change of chart in $\mathrm{Gr}^{\mathrm{ch}}$:

$$\phi^{ij}_{kl} : \mathcal{U}_{ij} \cap \mathcal{U}_{kl} \longrightarrow \mathcal{U}_{ij} \cap \mathcal{U}_{kl}\,,$$

where the submanifold in the domain is thought as an open supermanifold of $\mathcal{U}_{ij}$, while the submanifold in the codomain is thought as an open supermanifold of $\mathcal{U}_{kl}$. In other words we have

$$\phi^{ij}_{kl} : (U_{ij} \cap U_{kl}, \mathcal{O}_{ij}|_{U_{ij}}) \longrightarrow (U_{ij} \cap U_{kl}, \mathcal{O}_{kl}|_{U_{kl}})\,.$$

Once such morphisms are defined we will need to check the gluing conditions, namely

$$\phi^{kl}_{rs} \circ \phi^{ij}_{kl} = \phi^{ij}_{rs}, \qquad \text{in} \qquad \mathcal{U}_{ij} \cap \mathcal{U}_{kl} \cap \mathcal{U}_{rs}\,.$$

It is easier to see it with an example. Let us consider

$$\phi^{12}_{23} : \mathcal{U}_{12} \cap \mathcal{U}_{23} \longrightarrow \mathcal{U}_{12} \cap \mathcal{U}_{23}\,. \tag{4.42}$$

By the Chart Theorem 1.4.6, to give this morphism it is enough to specify four even and two odd global sections in $\mathcal{O}_{12}|_{U_{23}}$. The rule to determine such sections is as follows: first, we write the global sections of $\mathcal{U}_{12}$ in matrix form (see (4.41)) and then we multiply on the right such matrix by a 2×2 invertible matrix in such way that the result is in the form of global sections in $\mathcal{U}_{23}$ (again, see (4.41)). For the moment this is just a rule, but its geometrical meaning will appear evident when we introduce the notation of the functor of points. We have

$$\begin{pmatrix} 1 & 0 \\ 0 & 1 \\ u_{31} & u_{32} \\ u_{41} & u_{42} \\ \nu_{51} & \nu_{52} \end{pmatrix} \begin{pmatrix} -u_{32}u_{31}^{-1} & u_{31}^{-1} \\ 1 & 0 \end{pmatrix} = \begin{pmatrix} -u_{32}u_{31}^{-1} & u_{31}^{-1} \\ 1 & 0 \\ 0 & 1 \\ -u_{41}u_{32}u_{31}^{-1} + u_{42} & u_{41}u_{31}^{-1} \\ -\nu_{51}u_{32}u_{31}^{-1} + \nu_{52} & \nu_{51}u_{31}^{-1} \end{pmatrix}. \tag{4.43}$$

We have eliminated the superscripts in u_{kl}^{12} to lighten the notation. Notice that when we consider the even sections as sections on the *ordinary* $G(2, 4)$, the condition for sections in U_{12} to describe coordinates of points in $U_{12} \cap U_{23}$ is that the determinant of the 2×2 minor corresponding to rows $(2, 3)$ is invertible (see Section 2.5). Recalling that a super section is invertible if and only if its reduced part is so, we can safely assume that u_{31} is invertible.

Then, the morphism (4.42) is defined on the topological points using formula (4.43):

$$U_{12} \cap U_{23} \xrightarrow{|\phi_{23}^{12}|} U_{12} \cap U_{23}$$

$$\begin{pmatrix} 1 & 0 \\ 0 & 1 \\ u_{31} & u_{32} \\ u_{41} & u_{42} \\ 0 & 0 \end{pmatrix} \longrightarrow \begin{pmatrix} -u_{32}u_{31}^{-1} & u_{31}^{-1} \\ 1 & 0 \\ 0 & 1 \\ -u_{41}u_{32}u_{31}^{-1} + u_{42} & u_{41}u_{31}^{-1} \\ 0 & 0 \end{pmatrix}.$$

We have identified the points in U_{12} and U_{23} with matrices with complex entries (the u_{ij}'s) as in (4.41), as one does in the ordinary setting.

On the sheaves we use again (4.43):

$$\begin{array}{ccc} \mathcal{O}_{\mathcal{U}_{23}}|_{U_{12}} & \xrightarrow{(\phi_{23}^{12})^*} & (\phi_{23}^{12})^*(\mathcal{O}_{\mathcal{U}_{12}}|_{U_{23}}) \\ u_{11}^{23} & \longrightarrow & -u_{32}^{12}(u_{31}^{12})^{-1} \\ u_{12}^{23} & \longrightarrow & (u_{31}^{12})^{-1} \\ u_{41}^{23} & \longrightarrow & -u_{41}^{12}u_{32}^{12}(u_{31}^{12})^{-1} + u_{42}^{12} \\ u_{42}^{23} & \longrightarrow & u_{41}^{12}(u_{31}^{12})^{-1} \\ \nu_{51}^{23} & \longrightarrow & \nu_{51}^{12}u_{32}^{12}(u_{31}^{12})^{-1} + \nu_{52}^{12} \\ \nu_{52}^{23} & \longrightarrow & \nu_{51}^{12}(u_{31}^{12})^{-1}. \end{array}$$

It is now clear how to define the most general ϕ^{ij}_{kl}. We leave to the reader the straightforward but tedious verification that the gluing conditions are verified.

We now want to determine the functor of points of $\mathrm{Gr}^{\mathrm{ch}}$. We start by defining the functor $\mathcal{G} : (\text{smflds})^o \longrightarrow (\text{sets})$. On objects it is defined as follows: If T is a supermanifold then

$$\mathcal{G}(T) = \{\mathcal{F} \subset \mathcal{O}_T^{4|1} \,|\, \mathcal{F}\, \text{locally free subsheaf of rank } 2|0\},$$

where $\mathcal{O}_T^{4|1} := (\mathbb{C}^{4|1} \otimes \mathcal{O}_T)_0$. On morphisms, $\mathcal{G}$ is defined as follows: If $T \longrightarrow S$ is a morphism of supermanifolds, it induces a sheaf morphism $f^* : \mathcal{O}_S \longrightarrow f_*\mathcal{O}_T$. This sends subsheaves of $\mathcal{O}_S^{4|1}$ into subsheaves of $f_*(\mathcal{O}_T)^{4|1}$. A subsheaf of $f_*(\mathcal{O}_T)^{4|1}$ defines a subsheaf of $\mathcal{O}_T^{4|1}$.

We now prove that $\mathcal{G}$ is the functor of points of the Grassmannian supermanifold $\mathrm{Gr}^{\mathrm{ch}}$. We invite the reader to compare our discussion with the one in Chapter 2, relative to the ordinary setting.

Proposition 4.7.1. *$\mathcal{G}$ is the functor of points of the Grassmannian supermanifold* $\mathrm{Gr}^{\mathrm{ch}}$.

Proof. We show that there is a functorial correspondence between the sets $\mathcal{G}(T)$ and $\mathrm{Gr}^{\mathrm{ch}}(T) = \mathrm{Hom}(T, \mathrm{Gr}^{\mathrm{ch}})$, leaving to the reader all the routine checks.

If $\mathcal{F}_T \in \mathcal{G}(T)$, we want to construct a morphism $T \longrightarrow \mathrm{Gr}^{\mathrm{ch}}$ corresponding to it. We cover T with open sets V_i in such way that $\mathcal{F}_T(V_i)$ is free:

$$\mathcal{F}_T(V_i) = \mathrm{span}_{\mathcal{O}_T(V_i)} \left\{ \begin{pmatrix} t_{11} \\ t_{21} \\ t_{31} \\ t_{41} \\ \theta_{51} \end{pmatrix}, \begin{pmatrix} t_{12} \\ t_{22} \\ t_{32} \\ t_{42} \\ \theta_{52} \end{pmatrix} \right\},$$

where t_{ij}, θ_{kl} are, respectively, even and odd sections in $\mathcal{O}_T(V_i)$. Since the rank is $2|0$, it means that at least one of the 2×2 minors of the first four rows of the matrix obtained by the two generators is non degenerate. Hence we can right multiply by an element in $\mathrm{GL}_2(\mathcal{O}_T(V_i))$ to make this minor become the identity matrix. This amounts to a change of basis in $\mathcal{F}_T(V_i)$. Let us assume, for simplicity, that such minor corresponds to the first two rows:

$$\mathcal{F}_T(V_i) = \text{span}_{\mathcal{O}_T(V_i)} \left\{ \begin{pmatrix} 1 \\ 0 \\ s_{31} \\ s_{41} \\ \sigma_{51} \end{pmatrix}, \begin{pmatrix} 0 \\ 1 \\ s_{32} \\ s_{42} \\ \sigma_{52} \end{pmatrix} \right\}.$$

We have immediately a morphism

$$f_i : \mathcal{V}_i \longrightarrow \mathcal{U}_{12} \subset \text{Gr}^{\text{ch}},$$

where $\mathcal{V}_i$ is the supermanifold $(V_i, \mathcal{F}_T|_{V_i})$. In fact, by the Chart Theorem, the morphism f_i corresponds to the choice of 4 even and 2 odd sections in $\mathcal{O}(V_i)$, namely the s_{ij}'s and σ_{kl}'s.

One can check that the morphisms f_i's agree on $\mathcal{V}_i \cap \mathcal{V}_j$ to give a morphism $f : T \longrightarrow \text{Gr}^{\text{ch}}$. In fact the change of charts in Gr^{ch} are especially designed to make these f_i's glue.

Vice-versa, consider a morphism $f : T \longrightarrow \text{Gr}^{\text{ch}}$. Let $V_{ij} = |f|^{-1}(U_{ij})$ and let $f_{ij} = f|_{V_{ij}}$, so that

$$\mathcal{V}_{ij} \xrightarrow{f_{ij}} \mathcal{U}_{ij},$$

where $\mathcal{V}_{ij}, \mathcal{U}_{ij}$ are the open subsupermanifolds in T and Gr^{ch} corresponding to the open sets V_{ij} and U_{ij} respectively. By the Chart theorem we have that f_{ij} corresponds to the choice of four even and two odd sections in $\mathcal{O}_T(V_{ij})$. Let us call these six global sections as we did before

$$v_{k1}^{ij}, \quad v_{k2}^{ij}, \quad \eta_{51}^{ij}, \quad \eta_{52}^{ij}, \qquad k \neq i, j, \quad 1 \leq k \leq 4,$$

and organize them in a matrix (see (4.41)). We define $\mathcal{F}_T(V_{ij})$ as the span over $\mathcal{O}_T(V_{ij})$ of the columns of the given matrix. For example, for V_{12}:

$$\mathcal{F}_T(V_{12}) = \text{span}_{\mathcal{O}_T(V_{12})} \left\{ \begin{pmatrix} 1 \\ 0 \\ v_{31} \\ v_{41} \\ \eta_{51} \end{pmatrix}, \begin{pmatrix} 0 \\ 1 \\ v_{32} \\ v_{42} \\ \eta_{52} \end{pmatrix} \right\} \subset \mathcal{O}_T^{4|1}(V_{12}).$$

We can then define a sheaf $\mathcal{F}_T^{ij}$ over V_{ij} by suitably restricting the sections. For $V \subset V_{ij}$:

$$\mathcal{F}_T^{ij}(V) = \text{span}_{\mathcal{O}_T(V)} \left\{ \begin{pmatrix} 1 \\ 0 \\ v_{31}|_V \\ v_{41}|_V \\ \eta_{51}|_V \end{pmatrix}, \begin{pmatrix} 0 \\ 1 \\ v_{32}|_V \\ v_{42}|_V \\ \eta_{52}|_V \end{pmatrix} \right\}.$$

$\mathcal{F}_T^{ij}$ is a free subsheaf of $\mathcal{O}_T^{4|1}|_{V_{ij}}$ of rank $2|0$. As one can readily check the free sheaves $\mathcal{F}_T^{ij}$ glue together to give a locally free subsheaf of $\mathcal{O}_T^{4|1}$. We leave to the reader the check that the two maps that we have defined $\mathcal{G}(T) \longrightarrow \mathrm{Gr}^{\mathrm{ch}}(T)$ and $\mathrm{Gr}^{\mathrm{ch}}(T) \longrightarrow \mathcal{G}(T)$ are functorial in T and one the inverse of the other. ■

We now want to show that $\mathrm{Gr}^{\mathrm{ch}}$ is an homogeneous superspace for the supergroup $\mathrm{SL}(4|1)$. The supergroup $\mathrm{SL}(m|n)$ was studied in Section 1.8 in detail. We recall here the functor of points for $\mathrm{SL}(4|1)$ (Definition 1.8.8)

$$\begin{array}{ccc} (\text{smflds})^o & \xrightarrow{\mathrm{SL}(4|1)} & (\text{sets}) \\ T & \longrightarrow & \mathrm{SL}(4|1)(T), \end{array} \qquad \text{where}$$

$$\mathrm{SL}(4|1)(T) := \left\{ g = \begin{pmatrix} D & \tau \\ \gamma & d \end{pmatrix} \;\middle|\; D,\, d \text{ invertible, } \mathrm{Ber}(g) = 1 \right\}, \tag{4.44}$$

with D and d being 4×4 and 1×1 matrices, respectively, with entries in $\mathcal{O}(T)_0$ while α and β are 4×1 and 1×4 matrices, respectively, with entries in $\mathcal{O}(T)_1$.

It is easy to give the action of $\mathrm{SL}(4|1)$ on $\mathrm{Gr}^{\mathrm{ch}}$ in terms of the functor of points. For any supermanifold T we are going to define a morphism

$$\mathrm{SL}(4|1)(T) \times \mathrm{Gr}^{\mathrm{ch}}(T) \longrightarrow \mathrm{Gr}^{\mathrm{ch}}(T)$$

satisfying the usual properties of actions (see Chapter 1). Consider $\mathcal{F}_T \in \mathrm{Gr}^{\mathrm{ch}}(T)$. We want to define the locally free sheaf $g \cdot \mathcal{F}_T \subset \mathcal{O}_T^{4|1}$. Again we proceed in strict analogy with the ordinary setting (see Chapter 2). We cover T with open subsets V_i, such that $\mathcal{F}_T(V_i)$ is free. Then

$$\mathcal{F}_T(V_i) = \mathrm{span}_{\mathcal{O}_T(V_i)}\{u, v\} \subset \mathcal{O}_T(V_i)^{4|1}.$$

It makes then sense to define, for $g \in \mathrm{SL}(4|1)(V_i)$:

$$(g \cdot \mathcal{F}_T)(V_i) = \mathrm{span}_{\mathcal{O}_T(V_i)}\{g \cdot u, g \cdot v\}.$$

This is again a free $\mathcal{O}_T(V_i)$-module and, by restriction, it defines a free sheaf on V_i that we denote $g \cdot \mathcal{F}_T|_{V_i}$. The sheaves $g \cdot \mathcal{F}_T|_{V_i}$ glue to give a sheaf on T that we call $g \cdot \mathcal{F}_T$. It is important to remark that $g \cdot \mathcal{F}_T$ is only locally free, but not free in general.

Hence, we can define

$$\begin{array}{ccc} \mathrm{SL}(4|1)(T) \times \mathrm{Gr}^{\mathrm{ch}}(T) & \longrightarrow & \mathrm{Gr}^{\mathrm{ch}}(T) \\ g, \mathcal{F}_T & \longrightarrow & g \cdot \mathcal{F}_T\,. \end{array}$$

We leave to the reader the routine check that this actually defines an action of the complex supergroup $\mathrm{SL}(4|1)$ on the complex analytic supermanifold $\mathrm{Gr}^{\mathrm{ch}}$.

Let us consider the topological point of $\mathrm{Gr}^{\mathrm{ch}}$

$$\mathcal{F}_0 = \mathrm{span}_{\mathcal{O}_T}\{(1,0,0,0,0),(0,1,0,0,0)\} \in |\mathrm{Gr}^{\mathrm{ch}}| \subset \mathrm{Gr}^{\mathrm{ch}}(T).$$

$\mathcal{F}_0$ is a free sheaf (in particular, locally free). We compute the stabilizer,

$$\mathrm{Stab}(\mathcal{F}_0)(T) = \{g \in \mathrm{SL}(4|1)(T) \,|\, g\cdot\mathcal{F}_0 = \mathcal{F}_0\} \subset \mathrm{SL}(4|1)(T)\,.$$

It is immediate to verify that

$$\mathrm{Stab}(\mathcal{F}_0)(T) = \left\{\begin{pmatrix} g_{11} & g_{12} & g_{13} & g_{14} & \gamma_{15} \\ g_{21} & g_{22} & g_{23} & g_{24} & \gamma_{25} \\ 0 & 0 & g_{33} & g_{34} & \gamma_{35} \\ 0 & 0 & g_{43} & g_{44} & \gamma_{45} \\ 0 & 0 & \gamma_{53} & \gamma_{54} & g_{55} \end{pmatrix}\right\} \subset \mathrm{SL}(4|1)(T)\,.$$

The functor $T \mapsto \mathrm{Stab}(\mathcal{F}_0)(T)$ is the functor of points of a supergroup, that we call F_u^{ch}. We can now compare with the infinitesimal approach (4.7) and conclude that the stabilizer is a parabolic supergroup.

Proposition 4.7.2. *The Grassmannian supermanifold* $\mathrm{Gr}^{\mathrm{ch}}$ *is the quotient* $\mathrm{SL}(4|1)/F_u^{\mathrm{ch}}$.

Proof. We are going to use Proposition 1.9.7 in Section 1.9. We consider the morphism that we have defined above

$$\begin{array}{ccc} \mathrm{SL}(4|1)(T) & \xrightarrow{\pi} & \mathrm{Gr}^{\mathrm{ch}}(T) \\ g & \longrightarrow & g\cdot\mathcal{F}_0\,. \end{array}$$

In order to obtain our result, we need to show that π is a submersion at all topological points.

Let us assume $g\cdot\mathcal{F}_0 \in \mathcal{U}_{12}(T)$ (the reasoning when $g\cdot\mathcal{F}_0 \in \mathcal{U}_{ij}(T)$ is the same). Hence, locally we have

$$\begin{pmatrix} g_{11} & g_{12} & g_{13} & g_{14} & \gamma_{15} \\ g_{21} & g_{22} & g_{23} & g_{24} & \gamma_{25} \\ g_{31} & g_{32} & g_{33} & g_{34} & \gamma_{35} \\ g_{41} & g_{42} & g_{43} & g_{44} & \gamma_{45} \\ \gamma_{51} & \gamma_{52} & \gamma_{53} & \gamma_{54} & g_{55} \end{pmatrix} \longrightarrow \mathrm{span}\left\{\begin{pmatrix} g_{11} \\ g_{21} \\ g_{31} \\ g_{41} \\ \gamma_{51} \end{pmatrix}, \begin{pmatrix} g_{12} \\ g_{22} \\ g_{32} \\ g_{42} \\ \gamma_{52} \end{pmatrix}\right\}$$

$$= \mathrm{span}\left\{\begin{pmatrix} I \\ WZ^{-1} \\ \rho Z^{-1} \end{pmatrix}\right\},$$

where

$$\rho = \begin{pmatrix} \gamma_{51} \ \gamma_{52} \end{pmatrix}, \quad W = \begin{pmatrix} g_{31} & g_{32} \\ g_{41} & g_{42} \end{pmatrix}, \quad Z = \begin{pmatrix} g_{11} & g_{12} \\ g_{21} & g_{22} \end{pmatrix}.$$

Using the Chart theorem, the morphism of supermanifolds π is given by

$$g \longrightarrow \left(WZ^{-1}, \, \rho_1 Z^{-1}, \, (\tau_2 - WZ^{-1}\tau_1)d\right).$$

At this point one can compute the Jacobian and verify that at the identity π is surjective, so π is a submersion. We can repeat the same argument for all $\mathcal{U}_{ij}$ and then we obtain the result. ■

4.8 The super Plücker embedding of $\mathrm{Gr}^{\mathrm{ch}}$

We want to construct the super analogue of the Plücker embedding for the complex supermanifold $\mathrm{Gr}^{\mathrm{ch}}$ into the projective superspace $\mathbb{P}(E)$ (see Example 1.4.7) where

$$E = \wedge^2 \mathbb{C}^{4|1},$$

in analogy with the ordinary construction detailed in Chapter 2. It is important to remark that, in general, Grassmannian supermanifolds do not admit an embedding into a projective superspace (see Ref. [108] Chapter 3), however our Grassmannian $\mathrm{Gr}^{\mathrm{ch}}$ does, as we shall presently see.

We can then identify the complex supermanifold $\mathrm{Gr}^{\mathrm{ch}}$ with an algebraic subvariety in $\mathbb{P}(E)$; thus $\mathrm{Gr}^{\mathrm{ch}}$ acquires the additional structure of projective supervariety.

To construct the embedding of $\mathrm{Gr}^{\mathrm{ch}}$ into the projective superspace $\mathbb{P}(E)$ in terms of the functor of points, we need to give a natural transformation ϕ such that

$$\phi_T : \mathrm{Gr}^{\mathrm{ch}}(T) \longrightarrow \mathbb{P}(E)(T), \qquad T \in (\mathrm{smflds}).$$

Let $\mathcal{F}_T \in \mathrm{Gr}^{\mathrm{ch}}(T)$. We choose a cover of T by open subsets V_i such that $\mathcal{F}_T|_{V_i}$ is free and denote with $\mathcal{F}_i$ the restriction of $\mathcal{F}_T$ to V_i. We can then write

$$\mathcal{F}_i(V_i) = \mathrm{span}_{\mathcal{O}_T(V_i)} \{a, \, b\}, \qquad a = \begin{pmatrix} t_{11} \\ t_{21} \\ t_{31} \\ t_{41} \\ \theta_{51} \end{pmatrix}, \quad b = \begin{pmatrix} t_{12} \\ t_{22} \\ t_{32} \\ t_{42} \\ \theta_{52} \end{pmatrix}.$$

We define the morphism

$$\begin{array}{ccc} \mathrm{Gr}^{\mathrm{ch}}(V_i) & \xrightarrow{\phi_{V_i}} & \mathbb{P}(E)(V_i) \\ \mathrm{span}_{\mathcal{O}_T(V_i)}\{a,b\} & \longrightarrow & [a \wedge b]\,. \end{array}$$

where $[a \wedge b]$ is the equivalence class of the line generated by the vector $a \wedge b$ in $E(V_i)$, hence an element of $\mathbb{P}(E)(V_i)$ (see Section 1.7).

ϕ_{V_i} is well defined, since the choice of different basis elements for $\mathcal{F}_i$ will amount to multiply $a \wedge b$ by an invertible element in $\mathcal{O}_T(V_i)$, hence it will give the same element in $\mathbb{P}(E)(V_i)$. As one can readily check, when the $\mathcal{F}_i$ glue together to give $\mathcal{F}$, the corresponding images will glue according to the change of charts in $\mathbb{P}(E)$, and one obtains a well defined element $\phi_T(\mathcal{F}) \in \mathbb{P}(E)(T)$. The whole construction is manifestly functorial, hence the family of ϕ_T's defines a natural transformation between the functor of points of the Grassmannian $\mathrm{Gr}^{\mathrm{ch}}$ and the super projective space. By Yoneda's Lemma, such natural transformation corresponds to a supermanifold morphism $\phi : \mathrm{Gr}^{\mathrm{ch}} \longrightarrow \mathbb{P}(E)$, which we call the *super Plücker embedding* since its reduced part is indeed the Plücker embedding as detailed in Section 2.6.

We now want to determine the image of ϕ. In order to do this, it is better to introduce coordinates. Let us choose the (homogeneous) basis $(e_1, \dots, e_4, \epsilon_5)$ for $\mathbb{C}^{4|1}$ and consider the basis in $E = \wedge^2\mathbb{C}^{4|1}$:

$$e_i \wedge e_j, \quad 1 \le i < j \le 4, \qquad \epsilon_5 \wedge \epsilon_5\,, \qquad e_k \wedge \epsilon_5, \quad k = 1, \dots, 4\,,$$

which allows us to identify the super vector spaces $\wedge^2\mathbb{C}^{4|1} \cong \mathbb{C}^{7|4}$ and consequently to identify also $\mathbb{P}(E)$ and $\mathbb{P}^{6|4}$.

We have to find necessary and sufficient conditions for an element $Q \in \mathbb{P}^{6|4}(V_i)$ to be *decomposable* i.e. to be such that $Q = a \wedge b$ for suitable a and $b \in \mathcal{O}_T(V_i)^{4|1}$. Let us write such a generic Q as:

$$\begin{aligned} Q &= q + \lambda \wedge \epsilon_5 + a_{55}\epsilon_5 \wedge \epsilon_5\,, \quad \text{with} \\ q &= q_{12}e_1 \wedge e_2 + \cdots + q_{34}e_3 \wedge e_4\,, \quad q_{ij} \in \mathcal{O}_T(V_i)_0\,, \\ \lambda &= \lambda_1 e_1 + \cdots + \lambda_4 e_4, \quad \lambda_i \in \mathcal{O}_T(V_i)_1\,. \end{aligned} \tag{4.45}$$

Q is decomposable if and only if

$$\begin{aligned} Q &= (r + \xi\epsilon_5) \wedge (s + \theta\epsilon_5) \quad \text{with} \\ r &= r_1e_1 + \cdots + r_4e_4\,, \quad s = s_1e_1 + \cdots + s_4e_4\,, \end{aligned}$$

with $r_i, s_i \in \mathcal{O}_T(V_i)_0 \quad \xi, \theta \in \mathcal{O}_T(V_i)_1$. That is,

$$Q = r \wedge s + (\theta r - \xi s) \wedge \epsilon_5 + \xi\theta\epsilon_5 \wedge \epsilon_5\,,$$

which yields

$$q = r \wedge s, \quad \lambda = \theta r - \xi s, \quad a_{55} = \xi\theta\,.$$

These are equivalent to the following:

$$q \wedge q = 0, \qquad q \wedge \lambda = 0, \qquad \lambda \wedge \lambda = 2a_{55}q, \qquad \lambda a_{55} = 0\,.$$

Plugging (4.45) we obtain

$$\begin{aligned} &q_{12}q_{34} - q_{13}q_{24} + q_{14}q_{23} = 0, && \text{(classical Plücker relation)},\\ &q_{ij}\lambda_k - q_{ik}\lambda_j + q_{jk}\lambda_i = 0\,, && 1 \le i < j < k \le 4,\\ &\lambda_i\lambda_j = a_{55}q_{ij} && 1 \le i < j \le 4\,,\\ &\lambda_i a_{55} = 0\,. && \end{aligned} \tag{4.46}$$

These are the *super Plücker relations.* As one can readily check the differentials of such relations are linearly independent at each point, hence they define a closed analytic submanifold of $\mathbb{P}^{6|4}$ (see Chapter 1) that we call the *super Klein quadric* in $\mathbb{P}^{6|4}$. Notice that the Klein quadric can be also viewed as a superprojective variety, since the super Plücker relations are algebraic.

At this point, in order to show that the Plücker embedding realizes an isomorphism of $\mathrm{Gr}^{\mathrm{ch}}$ with the super Klein quadric, we need to show that the Plücker relations exhaust all of the relations among our coordinates. We are going to see the proof of this statement in the algebraic setting, since it is the one that we shall need in the sequel.

We need first to examine the algebraic supervariety structure of the Grassmannian $\mathrm{Gr}^{\mathrm{ch}}$. We defined the supermanifold structure of $\mathrm{Gr}^{\mathrm{ch}}$ using the open covering $\mathcal{U}_{ij}$, where $\mathcal{U}_{ij} \cong \mathbb{C}^{4|2}$ as analytic supermanifolds. In the same way, we can define a supervariety structure on $\mathrm{Gr}^{\mathrm{ch}}$ by taking the same topological open covering U_{ij} of the topological space $G(2,4)$, and defining

$$\mathcal{U}_{ij}^{\mathrm{alg}} = (U_{ij}, \mathcal{O}_{ij}^{\mathrm{alg}}), \qquad \text{where} \qquad \mathcal{O}_{ij}^{\mathrm{alg}} \cong \mathcal{O}_{\mathbb{A}^{4|2}}\,,$$

so that $\mathcal{U}_{ij}^{\mathrm{alg}}$ is isomorphic to the complex affine superspace $\mathbb{A}^{4|2}$. The sheaves $\mathcal{O}_{ij}^{\mathrm{alg}}$ glue together to give a supervariety structure on $G(2,4)$. We shall use the same symbol $\mathrm{Gr}^{\mathrm{ch}}$ to denote both the complex analytic supermanifold and the algebraic supervariety structures described above, the context making clear which one we are considering.

The functor of points of $\mathrm{Gr}^{\mathrm{ch}}$ as an algebraic supervariety is characterized exactly in the same way as for the analytic category, namely

$$\mathrm{Gr}^{\mathrm{ch}}(T) = \{\mathcal{F} \subset \mathcal{O}_T^{4|1} \mid \mathcal{F} \text{ locally free subsheaf of rank } 2|0\}, \qquad T \in (\mathrm{svar}),$$

where $\mathcal{O}_T^{4|1} := (\mathbb{C}^{4|1} \otimes \mathcal{O}_T)_0$ and $\mathcal{O}_T$, as usual, denotes the structural sheaf of the algebraic supervariety $T = (|T|, \mathcal{O}_T)$.

The supervariety $\mathrm{Gr}^{\mathrm{ch}}$ embeds into the projective superspace $\mathbb{P}^{6|4}$, via the Plücker embedding, which makes sense also in the algebraic category. We have the following proposition:

Proposition 4.8.1. *The Plücker embedding:*

$$\mathrm{Gr}^{\mathrm{ch}}(T) \xrightarrow{\phi_T} \mathbb{P}(E)(T), \qquad T \in (\mathrm{svar}),$$

defined on a suitable open cover $\{V_i\}$ *of* T *as*

$$\phi_{V_i}(\mathrm{span}\{a, b\}) = [a \wedge b]$$

realizes an isomorphism between the algebraic supervariety $\mathrm{Gr}^{\mathrm{ch}}$ *and the super Klein quadric, which is the projective algebraic subsupervariety of* $\mathbb{P}(E) \cong \mathbb{P}^{6|4}$ *defined by the homogeneous ideal* I_P *(see (4.46)) of the Plücker relations.*

Proof. The only thing that remains to be proven is that the Plücker relations exhaust all of the possible relations among the q_{ij}'s, λ_k, a_{55}. This can be done with the same argument we used for the ordinary setting (see Proposition 2.6.2). We are going to see an alternative proof in Proposition 4.9.5. ■

We define

$$\mathbb{C}[\mathrm{Gr}^{\mathrm{ch}}] := \mathbb{C}[q_{ij}, \lambda_k, a_{55}]/I_P \tag{4.47}$$

as the *coordinate superalgebra of* $\mathrm{Gr}^{\mathrm{ch}}$ *with respect to the Plücker embedding.* It is important to remark that such superalgebra is not intrinsecally associated to $\mathrm{Gr}^{\mathrm{ch}}$, but depends on the given projective embedding. In other words, different projective embeddings of $\mathrm{Gr}^{\mathrm{ch}}$ may give non isomorphic coordinate superalgebras, exactly as it happens in the ordinary setting.

4.9 The super Grassmannian $\mathrm{Gr}^{\mathrm{ch}}$ via invariant theory

We now want to examine the super Grassmannian $\mathrm{Gr}^{\mathrm{ch}}$ and its relation with super invariant theory (the reader can refer to Theorem 2.9.1 for the ordinary setting).

Consider the superalgebra of polynomials functions $\mathbb{C}[v_i, w_i, \xi_1, \xi_2]$ in the 8 even and 2 odd variables: v_i, w_i, ξ_1, ξ_2, $1 \leq i \leq 4$. Let us organize these variables in a (5×2) matrix form:

$$\begin{pmatrix} v & w \\ \xi_1 & \xi_2 \end{pmatrix} = \begin{pmatrix} v_1 & w_1 \\ \vdots & \vdots \\ v_4 & w_4 \\ \xi_1 & \xi_2 \end{pmatrix}.$$

The group $\mathrm{SL}_2(\mathbb{C})$ acts on the right on these matrices

$$\begin{pmatrix} v' & w' \\ \xi_1' & \xi_2' \end{pmatrix} = \begin{pmatrix} v & w \\ \xi_1 & \xi_2 \end{pmatrix} \cdot g, \qquad g \in \mathrm{SL}_2(\mathbb{C}).$$

Clearly this induces an action on $\mathbb{C}[v_i, w_i, \xi_1, \xi_2]$.

Definition 4.9.1. We say that $f \in \mathbb{C}[v_i, w_i, \xi_1, \xi_2]$ is *invariant* if

$$g \cdot f = f, \qquad \forall g \in \mathrm{SL}_2(\mathbb{C})$$

The set of all invariant polynomials forms a subring of $\mathbb{C}[v_i, w_i, \xi_1, \xi_2]$, which is called the *superring of invariants.* ■

It is immediate to verify that the following elements are invariant:

$$y_{ij} = v_i w_j - v_j w_i, \quad \theta_i = v_i \xi_2 - w_i \xi_1, \quad a = \xi_1 \xi_2 . \tag{4.48}$$

As one can readily check, they satisfy the *super Plücker relations* (4.46)

$$\begin{aligned} & y_{12}y_{34} - y_{13}y_{24} + y_{14}y_{23} = 0, && \text{(classical Plücker relation)} \\ & y_{ij}\lambda_k - y_{ik}\lambda_j + y_{jk}\lambda_i = 0, && 1 \leq i < j < k \leq 4, \\ & \lambda_i \lambda_j = a y_{ij} && 1 \leq i < j \leq 4, \\ & \lambda_i a = 0 . \end{aligned}$$

We want to show that the elements in (4.48) generate the superring of invariants and that (4.46) are all the relations among these generators, as it happens in the ordinary setting (see Chapter 2, Proposition 2.9.1).

Proposition 4.9.2. *(First fundamental theorem of super invariant theory for $\mathrm{SL}_2(\mathbb{C})$). Let*

$$f(v, w, \xi) = f_0(v, w) + \sum_i f_i(v, w)\xi_i + f_{12}(v, w)\xi_1\xi_2 \tag{4.49}$$

be an homogeneous invariant polynomial in $\mathbb{C}[v_i, w_i, \xi_1, \xi_2]$. Then $f_0(v, w)$ and $f_{12}(v, w)$ are ordinary invariants and

$$\sum_i f_i(v, w)\xi_i = \sum_i h_i(v, w)\theta_i i ,$$

with $h_i(v,w)$ also an ordinary invariant.

In other words

$$y_{ij} = v_i w_j - v_j w_i, \quad \theta_i = v_i \xi_2 - w_i \xi_1, \quad a = \xi_1 \xi_2 \tag{4.50}$$

generate the superring of invariants.

Proof. Let us take

$$g = \begin{pmatrix} a & b \\ c & d \end{pmatrix}, \quad \text{so} \quad \begin{pmatrix} v' & w' \\ \xi_1' & \xi_2' \end{pmatrix} = \begin{pmatrix} va + wc & vb + wd \\ \xi_1 a + \xi_2 c & \xi_1 b + \xi_2 d \end{pmatrix}.$$

Then we can see immediately that each term in (4.49) has to be an invariant so

$$f_0(v', w') = f_0(v, w), \qquad \sum_i f_i(v', w')\xi_i' = \sum_i f_i(v, w)\xi_i,$$
$$f_{12}(v', w')\xi_1'\xi_2' = f_{12}(v, w)\xi_1\xi_2 .$$

We have that f_0 is an ordinary invariant. Since $\xi_1'\xi_2' = \xi_1\xi_2 \det g$, then $f_{12}(v,w)$ is also an ordinary invariant. The odd terms θ_i are of the same form as the ordinary invariants y_{ij}, since the fact that ξ_i is odd plays no particular role here (recall that we are considering the action of the ordinary group $\mathrm{SL}_2(\mathbb{C})$). So by the same argument than in the ordinary case, there are no other odd invariants, besides those we have already found, that are linear in the odd variables ξ_1 and ξ_2. ∎

We now wish to give a result that describes completely the relations among the invariants, hence provides a description of the superring of invariants in terms of generators and relations.

Let us make the following change of the notation. Consider the polynomial superalgebra $\mathbb{C}[a_{ij}]$, $1 \leq i \leq 5$, $1 \leq j \leq 2$, with the parity defined as

$$p(a_{ij}) = p(i) + p(j), \quad \text{with } p(k) = 0 \text{ if } 0 \leq k \leq 4 \text{ and } p(5) = 1 .$$

Consider the following action of $\mathrm{SL}_2(\mathbb{C})$:

$$\begin{aligned} \mathbb{C}[a_{ij}] \times \mathrm{SL}_2(\mathbb{C}) &\longrightarrow \mathbb{C}[a_{ij}] \\ (a_{ij}, g) &\longrightarrow \textstyle\sum_k a_{ik} g_{kj} . \end{aligned}$$

In this notation, what we have just proven is that the following expressions:

$$d_{ij} = a_{i1}a_{j2} - a_{i2}a_{j1}, \quad 1 \leq i < j \leq 5,$$
$$d_{55} = a_{51}a_{52} ,$$

generate the superring of invariants.

We now need a generalization of the concept of tableau, introduced in Chapter 2, Definition 2.9.3.
Recall that a Young diagram is a finite collection of boxes or cells arranged in left-justified rows, with the row lengths weakly decreasing. Young diagrams with n boxes are in one to one correspondence with the partitions of the integer n.

Definition 4.9.3. Let (m, n) be a pair of integers and assign the following parity to the integers $1, \dots, m+n$: $p(i) = 0$ if $1 \leq i \leq m$ and $p(i) = 1$ if $m+1 \leq i \leq m+n$. A *super standard Young tableau*[4] (or *super standard tableau* for short) is obtained by filling in the boxes of the Young diagram with the first $m+n$ natural numbers in such a way that

- Each row consists of strictly increasing even parity integers and weakly increasing odd parity ones;
- Each column consists of weakly increasing even parity integers and strictly increasing odd parity ones.

■

We are interested in Young diagrams with all of the rows of the same length equal to 2 and filled with the numbers 1, 2, 3, 4, 5, where 5 is the only odd index.

Definition 4.9.4. Let $I_1 = (i_1, j_j), \dots, I_r = (i_r, j_r)$ be multiindices with $i_k, j_k = 1, \dots, 5$. A *standard monomial* in the indeterminates d_{ij}'s is a monomial $d_{I_1} \cdots d_{I_r}$ where the indices $I_1, \dots, I_r$ form a super standard tableau. ■

We have the following proposition:

Proposition 4.9.5. *(Second fundamental theorem of super invariant theory for* $\mathrm{SL}_2(\mathbb{C})$*). Let R be the subring of $\mathbb{C}[a_{ib}]$ generated by the determinants $d_{ij} = a_{i1}a_{j2} - a_{j1}a_{i2}$ and $d_{55} = a_{51}a_{52}$. Then $R \cong \mathbb{C}[a_{ib}]/I_P$, where I_P is the ideal of the super Plücker relations (4.46). In other words, I_P contains all the possible relations satisfied by d_{ij} and d_{55}.*

Proof. (Sketch). It is easy to verify that d_{ij} and d_{55} satisfy all of the above relations, the problem is to prove that these are the only relations.

[4]Some authors will call these *semistandard* or simply tableau without any adjective.

The proof of this fact is the same as in the classical setting. Let us briefly sketch it. Using the super Plücker relation one can verify that any monomial in R can be written as a linear combination of standard ones. This can be done directly or using the same argument for the classical case (see Ref. [71] pg 110 for more details). The standard monomials are also linearly independent, hence they form a basis for R as a complex super vector space. Again, this is done directly or with the same argument as in Ref. [71], pg 110. So given a relation in R, once we write each term as a standard monomial, we obtain that either the relation is identically zero (hence it is a relation in the Plücker ideal) or it gives a relation among the standard monomials. This would be a contradiction, so the result is proven. ■

This gives an alternative proof to the Proposition 4.8.1. The ring R in that case is $\mathbb{C}[\mathrm{Gr}^{\mathrm{ch}}]$.

4.10 The chiral Minkowski superspace

We now turn to the problem of the description of the big cell inside the Grassmannian $\mathrm{Gr}^{\mathrm{ch}}$. As we shall see in Section 4.14, it corresponds to the *chiral Minkowski superspace*, of great interest in physics.

We are mostly interested in the algebraic setting, since it is the most adequate to describe the quantum deformation that we shall examine in Chapter 5.

Definition 4.10.1. We define the *big cell* of the Grassmannian supervariety to be the open subvariety $\mathcal{U}_{12} \cong \mathbb{C}^{4|2}$. Its functor of points is

$$T \longrightarrow \mathcal{U}_{12}(T) = \mathrm{span}_{\mathcal{O}(T)} \left\{ \begin{pmatrix} 1 \\ 0 \\ t_{31} \\ t_{41} \\ \theta_{51} \end{pmatrix}, \begin{pmatrix} 0 \\ 1 \\ t_{32} \\ t_{42} \\ \theta_{52} \end{pmatrix} \right\} \subset \mathcal{O}(T)^{4|1},$$

where $\mathcal{O}(T)$ denotes as usual the global sections of the supervariety T. Notice that this definition makes sense also in the complex analytic category. ■

We shall denote an element $\mathcal{F}_T$ in $\mathcal{U}_{12}(T)$ as a 5×2 matrix

$$\begin{pmatrix} I \\ A \\ \alpha \end{pmatrix} = \begin{pmatrix} 1 & 0 \\ 0 & 1 \\ a_{31} & a_{32} \\ a_{41} & a_{42} \\ \alpha_{51} & \alpha_{52} \end{pmatrix}, \tag{4.51}$$

where each column corresponds to a generator of the free sheaf $\mathcal{F}_T$. As we know, such representation of $\mathcal{F}_T$ is unique.

We now want to understand how one can derive the superalgebra of algebraic functions on the affine supervariety $\mathcal{U}_{12}$ starting from the coordinate superalgebra $\mathcal{O}(\mathrm{Gr}^{\mathrm{ch}})$. We gave such a superalgebra in terms of generators and relations. The generators correspond to the coordinates in the Plücker embedding, which locally, in the functor of points notation, reads as

$$\begin{array}{ccc} \mathrm{Gr}^{\mathrm{ch}}(T) & \xrightarrow{\phi_T} & \mathbb{P}^{6|4}(T) \\ \mathrm{span}\{a,b\} & \longrightarrow & [d_{12}, d_{13}, \dots, d_{55}, d_{15}, \dots, d_{45}]\,, \end{array}$$

where T is suitably chosen so that the element in $\mathrm{Gr}^{\mathrm{ch}}(T)$ is free and not just locally free (a generic T that does not have this property can be covered by open sets with such a property). The topological space underlying $\mathcal{U}_{12}$ corresponds to the points in $|\mathrm{Gr}^{\mathrm{ch}}|$ with the corresponding d_{12} invertible. Consequently the coordinate superring of the affine open subvariety U_{12} of $\mathrm{Gr}^{\mathrm{ch}}$ corresponds to the *projective localization* of the Grassmannian superring at the coordinate d_{12}. In other words it consists of the elements of degree zero in

$$\mathbb{C}[d_{ij}d_{12}^{-1}, d_{55}d_{12}^{-1}] \subset \mathbb{C}[\mathrm{Gr}^{\mathrm{ch}}][d_{12}^{-1}].$$

As one can readily check, there are no relations among these generators so the big cell U_{12} of $\mathrm{Gr}^{\mathrm{ch}}$ is the affine superspace with coordinate ring

$$\mathcal{O}(U_{12}) = \mathbb{C}[x_{ij}, \xi_j] \approx \mathbb{C}^{4|2}\,, \tag{4.52}$$

where we set $x_{ij} = d_{ij}d_{12}^{-1}$, $x_{55} = d_{55}d_{12}^{-1}$ and $\xi_j = d_{j5}d_{12}^{-1}$.

We are now interested in the subsupergroup of $\mathrm{SL}(4|1)$ that preserves the big cell $\mathcal{U}_{12}$. Let $g \in \mathrm{SL}(4|1)(T)$; we want to determine the form of g such that

$$g \cdot \mathcal{U}_{12}(T) \subset \mathcal{U}_{12}(T)\,.$$

Working locally one can check that g must be of the form

$$g = \begin{pmatrix} x & 0 & 0 \\ tx & y & y\eta \\ d\tau & d\xi & d \end{pmatrix},$$

where x and y are even, invertible 2×2 matrices, t is an even, arbitrary 2×2 matrix, η a 2×1 odd matrix, τ, ξ are 1×2 odd matrices and d is an invertible even element.

We then define

$$F_l^{\mathrm{ch}}(T) = \{g \in \mathrm{SL}(4|1)(T) \mid g \cdot \mathcal{U}_{12}(T) \subset \mathcal{U}_{12}(T)\}.$$

Comparing with (4.7), we see that this is a parabolic subsupergroup of SL(4|1). The functor F_l^{ch} is the functor of points of an algebraic (also analytic) supergroup.

The action of the supergroup F_l^{ch} on the big cell $\mathcal{U}_{12}$ is as follows,

$$F_l^{\mathrm{ch}}(T) \times \mathcal{U}_{12}(T) \longrightarrow \mathcal{U}_{12}(T)$$

$$\left(\begin{pmatrix} x & 0 & 0 \\ tx & y & y\eta \\ d\tau & d\xi & d \end{pmatrix}, \begin{pmatrix} I \\ A \\ \alpha \end{pmatrix} \right) \longrightarrow \begin{pmatrix} I \\ A' \\ \alpha' \end{pmatrix},$$

where we use the notation (4.51). We compute the matrices A' and α' as follows. Since

$$\begin{pmatrix} x & 0 & 0 \\ tx & y & y\eta \\ d\tau & d\xi & d \end{pmatrix} \begin{pmatrix} I \\ A \\ \alpha \end{pmatrix} = \begin{pmatrix} x \\ tx + yA + y\eta\alpha \\ d\tau + d\xi A + d\alpha \end{pmatrix},$$

by right multiplying by the matrix x^{-1} (which amounts to a change of basis), we obtain a matrix corresponding to an element in $\mathcal{U}_{12}(T)$:

$$\begin{pmatrix} I \\ A' \\ \alpha' \end{pmatrix} = \begin{pmatrix} I \\ y(A + \eta\alpha)x^{-1} + t \\ d(\alpha + \tau + \xi A)x^{-1} \end{pmatrix}. \tag{4.53}$$

Notice that the subgroup of F_l^{ch} obtained by taking $\xi = 0$ corresponds, at the infinitesimal level, to the super Poincaré algebra plus dilations of Definition 4.6.1. In that case, we have the relation

$$d = \det x \det y.$$

In the next section we leave the chiral setting and go to the description of the full conformal superspace, the superflag $\mathcal{F}\ell(2|0, 2|1, 4|1)$.

4.11 The superflag $\mathcal{F}\ell$

In this section we want to construct the complex analytic flag supermanifold of $2|0$ and $2|1$ subspaces of the super vector space $\mathbb{C}^{4|1}$, $\mathcal{F}\ell(2|0, 2|1, 4|1)$.

As it turns out, the superflag is a quotient of the complex special linear supergroup $G := \mathrm{SL}(4|1)$ (whose Lie superalgebra is the Wess-Zumino superalgebra $\mathfrak{sl}(4|1)$ studied in Section 4.4) by a parabolic subgroup. G acts naturally on $\mathcal{F}\ell(2|0, 2|1, 4|1)$ and furthermore, $\mathcal{F}\ell(2|0, 2|1, 4|1)$ contains an open dense subsupermanifold isomorphic to $\mathbb{C}^{4|4}$. According to the argument in Section 4.6, we will define the *complex conformal superspace* as the superflag $\mathcal{F}\ell(2|0, 2|1, 4|1)$. The open, dense subsupermanifold called the *big cell*, will be the *complex Minkowski superspace*. We will then proceed as for the non super case $G(2, 4)$ or the chiral case $\mathrm{Gr}^{\mathrm{ch}}$, and show that the Poincaré supergroup times dilations, whose Lie superalgebra is the super Poincaré algebra plus dilations of Definition 4.6.1, leaves the big cell invariant.

Let us now procede to construct a supermanifold structure, that we shall denote $\mathcal{F}\ell$, on the topological space $G(2, 4)$, that admits an open cover by the six open subsets U_{ij}, $1 \leq i, j \leq 4$ described in Section 2.5. On each open set U_{ij} we will define a supermanifold structure, $\mathcal{U}_{ij} = (U_{ij}, \mathcal{O}_{ij})$, and we will then provide the morphisms $\phi_{ij}^{kl} : \mathcal{U}_{ij} \cap \mathcal{U}_{kl} \longrightarrow \mathcal{U}_{ij} \cap \mathcal{U}_{kl}$ of change of charts satisfying the gluing conditions

$$\phi_{rs}^{kl} \circ \phi_{kl}^{ij} = \phi_{rs}^{ij} \qquad \text{in} \qquad \mathcal{U}_{ij} \cap \mathcal{U}_{kl} \cap \mathcal{U}_{rs}\,.$$

In this way, we will then obtain a sheaf over $G(2, 4)$ and hence, a supermanifold structure $G(2, 4) = \cup_{i,j} U_{ij}$.

We take $\mathcal{U}_{ij} = (U_{ij}, \mathcal{O}_{ij})$ as the supermanifold $\mathbb{C}^{4|4}$ with global coordinates

$$u_{k1}^{ij}, \quad u_{k2}^{ij}, \quad \alpha_{51}^{ij}, \quad \alpha_{52}^{ij}, \quad \alpha_{k5}^{ij}, \qquad 1 \leq k \leq 4, \quad k \neq i, j\,.$$

As before, we organize the coordinates in matrix form (see (4.41)). This time we need two matrices, since we are parametrizing two subspaces. Also, one has to enforce the condition that the first subspace is inside the second one. This is done by introducing auxiliary coordinates

$$v_{k1}^{ij}, \quad v_{k2}^{ij}, \qquad 1 \leq k \leq 4, \quad k \neq i, j\,,$$

and then imposing certain constraints. Let us work out the example $(i, j) = (1, 2)$; then the procedure for arbitrary (i, j) will be clear. We have the following global coordinates on $\mathcal{U}_{12}$ (for clarity we omit the superindex $(1,2)$):

$$\begin{pmatrix} 1 & 0 \\ 0 & 1 \\ u_{31} & u_{32} \\ u_{41} & u_{42} \\ \alpha_{51} & \alpha_{52} \end{pmatrix}, \quad \begin{pmatrix} 1 & 0 & 0 \\ 0 & 1 & 0 \\ v_{31} & v_{32} & \alpha_{35} \\ v_{41} & v_{42} & \alpha_{45} \\ 0 & 0 & 1 \end{pmatrix}. \tag{4.54}$$

The auxiliary variables v_{rs} can be determined through the constraints

$$\begin{pmatrix} v_{31} & v_{32} \\ v_{41} & v_{42} \end{pmatrix} = \begin{pmatrix} u_{31} & u_{32} \\ u_{41} & u_{42} \end{pmatrix} - \begin{pmatrix} \alpha_{35} \\ \alpha_{45} \end{pmatrix} \begin{pmatrix} \alpha_{51} & \alpha_{52} \end{pmatrix}. \tag{4.55}$$

The significance of the v_{rs}'s and this relation will be clear when we go to the functor of points treatment of the superflag. For the moment, this is just a way to organize the global coordinate on a given open set.

Next, we need to specify how to go from one coordinate patch to the other. For example, we give the rule to go from $(1,2)$ to $(2,3)$:

$$\begin{pmatrix} 1 & 0 \\ 0 & 1 \\ u_{31} & u_{32} \\ u_{41} & u_{42} \\ \alpha_{51} & \alpha_{52} \end{pmatrix}, \quad \begin{pmatrix} 1 & 0 & 0 \\ 0 & 1 & 0 \\ v_{31} & v_{32} & \alpha_{35} \\ v_{41} & v_{42} & \alpha_{45} \\ 0 & 0 & 1 \end{pmatrix} \longrightarrow$$

$$\begin{pmatrix} 1 & 0 \\ 0 & 1 \\ u_{31} & u_{32} \\ u_{41} & u_{42} \\ \alpha_{51} & \alpha_{52} \end{pmatrix} \begin{pmatrix} 0 & 1 \\ u_{31} & u_{32} \end{pmatrix}^{-1}, \quad \begin{pmatrix} 1 & 0 & 0 \\ 0 & 1 & 0 \\ v_{31} & v_{32} & \alpha_{35} \\ v_{41} & v_{42} & \alpha_{45} \\ 0 & 0 & 1 \end{pmatrix} \begin{pmatrix} 0 & 1 & 0 \\ v_{31} & v_{32} & \alpha_{35} \\ 0 & 0 & 1 \end{pmatrix}^{-1}.$$

We leave to the reader the tedious check that the images of the u's and v's still satisfy the relation (4.55).

The matrix multiplication gives the sections that form a matrix in the prescribed form for $\mathcal{U}_{23}$. In general, to give $\phi_{ij}^{kl} : \mathcal{U}_{ij} \cap \mathcal{U}_{kl} \longrightarrow \mathcal{U}_{ij} \cap \mathcal{U}_{kl}$, by the Chart Theorem, it is enough to specify $4|4$ sections in $\mathcal{O}_{ij}$. If we organize the sections of $\mathcal{O}_{ij}$ in two matrices as above, we need to multiply the first matrix by the 2×2 inverse matrix corresponding to the minor with k, l rows, while the second matrix must be multiplied by the 3×3 inverse matrix corresponding to the minor with rows k, l, 5. We obtain then a matrix in the form prescribed for $\mathcal{U}_{kl}$ and the sections appearing in there are the ones needed to define ϕ_{ij}^{kl}. The conditions defining $\mathcal{U}_{ij}$ and $\mathcal{U}_{kl}$ make sure that such inverses are always well defined.

We leave to the reader the tedious verifications that the ϕ_{ij}^{kl}'s indeed define a supermanifold structure on the topological space $G(2,4)$. We call such supermanifold the *flag supermanifold* and denote with $\mathcal{F}\ell$.

We now go to the functor of points approach, that will shed some light on the cumbersome construction we have made of the supermanifold $\mathcal{F}\ell$.

We start by defining a functor from the category of analytic supermanifolds to the category of sets and then we prove that it is the functor of points of the analytic supermanifold $\mathcal{F}\ell$ that we have just defined.

Definition 4.11.1. We define the superflag functor $F : (\text{smflds})^o \longrightarrow$ (sets) on the objects as follows:

$$F(T) = \{\mathcal{F}_T \subset \mathcal{G}_T \subset \mathcal{O}_T^{4|1} \mid \mathcal{F}_T,\, \mathcal{G}_T \text{ locally free subsheaves of rank } 2|0,\, 2|1\},$$

where $\mathcal{O}_T^{4|1} := (\mathbb{C}^{4|1} \otimes \mathcal{O}_T)_0$. Since any morphism $\phi : T \longrightarrow S$ of supermanifolds corresponds contravariantly to a morphism of their structural sheaves, $F(\phi)$ is also defined. ■

Notice that when T is a point, $F(T)$ corresponds to the flags of $2|0$ and $2|1$ subspaces in $\mathbb{C}^{4|1}$, thus recovering the geometric meaning of the superflag of Definition 4.5.6. We urge the reader to go through the proof of the next proposition since it gives a geometrical interpretation of the change of charts that we have used to define the supermanifold $\mathcal{F}\ell$.

Proposition 4.11.2. *F is the functor of points of $\mathcal{F}\ell$.*

Proof. This proposition amounts essentially to show that there is a one-to-one functorial correspondence:

$$F(T) \longleftrightarrow \mathcal{F}\ell(T), \qquad T \in (\text{smflds}),$$

with $\mathcal{F}\ell(T) = \text{Hom}(T, \mathcal{F}\ell)$. We start with an element of $F(T)$, that is, two locally free sheaves $\mathcal{F}_T \subset \mathcal{G}_T \subset \mathcal{O}_T^{4|1}$ and we want to construct the corresponding morphism $f : T \longrightarrow \mathcal{F}\ell$. We choose a cover of T by open sets V_i's so that $\mathcal{F}_T$ and $\mathcal{G}_T$ are free:

$$\mathcal{F}_T(V_i) = \mathrm{span}_{\mathcal{O}_T(V_i)}\{u, v\}, \qquad \mathcal{G}_T(V_i) = \mathrm{span}_{\mathcal{O}_T(V_i)}\{u, v, w\},$$

$$u = \begin{pmatrix} t_{11} \\ t_{21} \\ t_{31} \\ t_{41} \\ \theta_{51} \end{pmatrix}, \qquad v = \begin{pmatrix} t_{12} \\ t_{22} \\ t_{32} \\ t_{42} \\ \theta_{52} \end{pmatrix}, \qquad w = \begin{pmatrix} \theta_{15} \\ \theta_{25} \\ \theta_{35} \\ \theta_{45} \\ t_{55} \end{pmatrix}. \tag{4.56}$$

The ranks of $\mathcal{F}_T(V_i)$ and $\mathcal{G}_T(V_i)$ are $2|0$ and $2|1$ respectively, so there is an invertible, order 2 minor in the matrix (u, v) and an invertible, order $2|1$ minor in the matrix (u, v, w). For the sake of clarity we may assume that such minors have rows $(1, 2)$ and $(1, 2, 5)$ respectively. In general they would have rows (i, j) and $(i, j, 5)$ (see the remark after the proof). By right multiplication by two suitable matrices we can write

$$\mathcal{F}_T(V_i) = \mathrm{span}_{\mathcal{O}_T(V_i)}\{a, b\}, \qquad \mathcal{G}_T(V_i) = \mathrm{span}_{\mathcal{O}_T(V_i)}\{c, d, e\},$$

$$a = \begin{pmatrix} 1 \\ 0 \\ s_{31} \\ s_{41} \\ \sigma_{51} \end{pmatrix}, \quad b = \begin{pmatrix} 0 \\ 1 \\ s_{32} \\ s_{42} \\ \sigma_{52} \end{pmatrix}, \quad c = \begin{pmatrix} 1 \\ 0 \\ w_{31} \\ w_{41} \\ 0 \end{pmatrix}, \quad d = \begin{pmatrix} 0 \\ 1 \\ w_{32} \\ w_{42} \\ 0 \end{pmatrix}, \quad e = \begin{pmatrix} 0 \\ 0 \\ \eta_{35} \\ \eta_{45} \\ 1 \end{pmatrix}. \tag{4.57}$$

Notice that this provides a *unique* way to express the generators of $\mathcal{F}_T(V_i)$ and $\mathcal{G}_T(V_i)$. The condition $\mathcal{F}_T(V_i) \subset \mathcal{G}_T(V_i)$ becomes

$$\mathrm{span}\{a, b, e\} = \mathrm{span}\{c, d, e\}.$$

If one right multiplies the matrix whose columns are a, b, e by the change of basis

$$\begin{pmatrix} 1 & 0 & 0 \\ 0 & 1 & 0 \\ \sigma_{51} & \sigma_{52} & 1 \end{pmatrix},$$

one sees right away that the w_{ij} must satisfy:

$$\begin{pmatrix} w_{31} & w_{32} \\ w_{41} & w_{42} \end{pmatrix} = \begin{pmatrix} s_{31} & s_{32} \\ s_{41} & s_{42} \end{pmatrix} - \begin{pmatrix} \sigma_{35} \\ \sigma_{45} \end{pmatrix} \begin{pmatrix} \eta_{51} & \eta_{52} \end{pmatrix}.$$

Then, using the Chart Theorem, we can define the morphism $f_i : \mathcal{V}_i \longrightarrow \mathcal{U}_{12}$ from the open submanifold $\mathcal{V}_i$ of T to the open submanifold $\mathcal{U}_{12}$ of $\mathcal{F}\ell$. It is clear that the changes of charts in the cumbersome definition of $\mathcal{F}\ell$ were especially deviced to make these f_i patch to give a well defined morphism from $T \longrightarrow \mathcal{F}\ell$. We leave to the reader all of the tedious verifications involved.

Conversely, assume that we have a morphism $f : T \longrightarrow \mathcal{F}\ell$. Since $|\mathcal{F}\ell| = U_{i,j}U_{ij}$, $1 \leq i < j \leq 4$, we define the open cover of T, $V_{ij} = f^{-1}(U_{ij})$ and $f_{ij} = f|_{V_{ij}}$. By the Chart's Theorem each f_{ij} corresponds to the choice of four even and two odd sections in $\mathcal{O}_T(V_{ij})$. We organize such sections, as usual, in matrix form. For example, for U_{12} we choose the sections u_{31}, u_{32}, u_{41}, u_{42}, α_{51}, α_{52} and we organize them in matrices as in (4.54). Notice that the v_{kl} in (4.54) are completely determined by Eq. (4.55) once we make our choice of sections.

We now define $\mathcal{F}_T(V_{12}) \subset \mathcal{G}_T(V_{12}) \subset \mathcal{O}_T(V_{12})$ as follows:

$$\mathcal{F}_T(V_{12}) = \text{span}_{\mathcal{O}_T(V_{12})}\{p, q\} \subset \mathcal{G}_T(V_{12}) = \text{span}_{\mathcal{O}_T(V_{12})}\{r, s, t\}$$

where p, q are the columns of the first matrix in the form (4.54), while r, s, t are the columns of the second matrix in there. Those vectors are precisely in the form (4.57). By restriction, we can define sheaves $\mathcal{F}_T(V_{12})$, $\mathcal{G}_T(V_{12})$ on V_{12} and one can repeat the same construction for all of the pairs of indices (i,j), $1 \leq i,j \leq 4$. The six pairs of sheaves so obtained glue together to give two sheaves $\mathcal{F}_T \subset \mathcal{G}_T \subset \mathcal{O}_T^{4|1}$, thus an element in $F(T)$. We leave to the reader all the verifications involved here. ∎

The proposition above allows us to express the functor of points of $\mathcal{F}\ell$ and of each of the submanifolds $\mathcal{U}_{ij}$ of the open covering $\mathcal{F}\ell$ quite explicitly. If $f \in \mathcal{F}\ell(T)$, f consists of a pair of a locally free subsheaves $\mathcal{F}_T \subset \mathcal{G}_T$ of $\mathcal{O}_T^{4|1}$ which locally (that is, on a suitable open cover of T) look like (refer to the notation as in Proposition 4.11.2 and (4.56)):

$$\mathcal{F}_T(V) = \text{span}_{\mathcal{O}_T(V)}\{u, v\} \subset \mathcal{G}_T(V) = \text{span}_{\mathcal{O}_T(V)}\{u, v, w\}$$

and have rank $2|0$, $2|1$. Such f belongs to $\mathcal{U}_{ij}(T)$ if and only if the minors corresponding respectively to rows i,j in the matrix (u,v), with columns the generators of $\mathcal{F}_T$, and to rows $i,j,5$ in the matrix (u,v,w), with columns the generators of $\mathcal{G}_T$, are non singular (refer to (4.56)). Notice that the condition on the minor in the matrix (u,v,w) relative to $\mathcal{G}_T$ is a consequence of the condition on the minor on the matrix (u,v) relative to $\mathcal{F}_T$. In fact, if the minor corresponding to rows i,j in (u,v) is invertible, the minor in (u,v,w) corresponding to rows $i,j,5$ is automatically invertible, since the element t_{55} has always to be invertible (the rank of $\mathcal{G}_T$ has to be $2|1$). Notice that the condition on the rank of $\mathcal{G}_T$ implies that at least one of the

$2|1$ minors in the matrix (u, v, w) is invertible, but one of such minors is invertible if and only if its diagonal blocks are invertible and one of such diagonal blocks is always t_{55}.

We have found six open submanifolds $\mathcal{U}_{ij}$ covering $\mathcal{F}\ell$ and a generic element $\mathcal{F}_T \subset \mathcal{G}_T$ in $\mathcal{F}\ell(T)$ belongs to one of them depending on whether one of the following of the six pairs of minors in the expression of $\mathcal{F}_T$ and $\mathcal{G}_T$ is invertible:

$$\mathcal{U}_{12} : (1,2) \quad (1,2,5) \qquad \mathcal{U}_{13} : (1,3) \quad (1,3,5) \qquad \mathcal{U}_{14} : (1,4) \quad (1,4,5)$$

$$\mathcal{U}_{23} : (2,3) \quad (2,3,5) \qquad \mathcal{U}_{24} : (2,4) \quad (2,4,5) \qquad \mathcal{U}_{34} : (3,4) \quad (3,4,5)$$

One should always keep in mind that the underlying topological space of $\mathcal{F}\ell$ is still $G(2,4)$ (as for $\mathrm{Gr}^{\mathrm{ch}}$) the Grassmannian of planes in $\mathbb{C}^4$, hence we have the same covering by open submanifolds studied in detail in Chapter 2 for the ordinary setting.

Now we come more in detail to the description of the cells $\mathcal{U}_{ij}$. An element $\mathcal{F}_T \subset \mathcal{G}_T$ in $\mathcal{U}_{ij}$ corresponds locally to a unique pair of matrices having as columns the generators of $\mathcal{F}_T$ and $\mathcal{G}_T$ expressed in the normalized form (for notation refer to (4.54)). This form is obtained from the generic generators as in (4.56) through a change of basis which allows us to write the identity in place of the minor corresponding to rows i, j for $\mathcal{F}_T$ and $i, k, 5$ for $\mathcal{G}_T$. So, for example, for $(i,j) = (1,2)$:

$$\mathcal{U}_{12}(T) \approx \left(\begin{pmatrix} I \\ A \\ \alpha \end{pmatrix}, \begin{pmatrix} I & 0 \\ B & \beta \\ 0 & 1 \end{pmatrix} \right), \qquad T \in (\text{smflds}), \tag{4.58}$$

where I is the identity, A and B are 2×2 matrices with even entries and $\alpha = (\alpha_1, \alpha_2)$, $\beta^t = (\beta_1, \beta_2)$ are rows with odd entries in $\mathcal{O}(T)$. As we saw in the proof of Proposition 4.11.2 the entries in A and B are not independent but they satisfy relations called *twistor relations*:

$$B = A - \beta\alpha \,. \tag{4.59}$$

This fact is of capital importance, since the *big cell* $\mathcal{U}_{12}$ will turn out to be the complex Minkowski superspace, as we are going to establish in the next section. The presence of the twistor relations marks a striking difference with the ordinary setting in which the big cell U_{12} of the Grassmannian $G(2,4)$ is an affine space and its entries are not subject to any relation. Here we have a relation which is non linear and adds an extra difficulty to the theory.

In the big cell $\mathcal{U}_{12}$ a flag is completely described by the triplet (A, α, β). Since there are no relations among A, α and β, we can take them as local coordinates for $\mathcal{U}_{12}$. We then have that $\mathcal{U}_{12}$ is an affine $4|4$ superspace.

We now turn to the problem of realizing the complex superflag as a quotient supermanifold.

4.12 The superflag as an homogeneous space

In Section 4.7 we defined, using the formalism of the functor of points, the action of the group $G = \mathrm{SL}(4|1)$ on the Grassmannian supermanifold $\mathrm{Gr}^{\mathrm{ch}}$. In particular, we defined $g \cdot \mathcal{F}_T$, with $g \in G(T)$ and $\mathcal{F}_T$ a rank $2|0$ locally free subsheaf of $\mathcal{O}_T^{4|1}$ (hence, an element of $\mathrm{Gr}^{\mathrm{ch}}(T)$). The same reasoning allows us to define $g \cdot \mathcal{G}_T$, where $\mathcal{G}_T$ is a locally free rank $2|1$ subsheaf of $\mathcal{O}_T^{4|1}$.

We can then define the following action of G on $\mathcal{F}\ell$:

$$\begin{array}{ccc} G(T) \times \mathcal{F}\ell(T) & \xrightarrow{\psi_T} & \mathcal{F}\ell(T) \\ (g, \mathcal{F}_T \subset \mathcal{G}_T) & \longrightarrow & g \cdot \mathcal{F}_T \subset g \cdot \mathcal{G}_T \,. \end{array}$$

We now look at the stabilizer of the topological point of $\mathcal{F}\ell$ corresponding to the pair of sheaves

$$\mathcal{F}_0 = \mathrm{span}\{e_1, e_2\} \subset \mathcal{G}_0 = \mathrm{span}\{e_1, e_2, \epsilon_5\} \in |\mathcal{F}\ell| \,,$$

where $(e_1, \dots, e_4, \epsilon_5)$ denotes, as usual, the canonical (homogeneous) basis in $\mathbb{C}^{4|1}$.

Let us compute the stabilizer:

$$\begin{aligned} \mathrm{Stab}(\mathcal{F}_0 \subset \mathcal{G}_0)(T) &= \{g \in \mathrm{SL}(4|1)(T) \,|\, g \cdot \mathcal{F}_0 = \mathcal{F}_0,\, g \cdot \mathcal{G}_0 = \mathcal{G}_0, \}, \\ \mathrm{Stab}(\mathcal{F}_0 \subset \mathcal{G}_0)(T) &\subset \mathrm{SL}(4|1)(T) \,. \end{aligned}$$

It is immediate to verify that

$$\mathrm{Stab}(\mathcal{F}_0 \subset \mathcal{G}_0)(T) = \left\{ \begin{pmatrix} g_{11} & g_{12} & g_{13} & g_{14} & \gamma_{15} \\ g_{21} & g_{22} & g_{23} & g_{24} & \gamma_{25} \\ 0 & 0 & g_{33} & g_{34} & 0 \\ 0 & 0 & g_{43} & g_{44} & 0 \\ 0 & 0 & \gamma_{53} & \gamma_{54} & g_{55} \end{pmatrix} \right\} \subset \mathrm{SL}(4|1)(T) \,.$$

The functor $T \mapsto \mathrm{Stab}(\mathcal{F}_0 \subset \mathcal{G}_0)(T)$ is the functor of points of a supergroup, that we call F_u. Comparing with Definition 4.6.1, we see that the Lie superalgebra associated to F_u is just the transposed version of $\mathfrak{f}^c$, the super Poincaré algebra plus dilations. These two superalgebras are conjugated and then, isomorphic.

Proposition 4.12.1. *The flag supermanifold* $\mathcal{F}\ell$ *is the quotient* $\mathrm{SL}(4|1)/F_u$.

Proof. We are going to use Proposition 1.9.7 in Section 1.9. The proof resembles very much the proof of Proposition 4.7.2, but we shall nevertheless briefly recapitulate it, given its importance in our treatment. We consider the morphism $\pi : G \longrightarrow \mathcal{F}\ell$, which, in the functor of points formalism, reads as:

$$\pi_T(g) = (g \cdot \mathcal{F}_0, g \cdot \mathcal{G}_0) .$$

We can express an element $g \in G(T)$ as

$$g = \begin{pmatrix} g_{ij} & \gamma_{i5} \\ \gamma_{5j} & g_{55} \end{pmatrix}, \qquad i, j = 1, \dots, 4 .$$

Let us assume that $g \cdot \mathcal{F}_0$, $g \cdot \mathcal{G}_0 \in \mathcal{U}_{12}(T) \subset F\ell(T)$. Hence, locally, we can write $g \cdot \mathcal{F}_0, g \cdot \mathcal{G}_0$ as:

$$g \cdot \mathcal{F}_0,\, g \cdot \mathcal{G}_0 \cong \begin{pmatrix} g_{11} & g_{12} \\ g_{21} & g_{22} \\ g_{31} & g_{32} \\ g_{41} & g_{42} \\ \gamma_{51} & \gamma_{52} \end{pmatrix}, \begin{pmatrix} g_{11} & g_{12} & \gamma_{15} \\ g_{21} & g_{22} & \gamma_{25} \\ g_{31} & g_{32} & \gamma_{35} \\ g_{41} & g_{42} & \gamma_{45} \\ \gamma_{51} & \gamma_{52} & g_{55} \end{pmatrix} = \begin{pmatrix} Z \\ W \\ \rho \end{pmatrix}, \begin{pmatrix} Z & \tau_1 \\ W & \tau_2 \\ \rho & a \end{pmatrix}, \tag{4.60}$$

where

$$Z = \begin{pmatrix} g_{11} & g_{12} \\ g_{21} & g_{22} \end{pmatrix}, \quad W = \begin{pmatrix} g_{31} & g_{32} \\ g_{41} & g_{42} \end{pmatrix}, \quad \rho = (\gamma_{51}, \gamma_{52}),$$

$$\tau_1 = \begin{pmatrix} \gamma_{15} \\ \gamma_{25} \end{pmatrix}, \quad \tau_2 = \begin{pmatrix} \gamma_{35} \\ \gamma_{45} \end{pmatrix}, \quad a = g_{55} .$$

We have to revert this into the standard form (4.58) by a change of basis. This amounts to multiply on the right the first factor with a matrix of $\mathrm{SL}_2(T)$ and the second factor by a matrix of $\mathrm{SL}(2|1)(T)$. These matrices are, respectively,

$$P = Z^{-1}, \qquad N = \begin{pmatrix} Q & \delta \\ \epsilon & d \end{pmatrix} = \begin{pmatrix} Z & \tau_1 \\ \rho & a \end{pmatrix}^{-1},$$

that is:

$$Q = (Z - \tau_1 \rho a^{-1})^{-1}, \qquad d = (a - \rho Z^{-1} \tau_1)^{-1},$$

$$\delta = -Z^{-1}\tau_1 (a - \rho Z^{-1}\tau_1)^{-1}, \qquad \epsilon = -\rho a^{-1} (Z - \tau_1 \rho a^{-1})^{-1} .$$

Then,

$$g \cdot \mathcal{F}_0,\, g \cdot \mathcal{G}_0 \cong \begin{pmatrix} \mathrm{id} \\ WZ^{-1} \\ \rho Z^{-1} \end{pmatrix}, \begin{pmatrix} \mathrm{id} & 0 \\ VY^{-1} & \tau \\ 0 & 1 \end{pmatrix},$$

where

$$V = W - a^{-1}\tau_2\rho, \quad Y = Z - a^{-1}\tau_2\rho, \quad \tau = (\tau_2 - WZ^{-1}\tau_1)(a - \rho Z^{-1}\tau_1)^{-1}.$$

The morphism π_T in these coordinates is given by

$$g \longmapsto \left(WZ^{-1}, \rho Z^{-1}, (\tau_2 - WZ^{-1}\tau_1)(a - \rho Z^{-1}\tau_1)^{-1}\right).$$

At this point one can compute the Jacobian and verify that, at the identity, $d\pi$ is surjective, so π is a submersion. We can repeat the same argument for all the $\mathcal{U}_{ij}$ and the result comes as an application of Proposition 1.9.7. ∎

Notice that the flag $\mathcal{F}_0 \subset \mathcal{G}_0$ corresponding to the identity in $\mathrm{SL}(4|1)/F_u$, in the coordinates (A, α, β) for the big cell, becomes:

$$\left(\begin{pmatrix} I \\ 0 \\ 0 \end{pmatrix}, \begin{pmatrix} I & 0 \\ 0 & 0 \\ 0 & 1 \end{pmatrix}\right) \approx (0, 0, 0).$$

We now look at the subgroup of G leaving the big cell $\mathcal{U}_{12}$ invariant. Such a subgroup is given, in the functor of points notation, by

$$\begin{array}{ccc} (\mathrm{smflds}) & \xrightarrow{F^c} & (\mathrm{sets}) \\ T & \longrightarrow & \begin{pmatrix} L & 0 & 0 \\ NL & R & R\chi \\ d\varphi & 0 & d \end{pmatrix}, \end{array} \tag{4.61}$$

with L, N, R being 2×2 even matrices, χ an odd, 1×2 matrix, φ an odd 2×1 matrix and d a scalar, all with entries in $\mathcal{O}(T)$. We have adapted the notation in order to have a simpler expression for the action of the supergroup later. Such functor corresponds to the functor of points of an analytic supergroup (we denote the supergroup with the same letter F^c than its functor of points, as usual). The Lie superalgebra of F^c is the Poincaré superalgebra plus dilations of Definition 4.6.1.

We are ready for a key proposition, which summarizes our discussion.

Proposition 4.12.2. *The flag supermanifold $\mathcal{F}\ell$, identified with the quotient* $\mathrm{SL}(4|1)/F_u$, *carries a natural action of the supergroup* $\mathrm{SL}(4|1)$. *The big cell* $\mathcal{U}_{12}$ *carries a natural action of the supergroup* $F^c \subset \mathrm{SL}(4|1)$, *the Poincaré supergroup times dilations. In the coordinates* (A, α, β) *introduced in (4.58, 4.59) and (4.61) such action reads*

$$\begin{aligned} A &\longrightarrow R(A + \chi\alpha)L^{-1} + N\,, \\ \alpha &\longrightarrow d(\alpha + \varphi)L^{-1}\,, \\ \beta &\longrightarrow d^{-1}R(\beta + \chi)\,. \end{aligned} \tag{4.62}$$

■

Definition 4.12.3. The flag supermanifold $\mathcal{F}\ell$ is the complex *conformal superspace* and its big cell $\mathcal{U}_{12}$ is the *complex Minkowski superspace.* ■

The superflag $\mathcal{F}\ell$ carries a natural action of the supergroup $\mathrm{SL}(4|1)$, which is also called the *complex conformal supergroup*, since its Lie superalgebra is $\mathfrak{sl}(4|1)$, the complex conformal superalgebra examined in detail in Section 4.4. The supergroup F^c described above, acting on $\mathcal{U}_{12}$, is called the *complex Poincaré supergroup times dilations* and its Lie superalgebra is the complex Poincaré superalgebra plus dilations of Definition 4.6.1.

Setting $L = R = \mathrm{id}$, $d = 1$ in (4.61) and (4.62) we obtain the *complex translation supergroup*, whose Lie superalgebra is (4.36) and whose action on the Minkowski space is

$$\begin{aligned} A &\longrightarrow A + N + \chi\alpha\,, \\ \alpha &\longrightarrow \alpha + \varphi\,, \\ \beta &\longrightarrow \beta + \chi\,. \end{aligned}$$

We see that φ and χ act just as odd translations in the same way than N is an even translation. The non trivial difference is the term $\chi\alpha$ which also acts as an even translation and it is due to the fact that the supergroup is non abelian.

If the odd part is zero, then the action of F^c on $\mathcal{U}_{12}$ reduces to the one of the classical Poincaré group times dilations on the ordinary Minkowski space.

We end by noticing that the superflag is the *superconformal compactification* of its big cell, the Minkowski superspace, just as the Grassmannian $G(2,4)$ was the *conformal compactification* of its big cell, the Minkowski space.

4.13 The real Minkowski superspace

In this section we want to construct the real form of the Minkowski superspace.

We start by explicitly computing the real form of SL(4|1) that corresponds to the real form $\mathfrak{su}(2,2|1)_\pm$ of the Lie superalgebra $\mathfrak{sl}(4|1)$ (see Section 4.4). We remind the reader that we use $j = \pm 1$ for the alternative signs appearing in $\mathfrak{su}(2,2|1)_\pm$. The conjugation $\sigma_\pm$ (4.17), whose set of fixed points is the real form of interest, is

$$\begin{array}{ccc} \mathfrak{sl}(4|1) & \xrightarrow{\sigma_\pm} & \mathfrak{sl}(4|1) \\ \begin{pmatrix} p & \alpha \\ \beta & d \end{pmatrix} & \longrightarrow & \begin{pmatrix} -Fp^\dagger F & j\mathrm{i}F\beta^\dagger \\ j\mathrm{i}\alpha^\dagger F & -\bar{d} \end{pmatrix}, \end{array} \tag{4.63}$$

where

$$F = \mathrm{i}\begin{pmatrix} 0 & -\mathrm{id} \\ \mathrm{id} & 0 \end{pmatrix}. \tag{4.64}$$

'id' denotes the 2×2 identity matrix.

We will use the functor of points approach. We refer to Section 1.10, Example 1.10.9.

Let $G = \mathrm{SL}(4|1,\mathbb{C})$ and let $G_\mathbb{R}$ be the complex supergroup G seen as a real supergroup (thus forgetting the complex structure). In order to obtain a real form of G we need an involutive morphism $\rho : G_\mathbb{R} \to G_\mathbb{R}$ that is antilinear with respect to the complex structure. Then, by Yoneda's lemma, this corresponds to a functorial family of maps:

$$G_\mathbb{R}(R) \xrightarrow{\rho_R} G_\mathbb{R}(R).$$

The procedure is similar to the one used in Example 1.10.9. We start by giving the following map

$$\begin{array}{ccc} G(R) & \xrightarrow{\theta_{R\pm}} & G(R) \\ g = \begin{pmatrix} D & \tau \\ \gamma & d \end{pmatrix} & \longrightarrow & g^{\theta_\pm} = \begin{pmatrix} D^\dagger & j\mathrm{i}\gamma^\dagger \\ j\mathrm{i}\tau^\dagger & \bar{d} \end{pmatrix}, \end{array}$$

where, as before, $j = \pm 1$. This map satisfies

$$(g^{\theta_\pm})^{\theta_\pm} = g, \qquad (hg)^{\theta_\pm} = g^{\theta_\pm} h^{\theta_\pm}. \tag{4.65}$$

One should have in mind two conventions: First, the one taken in (1.12) about the complex conjugation of a product of odd variables

$$\overline{\xi\xi'} = \bar{\xi}\bar{\xi}' \,.$$

Second, the convention for the transpose of the product of matrices A and B with odd entries (1.5)

$$(AB)^t = -B^t A^t \,.$$

The involution $\rho_\pm$, which gives the real form of G, is defined as

$$\begin{array}{ccc} G(R) & \xrightarrow{\rho_\pm} & G(R) \\ g & \longrightarrow & g^{\rho_\pm} = L(g^{\theta_\pm})^{-1}L, \end{array} \qquad \text{with} \quad L = \begin{pmatrix} F & 0 \\ 0 & 1 \end{pmatrix}$$

and F as in (4.64). We have that properties (4.65) also hold for $\rho_\pm$,

$$(g^{\rho_\pm})^{\rho_\pm} = g, \qquad (hg)^{\rho_\pm} = g^{\rho_\pm} h^{\rho_\pm} \,.$$

From this definition we obtain immediately the following proposition (see also Chapter 1, Section 1.10).

Proposition 4.13.1. *The map $g \to g^{\rho_\pm}$ is a natural transformation. It defines a ringed space involutive isomorphism $\rho : G_{\mathbb{R}} \to G_{\mathbb{R}}$ which is $\mathbb{C}$-antilinear with respect to the complex structure in G.*

The topological space $|G|^\rho$, consisting of the points fixed by $|\rho|$, has a real supermanifold structure and the supersheaf is composed of those sections $f \in \mathcal{O}_G$ such that $\rho^*(f) = f$ restricted to $|G|^\rho$. ■

We now have now the following proposition:

Proposition 4.13.2. *The involution $\rho_\pm$ of G defined above induces on* Lie(G) *the conjugation (4.63).*

Proof. We have to compute the tangent map at the identity, so we shall write

$$g = \begin{pmatrix} D & \tau \\ \gamma & d \end{pmatrix} \approx \begin{pmatrix} \mathrm{id} & 0 \\ 0 & 1 \end{pmatrix} + \varepsilon \begin{pmatrix} p & \alpha \\ \beta & c \end{pmatrix}, \qquad \begin{pmatrix} p & \alpha \\ \beta & c \end{pmatrix} \in \mathfrak{g} = \mathfrak{sl}(4|1) \,,$$

up to first order in ε (we can think on ε as an even variable with $\varepsilon^2 = 0$).

We now compute $g^{\rho_\pm}$ up to first order in ε:

$$g^{\rho_\pm} \approx \begin{pmatrix} \mathrm{id} & 0 \\ 0 & 1 \end{pmatrix} + \varepsilon L \begin{pmatrix} -p^\dagger & -ji\beta^\dagger \\ -ji\alpha^\dagger & -\bar{c} \end{pmatrix} L \,,$$

from which the result follows. ■

The complex super Poincaré group times dilations was defined in (4.61) and denoted as F^c. It is the subgroup of SL(4|1) whose T-points are of the form

$$g = \begin{pmatrix} L & 0 & 0 \\ M & R & R\varphi \\ d\chi & 0 & d \end{pmatrix},$$

with M, L and R 2×2 matrices, φ a 2×1 matrix, χ a 1×2 matrix and d a 1×1 matrix, all with entries in $\mathcal{O}(T)$. In the notation of (4.61), one has $M = NL$. We have defined the submatrices in a way that will be more convenient to express action of this group. The Lie superalgebra of F^c coincides with the one given in Definition 4.6.1 and denoted as $\mathfrak{f}^c$. As it happens at the infinitesimal level, the conjugation $\rho_\pm$ restricts to F^c and then it defines a real form of F^c, which is the *real Poincaré supergroup times dilations*. It will be denoted as F. We also recall that the two conjugations $\sigma_\pm$, at the Lie superalgebra level, give isomorphic real Poincaré superalgebras. Obviously, the same is true at the group level.

Let us compute this real form. First we notice that

$$g^{-1} = \begin{pmatrix} L^{-1} & 0 & 0 \\ -R^{-1}ML^{-1} + \varphi\chi L^{-1} & R^{-1} & -\varphi d^{-1} \\ -\chi L^{-1} & 0 & d^{-1} \end{pmatrix},$$

from which

$$g^{\rho_\pm} = \begin{pmatrix} {R^\dagger}^{-1} & 0 & 0 \\ {L^\dagger}^{-1}M^\dagger {R^\dagger}^{-1} + {L^\dagger}^{-1}\chi^\dagger\varphi^\dagger & {L^\dagger}^{-1} & j{L^\dagger}^{-1}\chi^\dagger \\ j\bar{d}^{-1}\varphi^\dagger & 0 & \bar{d}^{-1} \end{pmatrix}.$$

It follows that the fixed points are those that satisfy the conditions

$$L = {R^\dagger}^{-1}, \qquad \varphi = j\chi^\dagger, \qquad ML^{-1} = (ML^{-1})^\dagger + j{L^\dagger}^{-1}\chi^\dagger\chi L^{-1}. \quad (4.66)$$

The condition $d = \bar{d}^{-1}$ follows from the determinant condition and (4.66).

To get a more familiar form for the reality conditions, let us define $N = ML^{-1}$ and

$$T := N - \frac{1}{2}j{L^\dagger}^{-1}\chi^\dagger\chi L^{-1}.$$

Then, (4.66) becomes

$$L = {R^\dagger}^{-1}, \qquad \varphi = j\chi^\dagger, \qquad T = T^\dagger. \quad (4.67)$$

We want now to compute the real form of the superflag and its big cell. We recall that the flag supermanifold $\mathcal{F}\ell = \mathcal{F}\ell(2|0, 2|1, 4|1)$ could be seen as an homogeneous space of the supergroup SL(4|1) (see Section 4.12) and that its stabilizer was the parabolic subsupergroup with points

$$\begin{pmatrix} g_{11} & g_{12} & g_{13} & g_{14} & \gamma_{15} \\ g_{21} & g_{22} & g_{23} & g_{24} & \gamma_{25} \\ 0 & 0 & g_{33} & g_{34} & 0 \\ 0 & 0 & g_{43} & g_{44} & 0 \\ 0 & 0 & \gamma_{53} & \gamma_{54} & g_{55} \end{pmatrix} = \begin{pmatrix} P & Q & \gamma \\ 0 & S & 0 \\ 0 & \gamma' & c \end{pmatrix} .$$

This supergroup was denoted as F_u. By the discussion in Section 4.5, we know that this supergroup is isomorphic to the complex Poincaré supergroup times dilations, F^c. In fact, by the same arguments used there, the conjugation $\rho_\pm$ restricts to F_u and then, a conjugation is induced on the quotient space $\mathcal{F}\ell = \mathrm{SL}(4|1)/F_u$.

The easiest way of seeing this conjugation and its associated real form explicitly is to work on the big cell and to consider the Poincaré group F^c, in the functor of points notation, as the group of T-points preserving the big cell.

Let (A, α, β) be the coordinates of a point in the big cell (4.58). Since the big cell is a subset of $(\mathrm{SL}(4|1)/F_u)(T)$, we can choose a representative of the point with coordinates (A, α, β) in the group $\mathrm{SL}(4|1)(T)$

$$g = \begin{pmatrix} \mathrm{id} & 0 & 0 \\ A & \mathrm{id} & \beta \\ \alpha & 0 & 1 \end{pmatrix} , \qquad gF_u(T) \in (\mathrm{SL}(4|1)/F_u)(T) , \tag{4.68}$$

for T a supermanifold (we may have to replace T by an open subset belonging to a suitable cover of T, see the observations at the end of Section 2.10). Notice that the chosen representative belongs to $F^c(T)$. g is the representative of (A, α, β) because

$$g \begin{pmatrix} \mathrm{id} \\ 0 \\ 0 \end{pmatrix} = \begin{pmatrix} \mathrm{id} \\ A \\ \alpha \end{pmatrix} , \qquad g \begin{pmatrix} \mathrm{id} & 0 \\ 0 & 0 \\ 0 & 1 \end{pmatrix} = \begin{pmatrix} \mathrm{id} & 0 \\ A & \beta \\ \alpha & 1 \end{pmatrix} \approx \begin{pmatrix} \mathrm{id} & 0 \\ A - \beta\alpha & \beta \\ 0 & 1 \end{pmatrix} .$$

We first compute the inverse,

$$g^{-1} = \begin{pmatrix} \mathrm{id} & 0 & 0 \\ -A + \beta\alpha & \mathrm{id} & -\beta \\ -\alpha & 0 & 1 \end{pmatrix} ,$$

and then $g^{\rho_\pm}$

$$g^{\rho_\pm} = \begin{pmatrix} \mathrm{id} & 0 & 0 \\ A^\dagger + \alpha^\dagger \beta^\dagger & \mathrm{id} & j\alpha^\dagger \\ j\beta^\dagger & 0 & 1 \end{pmatrix} .$$

From this, the points that are fixed by $\rho_\pm$ must satisfy

$$A = A^\dagger + j\alpha^\dagger \alpha, \qquad \beta = j\alpha^\dagger .$$

As before, we can make a convenient change of coordinates,

$$C := A - \frac{1}{2} j\alpha^\dagger \alpha , \tag{4.69}$$

so the reality condition is

$$C = C^\dagger, \qquad \beta = j\alpha^\dagger . \tag{4.70}$$

The reality condition on the ordinary Minkowski space then reduces to the one used in Chapter 2.

Remark 4.13.3. In all of these calculations we have assumed to be in a situation in which: $(\mathrm{SL}(4|1)/F_u)(T) = \mathrm{SL}(4|1)(T)/F_u(T)$. This is not a restrictive assumption, since if this is not the case, we can always argue by taking a suitable open cover of the supermanifold T for which our assumption holds and then glue the open subsets using a technique we described in detail in Section 4.11. For the sake of clarity we decided not be bothered by these technicalities, which are easily overcome. ■

4.14 Complex superfields

In this and the next three sections we intend to make contact with the way in which the Minkowski superspace appears in physics. We start from the definition of the complex Minkowski superspace as the big cell U_{12} of the superflag $\mathcal{F}\ell$ (Definition 4.12.3).

We will denote the complex Minkowski superspace as $\mathbf{M}^c := U_{12} \cong \mathbb{C}^{4|4}$, and we will consider it as a complex analytic supermanifold. In all the developments the functor of points notation is implicit (see also Remark 4.13.3). We have on $\mathbf{M}^c$ global coordinates (A, α, β) (4.58, 4.59). We adopt here the coordinates

$$C = A - \frac{1}{2}\beta\alpha ,$$

so the reality condition (4.70) is easier to impose. We also rename the odd coordinates $(\beta, \alpha) \to (\theta, \bar{\theta}^t)$ to comply with the notation in the physics literature. At this point, $\bar{\theta}$ is just a pair of odd variables different from θ. We will see later that $\bar{\theta}$ can be thought as the 'complex conjugate' of θ.

The functor of points of the big cell is given, for $T \in$ (smflds), as (see (4.58))

$$\mathcal{U}_{12}(T) \approx \left(\begin{pmatrix} \mathrm{id} \\ A \\ \alpha \end{pmatrix}, \begin{pmatrix} \mathrm{id} & 0 \\ B & \beta \\ 0 & 1 \end{pmatrix} \right) \longrightarrow \left(\begin{pmatrix} \mathrm{id} \\ C + \frac{1}{2}\theta\bar{\theta}^t \\ \bar{\theta}^t \end{pmatrix}, \begin{pmatrix} \mathrm{id} & 0 \\ C - \frac{1}{2}\theta\bar{\theta}^t & \theta \\ 0 & 1 \end{pmatrix} \right), \tag{4.71}$$

which satisfy the twistor relation (4.59) $B = A - \theta\bar{\theta}^t$.

We also make a change of variables in the Poincaré supergroup times dilations with respect to (4.61). The supergroup, in the functor of points notation, becomes

$$\begin{pmatrix} L & 0 & 0 \\ TL + \frac{1}{2}R\varphi\bar{\varphi}^t & R & R\varphi \\ d\bar{\varphi}^t & 0 & d \end{pmatrix}, \qquad T = N - \frac{1}{2}R\varphi\bar{\varphi}^t L^{-1}. \tag{4.72}$$

Its action on $\mathbf{M}^c$ is then

$$\begin{aligned} C &\longrightarrow R(C + \frac{1}{2}\varphi\bar{\theta}^t - \frac{1}{2}\theta\bar{\varphi}^t)L^{-1} + T, \\ \theta &\longrightarrow d^{-1}R(\theta + \varphi), \\ \bar{\theta} &\longrightarrow dL^{-1t}(\bar{\theta} + \bar{\varphi}). \end{aligned} \tag{4.73}$$

We may want to use directly the coordinates x^μ of the Minkowski space. These are defined in terms of the Pauli matrices as in Chapter 2 (2.54)

$$\sigma_0 = \begin{pmatrix} 1 & 0 \\ 0 & 1 \end{pmatrix}, \quad \sigma_1 = \begin{pmatrix} 0 & 1 \\ 1 & 0 \end{pmatrix}, \quad \sigma_2 = \begin{pmatrix} 0 & -i \\ i & 0 \end{pmatrix}, \quad \sigma_3 = \begin{pmatrix} 1 & 0 \\ 0 & -1 \end{pmatrix}, \tag{4.74}$$

which satisfy

$$\sigma_\mu^\dagger = \sigma_\mu, \quad \sigma_a\sigma_b = \delta_{ab}\mathrm{id} + \mathrm{i}\epsilon_{abc}\sigma_c, \quad \mathrm{tr}\,\sigma_a = 0, \quad a, b, c = 1, 2, 3.$$

Then we define

$$C = \sum_{\mu=0}^{3} x^\mu \sigma_\mu = \begin{pmatrix} x^0 + x^3 & x^1 - \mathrm{i}x^2 \\ x^1 + \mathrm{i}x^2 & x^0 - x^3 \end{pmatrix}, \tag{4.75}$$

and one can check that

$$x^\mu = \frac{1}{2}\mathrm{tr}(C\sigma_\mu). \tag{4.76}$$

For the moment being, $x^\mu, \mu = 0, \ldots, 3$, are even, complex, global coordinates, but it is apparent that when we impose the reality condition they will become real coordinates.

The following notation for the odd coordinates is also customary in physics,

$$\theta = \begin{pmatrix} \theta^1 \\ \theta^2 \end{pmatrix}, \qquad \theta^\alpha, \quad \alpha = 1, 2,$$

$$\bar{\theta} = \begin{pmatrix} \bar{\theta}^1 \\ \bar{\theta}^2 \end{pmatrix}, \qquad \bar{\theta}^{\dot{\alpha}}, \quad \dot{\alpha} = 1, 2. \tag{4.77}$$

This helps to distinguish, in the explicit formulas, the representation of the Lorentz group carried by the odd variables. We recall that the (spin group of the) complex Lorentz group is

$$\mathbf{L}^c = (\mathrm{SL}_2(\mathbb{C})_\mathbb{R})^c = \mathrm{SL}_2(\mathbb{C}) \times \mathrm{SL}_2(\mathbb{C}).$$

Then θ is in the representation $S^+ = D(1/2, 0)$ of this group while $\bar{\theta}$ is in the (contragredient) representation $S^- = D(0, 1/2)$ (see (4.14) and comments after). In fact, we get

$$\begin{aligned} x^\mu &\longrightarrow \Lambda^\mu{}_\nu x^\nu, & \Lambda &\in \mathrm{SO}(1,3)^c, \\ \theta^\alpha &\longrightarrow R^\alpha{}_\beta \theta^\beta, & R &\in \mathrm{SL}_2(\mathbb{C}), \\ \bar{\theta}^{\dot{\alpha}} &\longrightarrow L^{-1t\dot{\alpha}}{}_{\dot{\beta}} \bar{\theta}^{\dot{\beta}} & L &\in \mathrm{SL}_2(\mathbb{C}). \end{aligned} \tag{4.78}$$

The matrix Λ can be determined from R and L through (4.75).

Remark 4.14.1. *Conventions on indices.* Generically, when a matrix is written as $A^a{}_b$, the indices more to the left indicate the row and the indices to the right indicate the column. Then the transpose matrix is written as

$$A^t{}_b{}^a = A^a{}_b,$$

that is, interchanging rows with columns.

If the coordinates of a vector space are numerated with a superindex (that is an index appearing as a superscript), then the coordinates in its dual vector space are enumerated with a subindex (and vice versa).

We use a superindex α to enumerate coordinates in a vector space that carries the representation S^+ and a superindex $\dot{\alpha}$ for coordinates in a vector space that carries the contragredient representation of S^-.

Additionally, 'vector' indices, that is, indices that run over the fundamental representation of $\mathrm{SO}(1,3)$ (or its complexification), are denoted with letters $\mu, \nu, \rho, \ldots$, while spinor indices are denoted with letters $\alpha, \beta, \gamma, \ldots$ or the dotted version of them.

These conventions may appear as unnecessary or vague, but they work quite well in calculations. ■

In particular, looking at (4.78), it is clear that C has indices $C^{\alpha\dot{\alpha}}$, so also the sigma matrices are

$$C^{\alpha\dot{\alpha}} = x^{\mu}\sigma_{\mu}^{\alpha\dot{\alpha}} .$$

We have that

$$R^{\alpha}{}_{\beta}\sigma_{\mu}^{\beta\dot{\beta}}L^{-1}{}_{\dot{\beta}}{}^{\dot{\alpha}} = \Lambda^{t}{}_{\mu}{}^{\nu}\sigma_{\nu}^{\alpha\dot{\alpha}}, \qquad R\sigma_{\mu}L^{-1} = \Lambda^{t}{}_{\mu}{}^{\nu}\sigma_{\nu} , \tag{4.79}$$

which defines Λ^t, and then, Λ. The sigma matrices are the intertwiners between the representation $S^{-} \otimes S^{+}$ and the fundamental representation of $\mathrm{SO}(1,3)$.

Once the notation and conventions are stated, we want to describe the superalgebra of global sections of $\mathbf{M}^c$, that is

$$\mathcal{O}(\mathbf{M}^c) = \mathcal{H}(\mathbb{C}^4) \otimes \wedge[\theta^1, \theta^2, \bar{\theta}^{\dot{1}}, \bar{\theta}^{\dot{2}}] ,$$

where $\mathcal{H}$ stands for holomorphic functions. We will simply write $\mathcal{H}(\mathbb{C}^4)[\theta, \bar{\theta}]$. An element $\Psi \in \mathcal{O}(\mathbf{M}^c)$ is called in physics a (holomorphic) *complex superfield*. We can write it explicitly in a basis of $\wedge[\theta, \bar{\theta}]$ as

$$\begin{aligned}\Psi(x) = \sum_I \psi_I \xi^I &= \psi_0(x) + \psi_{\alpha}(x)\theta^{\alpha} + \psi_{\dot{\alpha}}(x)\bar{\theta}^{\dot{\alpha}} + \psi_{\alpha\beta}(x)\theta^{\alpha}\theta^{\beta} \\ &+ \psi_{\alpha\dot{\beta}}(x)\theta^{\alpha}\bar{\theta}^{\dot{\beta}} + \psi'_{\dot{\alpha}\dot{\beta}}(x)\bar{\theta}^{\dot{\alpha}}\bar{\theta}^{\dot{\beta}} + \psi_{\alpha\beta\dot{\gamma}}(x)\theta^{\alpha}\theta^{\beta}\bar{\theta}^{\dot{\gamma}} \\ &+ \psi_{\alpha\dot{\beta}\dot{\gamma}}(x)\theta^{\alpha}\bar{\theta}^{\dot{\beta}}\bar{\theta}^{\dot{\gamma}} + \psi_{\alpha\beta\dot{\gamma}\dot{\delta}}(x)\theta^{\alpha}\theta^{\beta}\bar{\theta}^{\dot{\gamma}}\bar{\theta}^{\dot{\delta}} .\end{aligned} \tag{4.80}$$

We have denoted the odd variables generically as ξ. I is a multiindex $(\alpha, \beta, \dots, \dot{\alpha}, \dot{\beta}, \dots)$ and sum over repeated indices is understood. The holomorphic functions appearing as coefficients in the odd variables expansion are called *component fields*. Notice that because the odd variables anticommute, the component fields can be chosen antisymmetric in all their indices. We can see now the utility of the convention (4.77): we have attached to the component fields indices that unequivocally indicate the term of the expansion to which they belong. But this encodes more information, as we are going to see.

In physics, it is also common to denote $\Psi(x)$ as $\Psi(x, \theta, \bar{\theta})$. In this way one keeps thinking on Ψ as a 'function' of the even and odd variables. We know that this interpretation is not correct, but still, the notation can be useful in many situations.

Let us see an instance of this. The action of the Poincaré supergroup on $\mathbf{M}^c$ (4.62) gives rise to a coaction of the Hopf algebra (See Definition (1.11.8)) $\mathcal{O}(P^c)$ on $\mathcal{O}(\mathbf{M}^c)$. We will focus only on the action of the Lorentz group $\mathbf{L}^c \subset P^c$, although everything we say can be extended to the full Poincaré supergroup. The coaction is then

$$\begin{array}{ccc} \mathcal{O}(\mathbf{M}^c) & \xrightarrow{\Delta} & \mathcal{O}(\mathbf{L}^c) \otimes \mathcal{O}(\mathbf{M}^c) \\ \Psi & \longrightarrow & \Delta\Psi\,. \end{array} \tag{4.81}$$

Notice that $\mathcal{O}(\mathbf{L}^c) \otimes \mathcal{O}(\mathbf{M}^c) \subset \mathcal{O}(\mathbf{L}^c \times \mathbf{M}^c)$. As an element of $\mathcal{O}(\mathbf{L}^c \times \mathbf{M}^c)$, the coaction takes this simple form

$$\Delta\Psi(x, \theta, \bar{\theta}) = \Psi(\Lambda^{-1}x, L\theta, R^{-1}\bar{\theta})\,. \tag{4.82}$$

The notation is reminiscent of the interpretation in the purely even case: the action of a group on a manifold induces a representation on the space of functions of such manifold with the formula

$$(g, m) \to gm, \qquad f \to R_g(f) \mid R_g(f)(x) = f(g^{-1}x)\,.$$

One can now attach to the indices in $\psi_{\alpha\dot{\beta}\dots}$ a true meaning and define the transformations

$$\begin{aligned} &\psi_0'(x) = \psi_0(\Lambda^{-1}x), \\ &\psi_\alpha'(x) = {R^{-1t}}_\alpha{}^\beta \psi_\beta(\Lambda^{-1}x), && {R^{-1t}}_\alpha{}^\beta := {R^{-1}}^\beta{}_\alpha \\ &\psi_{\alpha,\dot{\alpha}}'(x) = L_{\dot{\alpha}}{}^{\dot{\beta}} {R^{-1t}}_\alpha{}^\beta \cdots \psi_{\beta\dot{\beta}}(\Lambda^{-1}x), && L_{\dot{\alpha}}{}^{\dot{\beta}} := {L^t}^{\dot{\beta}}{}_{\dot{\alpha}}, \\ &\vdots \end{aligned} \tag{4.83}$$

The component fields are then organized as sections of a (trivial) bundle over $\mathbf{M}^c$ whose structure group is the (complex) Lorentz group and the fiber carries the corresponding representation of it, that is, antisymmetrized tensors products of copies of the representations $S^\pm$. The component fields are the ordinary *fields* of (Classical and) Quantum Field Theory, that is, the objects appearing in the wave equations (see the end of Section 3.2) that are associated to unitary representations of the Poincaré group.

Let us now consider the bundle corresponding to a component field, say $\psi_{\alpha\dot{\alpha}\dots}$. We can think of it as the direct product of the basis $\mathbf{M}^c$ times the representation space of the Lorentz group, $S^+ \wedge S^- \wedge \cdots$. Let S_g be such representation for an element of the group $g = (R, L)$. Then we have, generically

$$(S_g\psi)_{\alpha\dot{\alpha}\dots}(x) = L_{\dot{\alpha}}{}^{\dot{\beta}} {R^{-1t}}_\alpha{}^\beta \cdots \psi_{\beta\dot{\beta}\dots}(x)\,,$$

and

$$S_g\Psi(x) = \sum_I (S_g\psi(x))_I \xi^I\,.$$

S_g is just the transformation on the fiber. Combining (4.82) with (4.83) we can define

$$\Psi'(x,\theta,\bar{\theta}) = S_g \Psi(\Lambda^{-1}x, L\theta, R^{-1}\bar{\theta}) .$$

With this definition, it is clear that

$$\Psi'(x,\theta,\bar{\theta}) = \Psi(x,\theta,\bar{\theta}) .$$

Physicists say that Ψ is a *scalar superfield.*

Superfields other than the scalar ones are also used in physics. For example, one can think on a spinor superfield, say Ψ_α, and this index is added to all the component fields.

4.15 Real superfields

In Section 4.13 we discussed the real form of the Minkowski space in terms of the functor of points. We want to discuss it here with some detail in terms of the sheaf (Definitions 1.10.5 and 1.10.6), which will produce a reality condition on $\mathcal{O}(\mathbf{M}^c)$. We proceed as in Example 1.10.3. Here the real form $\mathbf{M}$ has to be chosen in such way that the action of F^c on $\mathbf{M}^c$ reduces to the action of F on $\mathbf{M}$, where F is the real form of F^c. As before, we will only be concerned with the Lorentz group action. The reader can complete the details for the full supergroup F.

Let us introduce, in the first place, the complex conjugate manifold of $\mathbf{M}^c$, $\mathbb{M}^c$ (1.11). The topological space is the same, $|\mathbf{M}^c| = |\mathbb{M}^c| \cong \mathbb{C}^4$, but the sheaf is

$$\mathcal{O}(\mathbb{M}^c) = \bar{\mathcal{H}}(\mathbb{C}^4)[\eta,\bar{\eta}] ,$$

where $\bar{\mathcal{H}}$ stands for antiholomorphic functions and $\eta, \bar{\eta}$ are new variables, not related, a priori, with θ and $\bar{\theta}$. We have to give also an *antilinear* morphism of sheaves

$$\begin{array}{ccc} \mathcal{O}(\mathbf{M}^c) & \xrightarrow{\sigma} & \mathcal{O}(\mathbb{M}^c) \\ \psi(x),\theta,\bar{\theta} & \longrightarrow & \overline{\psi(x)},\bar{\eta},\eta\,, \end{array}$$

where $\overline{\psi(x)}$ means complex conjugation in $\mathbb{C}$. We need now a *linear* isomorphism of ringed spaces $\hat{\tau} = (|\tau|, \tau^*)$

$$\begin{array}{ccc} \mathbb{C}^4 & \xrightarrow{|\tau|} & \mathbb{C}^4 \\ x & \longrightarrow & \bar{x}, \end{array} \qquad \begin{array}{ccc} \mathcal{O}(\mathbf{M}^c) & \xrightarrow{\tau^*} & \mathcal{O}(\mathbb{M}^c) \\ \psi(x),\theta,\bar{\theta} & \longrightarrow & \psi(\bar{x}),\eta,\bar{\eta}\,. \end{array}$$

Finally, we get the *antilinear* morphism of ringed spaces $\rho = (|\rho|, \rho^*)$ where $|\rho| = |\tau|$ and $\rho^* = \sigma^{-1} \circ \tau^*$

$$\begin{array}{ccc} \mathcal{O}(\mathbf{M}^c) & \xrightarrow{\rho^*} & \mathcal{O}(\mathbf{M}^c) \\ \psi(x), \theta, \bar{\theta} & \longrightarrow & \overline{\psi(\bar{x})}, \bar{\theta}, \theta \,. \end{array}$$

The morphism ρ is involutive and antilinear and it is going to define the real form. The topological space is

$$|\mathbf{M}| = \{x \in |\mathbf{M}^c| \mid x = \bar{x}\}$$

and the sheaf is

$$\mathcal{O}_{\mathbf{M}} = \mathcal{O}_{\mathbf{M}^c}\big|_{\mathbf{M}}, \quad \text{where} \quad \mathcal{O}_{\mathbf{M}^c_\rho}(U) = \{\Psi \in \mathcal{O}_{\mathbf{M}^c}(U) \mid \rho^*(\Psi) = \Psi\} \tag{4.84}$$

for U open in $|\mathbf{M}|$. The variables $\eta, \bar{\eta}$ introduced as the complex conjugates disappear in the final formulation and we can indeed interpret $\bar{\theta}$ as the complex conjugate of θ.

At the level of field components, (4.84) reads

$$\begin{aligned} &\bar{\psi}_0 = \psi_0, \qquad \bar{\psi}_\alpha = \psi_{\dot{\alpha}}, \qquad \bar{\psi}_{\alpha\beta} = \psi_{\dot{\alpha}\dot{\beta}} \qquad \bar{\psi}_{\alpha\dot{\beta}} = \psi_{\dot{\alpha}\beta} = -\psi_{\beta\dot{\alpha}}, \\ &\bar{\psi}_{\alpha\beta\dot{\gamma}} = \psi_{\gamma\dot{\alpha}\dot{\beta}}, \qquad \bar{\psi}_{\alpha\beta\dot{\gamma}\dot{\delta}} = \psi_{\gamma\delta\dot{\alpha}\dot{\beta}} \,. \end{aligned} \tag{4.85}$$

(We recall again the convention (1.12): the complex conjugation of a product of odd variables does not affect the order.) A superfield satisfying these conditions is called a *real superfield.*

It is now straightforward to check that the coaction of the real Lorentz group is well defined on $\mathcal{O}_M$. The real Lorentz group is obtained from (4.67) by imposing

$$L^{-1\dagger} = R, \quad \Leftrightarrow \quad L^{-1t} = \bar{R}\,,$$

where $\bar{R}$ is the complex conjugate of the matrix R. By simple inspection of (4.85), one can see that the real form $\mathrm{SL}_2(\mathbb{C})_\mathbb{R}$ acts on the sheaf of real superfields.

Let us analyze the component fields of the real, scalar superfield. In the first place we just have to find what is the representation of the Lorentz group that corresponds to the spinorial representations carried by the component fields. These, we can identify for their indices, remembering that they are antisymmetrized and that we have the reality conditions (4.85). The Lorentz representation of the field is associated to a wave equation which, in turn, corresponds to a unitary representation of the Poincaré group. These are labeled by two numbers, the mass and the spin (or helicity, for 0 mass). (See Section 3.2.)

By inspection, there are two real scalar fields ψ_0 and $\psi_{\alpha\beta\dot\gamma\dot\delta}$, one complex scalar field $\psi_{\alpha\beta}$, two spinor fields $\psi_\alpha(x)$ and $\psi_{\alpha\dot\beta\dot\gamma}$ and one vector field $\psi_{\alpha\dot\beta}$ (in the fundamental representation of SO(1, 3)). The rest of the field components are not independent since they are the complex conjugates of the ones that we have just listed.

The vector field is used, for example, in Maxwell theory, being identified with the electromagnetic potential. The Maxwell equations correspond to a massless particle of helicity 1 (the photon). There is also an equivalent wave equation for the massive spin 1 particle. The spinor fields, together with the Dirac equation, correspond to spin 1/2 particles and the scalar fields, complex or real, when satisfying the Klein-Gordon equation, are spin 0 particles. The mass, we cannot learn from here, and one should go to the specific theory to compute the wave equations and see if they have a mass term. We can only say that all the particles appearing in the same superfield in a supersymmetric field theory have the same mass[5].

The real, scalar superfield can be used in a supersymmetric generalization of the Maxwell theory (for zero mass). Then, half of the scalar and spinor fields become spurious variables that can be eliminated by symmetries, and we are left with one helicity 1, one helicity 1/2 and two helicity 0 particles which satisfy the adequate wave equations. An irreducible unitary representation of the Poincaré supergroup arises in this way, as direct sum of representations of different helicity (or spin) and the same mass. The representation that we have described very briefly here is called in physics the *massless vector multiplet* and it is denoted by its helicity content as $[2(0), 1(1/2), 1(1)]$.

This is the way in which particles with different spins appear as components of the same object in supersymmetric theories.

4.16 Chiral superfields and covariant derivatives

The complex superfield is sometimes too large (has too many field components) and one has to impose some constraints on it. The constraints have to be compatible with the Poincaré supersymmetry, though. We saw one possibility, that is, imposing the reality of the superfield, but there are

[5]We have only spoken about free wave equations, but the concept of particle can still be maintained when the interactions can be treated perturbatively.

other possibilities that also arise in physical theories. From this point of view, we are going to define *chiral superfields* and, of course, we will find out how this definition is related to the chiral Minkowski superspace (see Section 4.10).

We recall here the concept of derivation of an associative superalgebra $\mathcal{A}$ (Definition 1.5.1). A linear map $D^{L,R} : \mathcal{A} \to \mathcal{A}$ satisfying either

- $D^L(a \cdot b) = D^L(a) \cdot b + (-1)^{m|a|} a \cdot D^L(b)$ or
- $D^R(a \cdot b) = (-1)^{m|b|} D^R(a) \cdot b + a \cdot D^R(b) \qquad \forall\, a, b \in \mathcal{A},\ m = 0, 1\,,$

is said to be a left or right derivation, respectively, m denoting the parity of the derivation $D^{L,R}$. The concepts of left and right derivations are not independent, since for any left derivation one can define a right derivation as

$$D^R \Psi = (-1)^{m(p_\Psi + 1)} D^L \Psi\,.$$

They can then be used alternatively and we choose to use left derivations. We will suppress the index L in what follows.

The basic derivations of the complex superalgebra $\mathcal{O}(\mathbf{M}^c)$ are

$$\begin{array}{llllll} \partial_\mu & \text{such that} & \partial_\mu \theta^\alpha = 0, \,, & \partial_\mu \bar{\theta}^{\dot\alpha} = 0 & \partial_\mu \psi_{\beta\dot\beta\ldots} = \partial \psi_{\beta\dot\beta\ldots}/\partial x^\mu, \\ \nabla_\alpha & \text{such that} & \nabla_\alpha \theta^\beta = \delta^\beta_\alpha, & \nabla_\alpha \theta^{\dot\beta} = 0, & \nabla_\alpha \psi_{\beta\dot\beta\ldots} = 0, \\ \bar\nabla_{\dot\alpha} & \text{such that} & \bar\nabla_{\dot\alpha} \theta^\beta = 0, & \bar\nabla_{\dot\alpha} \theta^{\dot\beta} = \delta^{\dot\beta}_{\dot\alpha}, & \bar\nabla_{\dot\alpha} \psi_{\beta\dot\beta\ldots} = 0\,. \end{array} \tag{4.86}$$

Then we extend them to an arbitrary element Ψ using the property of being a left derivation. ∂_μ is an even derivation ($m = 0$) and the others are odd derivations, ($m = 1$). In physics it is usual to denote ∇_α and $\bar\nabla_{\dot\alpha}$ as ∂_α and $\bar\partial_{\dot\alpha}$, but we prefer to avoid ambiguities when $\mu = 1, 2$ and $\alpha = 1, 2$.

Analogously to the ordinary case, a super vector field on a supermanifold is a derivation of the algebra of (local) sections (see Definition 1.5.3). For example, the derivations defined in (4.86) are super vector fields on $\mathbf{M}^c$.

A tangent vector at a topological point m is a map from the stalk at m to the complex field satisfying also the derivation property (see Definition 1.5.2).

We want to compute the super vector fields associated to the *infinitesimal action* of the Poincaré supergroup on the Minkowski superspace, analogously to the way we computed the vector fields of the action of the conformal group in Chapter 2, Section 2.3. Let us first define properly these super vector fields in a general context (see Section 8.2 in Ref. [25]).

Let G be a Lie supergroup with Lie superalgebra $\mathfrak{g}$, that we will identify here with the tangent space at the identity of G, T_eG and let $\mathcal{M}$ be a supermanifold. Let

$$G \times \mathcal{M} \xrightarrow{\ a\ } \mathcal{M}$$

be an action of G on $\mathcal{M}$ (see Definition 1.9.1). Let $v \in T_eG$ and let U_e be a neighborhood of e. Let U be a suitable open set $U \subset \mathcal{M}$. Then the composition $\rho_a(v) = (v \otimes \mathrm{id}) \circ a^*$,

$$\mathcal{O}_{\mathcal{M}}(U) \xrightarrow{\ a^*\ } \mathcal{O}_{G\times\mathcal{M}}(U_e \times U) \xrightarrow{\ v\otimes \mathrm{id}\ } \mathcal{O}_{\mathcal{M}}(U) \tag{4.87}$$

is a derivation of $\mathcal{O}_{\mathcal{M}}$. This establishes a morphism

$$\begin{aligned} \mathfrak{g} &\xrightarrow{\ \rho_a\ } \mathrm{Vec}_{\mathcal{M}} \\ v &\longrightarrow \rho_a(v)\,. \end{aligned} \tag{4.88}$$

The super vector fields $\rho_a(v)$ give the infinitesimal action of $\mathfrak{g}$ on $\mathcal{M}$.

Notice that when G is the Poincaré supergroup and $\mathcal{M}$ the Minkowski superspace, the map a^* acting on global sections is just the coaction (4.81).

We are going to compute the vector fields associated to the translation superalgebra,

$$\begin{pmatrix} \mathrm{id} & 0 & 0 \\ T + \frac{1}{2}\varphi\bar{\varphi}^t & \mathrm{id} & \varphi \\ \bar{\varphi}^t & 0 & 1 \end{pmatrix},$$

which acts on $\mathbf{M}^c$ as

$$\begin{aligned} C &\longrightarrow C + \frac{1}{2}\varphi\bar{\theta}^t - \frac{1}{2}\theta\bar{\varphi}^t + T, \\ \theta &\longrightarrow \theta + \varphi, \\ \bar{\theta} &\longrightarrow \bar{\theta} + \bar{\varphi}\,. \end{aligned} \tag{4.89}$$

The identity e in the translation supergroup has coordinates

$$\varphi = 0, \quad \bar{\varphi} = 0, \qquad T = t^{\mu}\sigma_{\mu} = 0\,.$$

The tangent space at the identity is spanned by the derivations

$$\left.\frac{\partial}{\partial\varphi^{\alpha}}\right|_e, \quad \left.\frac{\partial}{\partial\bar{\varphi}^{\dot{\alpha}}}\right|_e, \quad \left.\frac{\partial}{\partial t^{\mu}}\right|_e.$$

The odd derivations are defined analogously to the derivations ∇_{α}, $\bar{\nabla}_{\dot{\alpha}}$.

We can now use the same format that we used to write the Lorentz coaction, and it will be enough to assume now that t^μ, φ and $\bar{\varphi}$ are infinitesimal parameters (no quadratic terms can contribute to ρ_a in (4.88)). This does not change (4.89), since it is linear. We then have

$$\Delta\Psi(x,\theta,\bar{\theta}) = \Delta x^\mu \partial_\mu \Psi(x,\theta,\bar{\theta}) + \Delta\theta^\alpha \nabla_\alpha \Psi(x,\theta,\bar{\theta}) + \Delta\bar{\theta}^{\dot{\alpha}} \bar{\nabla}_{\dot{\alpha}} \Psi(x,\theta,\bar{\theta}) . \tag{4.90}$$

The reader can check that the order of the factors in each term in the above equation is the right one for $\nabla, \bar{\nabla}$ being left derivations. If they were right derivations we would have to write

$$\nabla_\alpha \Psi(x,\theta,\bar{\theta})\Delta\theta^\alpha + \nabla_{\dot{\alpha}} \Psi(x,\theta,\bar{\theta})\Delta\bar{\theta}^{\dot{\alpha}} .$$

This describes the map a^* in (4.87) on global sections. We can now write it as

$$\Delta\Psi(x,\theta,\bar{\theta}) = \big(\varphi^\alpha Q_\alpha + \bar{\varphi}^{\dot{\alpha}} \bar{Q}_{\dot{\alpha}} + t^\mu P_\mu\big)\Psi(x,\theta,\bar{\theta}) .$$

The coefficients can be then identified as

$$\rho_a\left(\frac{\partial}{\partial\varphi^\alpha}\bigg|_e\right) = Q_\alpha, \qquad \rho_a\left(\frac{\partial}{\partial\varphi^{\dot{\alpha}}}\bigg|_e\right) = \bar{Q}_{\dot{\alpha}}, \qquad \rho_a\left(\frac{\partial}{\partial t^\mu}\bigg|_e\right) = P_\mu .$$

After some calculations we get

$$P_\mu = \partial_\mu, \qquad Q_\alpha = \nabla_\alpha + \frac{1}{4}\bar{\theta}^{\dot{\alpha}} \bar{\sigma}^\mu_{\dot{\alpha}\alpha}\partial_\mu, \qquad \bar{Q}_{\dot{\alpha}} = \bar{\nabla}_{\dot{\alpha}} + \frac{1}{4}\bar{\sigma}^\mu_{\dot{\alpha}\alpha}\theta^\alpha\partial_\mu . \tag{4.91}$$

Here we have used the definition

$$\bar{\sigma}^\mu_{\dot{\alpha}\alpha} := \bar{\sigma}^{\alpha\dot{\alpha}}_\mu . \tag{4.92}$$

We can check that this definition is consistent. Let us denote with A the transformation

$$L\bar{\sigma}^\mu R^{-1} = A^\mu{}_\nu \bar{\sigma}^\nu .$$

One has to prove that $A^\mu{}_\nu = \Lambda^\mu{}_\nu$. Using (4.79) and (4.92) it is not difficult to actually prove the statement from direct calculation.

Notice that with this definition, equation (4.76) can be expressed more aesthetically:

$$x^\mu = \frac{1}{2}\mathrm{tr}(C\bar{\sigma}^\mu) = \frac{1}{2}\, C^{\alpha\dot{\alpha}}\, \bar{\sigma}^\mu_{\dot{\alpha}\alpha} .$$

The three vector fields (4.91) satisfy the commutation rules of the translation superalgebra

$$[Q_\alpha, \bar{Q}_{\dot{\alpha}}]_+ = \frac{1}{2}\bar{\sigma}^\mu_{\dot{\alpha}\alpha} P_\mu, \qquad \text{and the rest zero} .$$

We have specified here that Q_α and $\bar{Q}_{\dot\alpha}$ satisfy anticommutation relations with the subindex '$+$', as it is common in physics. In comparing with the physics literature, one has to take into account that we can reabsorb numerical factors by redefining

$$\frac{1}{2}(\theta, \bar\theta) \longrightarrow (\theta, \bar\theta)\,,$$

and that we do not expect the usual 'i' factors that appear as a result of the non natural convention chosen in physics for which

$$\overline{\xi\xi'} = \bar\xi'\bar\xi\,.$$

The derivations Q_α and $\bar{Q}_{\dot\alpha}$ are called the *supersymmetry charges*: they are the generators of the odd translations on the Minkowski superspace. The generator of the ordinary translations is the *momentum operator* P_μ.

The Minkowski superspace can be identified with the translation supergroup by means of the map (4.68). We already saw this fact at the infinitesimal level in (4.38). This means that, other than the left action that we have studied up to now, we also have the right action of the supergroup on itself, or, using the identification, on the Minkowski superspace. In the functor of points notation, the action is

$$\begin{pmatrix} \mathrm{id} & 0 & 0 \\ C + \frac{1}{2}\theta\bar\theta^t & \mathrm{id} & \theta \\ \bar\theta^t & 0 & 1 \end{pmatrix} \longrightarrow \begin{pmatrix} \mathrm{id} & 0 & 0 \\ C + \frac{1}{2}\theta\bar\theta^t & \mathrm{id} & \theta \\ \bar\theta^t & 0 & 1 \end{pmatrix} \begin{pmatrix} L & 0 & 0 \\ TL + \frac{1}{2}R\varphi\bar\varphi^t & R & R\varphi \\ d\bar\varphi^t & 0 & d \end{pmatrix},$$

which in coordinates becomes

$$\begin{aligned} C &\longrightarrow C - \frac{1}{2}\varphi\bar\theta^t + \frac{1}{2}\theta\bar\varphi^t + T, \\ \theta &\longrightarrow \theta + \varphi, \\ \bar\theta &\longrightarrow \bar\theta + \bar\varphi\,. \end{aligned}$$

The infinitesimal generators of this action are the (left) derivations

$$P_\mu = \partial_\mu, \qquad D_\alpha = \nabla_\alpha - \frac{1}{4}\bar\theta^{\dot\alpha}\bar\sigma^\mu_{\dot\alpha\alpha}\partial_\mu, \qquad \bar{D}_{\dot\alpha} = \bar\nabla_{\dot\alpha} - \frac{1}{4}\bar\sigma^\mu_{\dot\alpha\alpha}\theta^\alpha\partial_\mu\,, \quad (4.93)$$

whose commutation relations are

$$[D_\alpha, \bar{D}_{\dot\alpha}]_+ = -\frac{1}{2}\bar\sigma^\mu_{\dot\alpha\alpha}P_\mu, \qquad \text{and the rest zero}\,.$$

This algebra is isomorphic to the previous one (and then to the translation superalgebra), through the map $P_\mu \to -P_\mu$. It is easy to see that the supercharges anticommute with the derivations D and $\bar{D}$ (as it should be, since left and right translations in the supergroup commute)

$$[D_\alpha, Q_\beta]_+ = [D_\alpha, \bar{Q}_{\dot{\beta}}]_+ = [\bar{D}_{\dot{\alpha}}, Q_\beta]_+ = [\bar{D}_{\dot{\alpha}}, \bar{Q}_{\dot{\beta}}]_+ = 0\,.$$

It is for this reason that physicists call D and $\bar{D}$ *supersymmetric covariant derivatives* or simply *covariant derivatives*, although they are not related to any connection form.

We go now to an important definition.

Definition 4.16.1. A *chiral superfield* is a superfield Φ such that

$$\bar{D}_{\dot{\alpha}}\Phi = 0\,. \tag{4.94}$$

■

Because of the anticommuting properties of the D's and Q's, we have that

$$\bar{D}_{\dot{\alpha}}\Phi = 0 \quad \Rightarrow \quad \bar{D}_{\dot{\alpha}}(Q_\beta\Phi) = 0, \quad \bar{D}_{\dot{\alpha}}(\bar{Q}_{\dot{\beta}}\Phi) = 0.$$

This means that the supertranslation algebra acts on the space of chiral superfields (chiral subspace). In fact, it is not difficult to see that the translation supergroup itself acts on the chiral subspace.

Moreover, the Lorentz group also acts on the chiral subspace (in physics one says that the constraint is Lorentz covariant), and in fact, the whole Poincaré supergroup does act on it.

On the other hand, due to the derivation property,

$$\bar{D}_{\dot{\alpha}}(\Phi\Psi) = \bar{D}_{\dot{\alpha}}(\Phi)\Psi + (-1)^{p(\Phi)}\Phi\bar{D}_{\dot{\alpha}}(\Psi),$$

we have that the product of two chiral superfields is again a chiral superfield. Chiral superfields form a superalgebra.

The condition $\bar{D}_{\dot{\alpha}}\Phi = 0$ defining the chiral superfield can be reinterpreted by performing a change of coordinates. We first notice that the quantities

$$y^\mu = x^\mu + \frac{1}{4}\theta^\alpha\bar{\sigma}^\mu_{\alpha\dot{\alpha}}\theta^{\dot{\alpha}} \qquad \text{and} \qquad \theta^\alpha \tag{4.95}$$

satisfy

$$\bar{D}_{\dot{\alpha}}y^\mu = 0, \qquad \bar{D}_{\dot{\alpha}}\theta^\alpha = 0.$$

Using the derivation property, any superfield that is (formally) of the form

$$\Phi(y^\mu, \theta) \qquad \text{satisfies} \quad \bar{D}^L_{\dot{\alpha}}\Phi = 0\,,$$

and so it is a chiral superfield. This provides in fact the general solution of the constraint $\bar{D}^L_{\dot\alpha}\Phi = 0$.

One can now define the change of coordinates

$$x^\mu,\ \theta^\alpha,\ \bar\theta^{\dot\alpha}\ \longrightarrow\ y^\mu = x^\mu + \frac{1}{4}\theta^\alpha\bar\sigma^\mu_{\alpha\dot\alpha}\bar\theta^{\dot\alpha},\ \theta^\alpha,\ \bar\theta^{\dot\alpha}.$$

A superfield may be expressed in both coordinate systems

$$\Phi(x,\theta,\bar\theta) = \Phi'(y,\theta,\bar\theta)\,,$$

and so can the covariant derivatives and supersymmetry charges, which take the form

$$D_\alpha\Phi' = \nabla_\alpha - \frac{1}{2}\bar\theta^{\dot\alpha}\bar\sigma^\mu_{\dot\alpha\alpha}\partial_\mu, \qquad \bar{D}_{\dot\alpha}\Phi' = \bar\nabla_{\dot\alpha},$$

$$\bar{Q}_{\dot\alpha}\Phi' = \bar\nabla_{\dot\alpha} + \frac{1}{2}\bar\sigma^\mu_{\dot\alpha\alpha}\theta^\alpha\partial_\mu, \qquad Q_\alpha\Phi' = \nabla_\alpha\,.$$

In the new coordinate system the chirality condition is simply

$$\bar\nabla_{\dot\alpha}\Phi' = 0\,.$$

This shows that chiral scalar superfields are global sections of the supermanifold $\mathbb{C}^{4|2}$, that is, elements of

$$\mathcal{O}(\mathbb{C}^{4|2}) = \mathcal{H}(\mathbb{C}^4)\otimes\wedge[\theta^1,\theta^2]\,.$$

By reversing the role of θ and $\bar\theta$ one could define also the *antichiral superfields*, that is, superfields satisfying

$$D_\alpha\Phi = 0\,.$$

There is then a different change of coordinates

$$x^\mu,\ \theta^\alpha,\ \bar\theta^{\dot\alpha}\ \longrightarrow\ z^\mu = x^\mu - \frac{1}{4}\bar\theta^{\dot\alpha}\bar\sigma^\mu_{\dot\alpha\alpha}\theta^\alpha,\ \theta^\alpha,\ \bar\theta^{\dot\alpha}\,.$$

Obviously, antichiral superfields enjoy the same properties as chiral superfields.

We are ready now to interpret the chiral and antichiral fields in geometric terms. A straightforward calculation shows that the coordinates y^μ and z^μ correspond, respectively, to

$$y^\mu = \frac{1}{2}\mathrm{tr}\big(C - \frac{1}{2}\theta\bar\theta^t\big) = \frac{1}{2}\mathrm{tr}\,B,$$

$$z^\mu = \frac{1}{2}\mathrm{tr}\big(C + \frac{1}{2}\theta\bar\theta^t\big) = \frac{1}{2}\mathrm{tr}\,A\,,$$

for which we refer to (4.71).

The big cell of the superflag can be seen as the subset of the product

$$\mathcal{U}_{12} \subset \mathcal{V}_{12} \times \mathcal{V}'_{12}, \qquad \mathcal{V}_{12} \approx \begin{pmatrix} \mathrm{id} \\ A \\ \bar{\theta} \end{pmatrix}, \quad \mathcal{V}'_{12} \approx \begin{pmatrix} \mathrm{id} & 0 \\ B & \theta \\ 0 & 1 \end{pmatrix}$$

satisfying the twistor relations $B = A - \theta\bar{\theta}^t$. Also, $\mathcal{V}_{12}$ and $\mathcal{V}'_{12}$ can be interpreted as big cells of the super Grassmannians

$$\mathcal{V}_{12} \subset G(2|0,4|1), \qquad \mathcal{V}'_{12} \subset G(2|1,4|1)\,.$$

These two Grassmannians are equivalent, and we have described them in detail Sections 4.7, 4.8, 4.9 and 4.10. In fact, we have called $\mathrm{Gr}^{\mathrm{ch}} = G(2|0,4|1)$ the *chiral conformal superspace* (strictly, it should be called the *antichiral conformal superspace*). These Grassmannians are related to the parabolic subalgebras $\mathfrak{f}_2$, for $G(2|0,4|1)$, and $\mathfrak{f}_4$, for $G(2|1,4|1)$, in (4.32). Their intersection

$$\mathfrak{f}_2 \cap \mathfrak{f}_4 = \mathfrak{f}_3 = \mathfrak{f}^c$$

is the complex Poincaré superalgebra plus dilations. They both have an action of the complex conformal supergroup $\mathrm{SL}(4|1)$.

We have then completed the interpretation of the Grassmannians $G(2|1,4|1)$ and $G(2|0,4|1)$ as the chiral and antichiral conformal superspaces.

We want to end by giving the field content of a chiral superfield. Notice that the expansion (4.80) now reduces to

$$\Phi(z,\theta) = \sum_I \phi_I \xi^I = \phi_0(z) + \phi_\alpha(z)\theta^\alpha + \phi_{\alpha\beta}(z)\theta^\alpha\theta^\beta\,.$$

The component fields ϕ_0 and $\phi_{\alpha\beta}$ are spin 0 fields, while ϕ_α is a spin 1/2 field. The complex conjugate of this field, $\bar{\Phi}(y,\bar{\theta})$ is an antichiral superfield, and the field content is the same. This irreducible representation of the Poincaré supergroup is called the *chiral multiplet*, it can be massive or massles and it is denoted by its spin or helicity content as $[2(0), 1(1/2)]$.

4.17 References

The spin-statistics theorem was first formulated in 1939 by M. Fierz, and later reworked out by W. Pauli. Later on, both J. Schwinger and R. Feyman have given their contribution and we refer the reader to [51] for a complete treatment. For the spinors and spin representations our main reference is the paper by D'Auria et al. [36] and the book by V. S. Varadarajan [147] where the material is treated more extensively. The infinitesimal treatment of Minkowski superspace appeared in [108] by Manin and also in [147]. The Minkowski and conformal superspaces as homogeneous spaces and their real forms appeared in [66], [26].

Chapter 5

Deformations of the Minkowski and Conformal Superspaces

In this chapter we construct a quantum deformation of the Minkowski and conformal spaces, together with the natural actions of their groups of symmetries, the Poincaré and conformal groups. The deformation is also constructed for the chiral Minkowski and conformal superspaces, according to the philosophy stated in the Introduction.

As we have explained in the introduction to Chapter 3, the current theory of gravity, General Relativity, is a theory that is not congruent with quantum physics. One can formulate it as a classical field theory, but then its quantization, using the ordinary methods of QFT, fails clamorously. One could not expect perhaps such 'boring' behaviour from a theory whose dynamical variable is the metric of spacetime itself and that is highly non linear. Gravity is extremely weak and it has not been tested at very short distances, the realm of the quantum world. The regime where gravitational interaction and quantum effects become simultaneously relevant is not accessible to experiments in our labs, so the mathematical consistency is by now the best guide that we have in the search for such theory.

There is, though, a quantum gravity theoretical 'lab' where different ideas on the short distance structure of spacetime can be tested. These are *black holes*. The history starts with the spherically symmetric solution of Einstein's equations of GR, the Schwarzschild metric. It describes the metric of spacetime created by a spherically symmetric matter distribution in the vacuum outside it. The solution presents a singularity at the origin, but it also presents a spherical surface surrounding the singularity that has bizarre properties: it is called the *event horizon*, and depends only on the total mass of the matter distribution. When all the matter is contained inside the event horizon, one says that the matter has undergone gravitational collapse, and the event horizon becomes a surface that can be reached (in principle) by an observer. One can, for example, send a probe to test its behaviour. The causal structure of spacetime in the event horizon is such that, once the probe reaches it, it cannot return anymore. Objects are virtually 'swallowed' inside the horizon, and even light cannot escape.

This is why matter that has collapsed inside its event horizon is said to form a black hole. In fact, it is possible to show that the matter will keep collapsing into the singularity. When Schwarzschild computed the solution (1916), even Einstein thought that it was unphysical. Nowadays, recent observations show that, very likely, there is a supermassive black hole at the center of most galaxies.

Black holes, then, show all the characteristics of a quantum gravity lab: the gravitational field is very intense and matter is compressed at very short distances. Under such conditions we cannot presume GR to be valid, and in fact we can regard the singularity as a crash of GR. Quantum gravity is supposed to cure these problems in a still unknown way.

There has been a lot of research around black holes. One may think on having a quantum field near the horizon of a black hole, which opens a fascinating window to observe the behaviour of the quantum theory under extreme gravitational fields. This is not yet quantum gravity, but even at this level, black holes show very intriguing properties. They are thermal objects, emitting black body radiation (Hawking radiation [86])[1] and one can attach to them a quantity with the properties of the entropy, the *Bekenstein-Hawking entropy* [10, 13, 86], which depends on the area of the event horizon. Thermodynamics reflects the macroscopic, statistical behaviour of a system of many bodies obeying microscopic laws that govern the individual interactions. The fact that black holes present thermodynamic properties implies that there should be an underlying microscopic theory that cannot be anything else but quantum gravity. In fact, the interpretation of the Bekenstein-Hawking entropy in terms of string theory is an active area of research with many promising results.

Another place where physicists look for quantum gravity effects is in cosmological observations. The discovery by E. Hubble in 1929 that galaxies are receding from each other marked the beginning of a cosmological theory that states that the Universe started in a singularity and that it is constantly expanding. The initial singularity, known as the *Big Bang*, is in a way similar to a black hole singularity. Short time after the Big Bang, an

[1] So, after all they are not totally black!

hypothetical observer would be in an ideal scenario to measure quantum gravity effects. If observing the behaviour of matter and spacetime near a black hole looks like a difficult task, the idea of, somehow, reaching the first instants of the Universe seems even more implausible. Nevertheless, there is a faint echo of such violent expansion that can be observed nowadays. It is the *Cosmic Microwave Background (CMB)*, an electromagnetic radiation formed about 380.000 years after the Big Bang (the age of the Universe is estimated in 13.8 billion years) that impregnates the Universe. This radiation, filling the Universe, is the same in every direction (with only very tiny fluctuations) and it was discovered in 1964 by A. Penzias and R. Wilson (they were awarded the Nobel prize in 1978). Because of the Universe expansion, its wavelength increased with time as to reach the wavelength that is measured today, of the order of 2mm.

In the current cosmological model, the Universe went through a period of very fast expansion (known as *inflation*) very shortly after the Big Bang. This period of spectacular expansion lasted very little (much less than one second), but under those extreme conditions, gravitons, the quantum particles associated to gravity, were produced with a characteristic spectrum. The CMB was produced much later, but an imprint of the gravitons of the inflationary period should have been left on the electromagnetic radiation. It is this imprint what seems to have been measured very recently (announced in March 2014) with the microwave telescope BICEP2 of the Harvard-Smithsonian Center installed in the Antarctic base Amundsen-Scott. If confirmed, this could be the first experiment directly measuring gravitational particle production and quantum gravity effects.

Going back to theoretical arguments, there is an interesting estimation of the smallest measurable longitude. It is obtained by combining arguments from quantum physics and GR. If we want to resolve an object we need a wave (for example, a light wave) with wavelength of the same order of magnitude than the size of the object. If we intend to resolve a very small length, then the wavelength is also very small and the frequency of the wave has to be very large. Given the relation (3.3) between frequency and energy, this means that, at some point, we will have a big amount of energy focussed in a very small region of space. Such energy will then start a process of gravitational collapse and form a black hole. This poses a lower limit to length measurements which is estimated to be of the order of Planck's length, ℓ_{P}:

$$\ell_{\mathrm{P}} = \sqrt{\frac{\hbar G}{c^3}} = 1.616\,199(97) \times 10^{-35}\mathrm{m}, \qquad r_{\mathrm{p}} = 0.8775(51) \times 10^{-15}\mathrm{m}\,,$$

where c is the speed of light in a vacuum, G is the gravitational constant, and $\hbar$ is the reduced Planck constant. We also give the radius of a proton, r_p, which is 20 orders of magnitude bigger. So Planck's length is extraordinarily small.

Beyond that limit we are not allowed to extrapolate our concept of spacetime as a smooth manifold. The limit in length measurements reminds us of the uncertainty principle of quantum mechanics: if two observables have a non zero commutator, then there are no states that are eigenvectors of the two observables simultaneously. In the lab, this is manifested in the impossibility of measuring the two observables with arbitrary precision simultaneously. It may happen that the impossibility of measuring lengths with arbitrary precision is due to a non commutative product among spacetime coordinates. So it is conceivable that spacetime at such small distances shows the properties of a noncommutative space.

It is then interesting to study the deformations of the algebra of functions on spacetime to a noncommutative algebra, in the same way that the quantization of a physical system could be regarded as a deformation of the algebra of functions on phase space. The parameter of the deformation should then be related with Planck's length.

This is probably the leitmotif of an abundant research on deformations of Minkowski space. It would be impossible to give here an exhaustive account of references on the subject. Instead, we mention a few of the original references, Refs. [7,42,103,106,117], which can introduce the interested reader in the area.

We want to mention, though, that deforming Minkowski space is still very far from any quantum gravity considerations, since Minkowski space is a flat manifold, just the case when gravity is absent. Still, consistency of quantum Minkowski space seems to us a basic conceptual problem, as well as a more practical one. One may think, for example, on a regime where gravity could be added perturbatively to a background quantum Minkowski space.

There are plenty of deformations of Minkowski space when one sees it simply as $\mathbb{R}^4$. For example, for differential deformations, one could use Kontsevich's classification [95]. Nevertheless, the most interesting are the ones that somehow preserve the action of the Poincaré group or some quantum version of it.

The deformation that we propose here goes a bit further than previous ones because it also quantizes the conformal space. It was developed in Refs. [54–56]. We shall give such deformation using the concept of *quantum group* and *quantum homogeneous space*.

Let us recall briefly the philosophy exposed in the Introduction, where the commutative diagram (0.4)

$$\begin{array}{ccc} \mathcal{C} \times \mathbf{C} & \longrightarrow & \mathbf{C} \\ \uparrow & & \uparrow \\ \mathcal{P} \times \mathbf{M} & \longrightarrow & \mathbf{M} \end{array} \tag{5.1}$$

stated the relation between Minkowski and conformal spaces and Poincaré and conformal groups. This relation has been explained at length in Chapter 2. In Chapter 4 we showed how it could be carried over to the super setting, by enlarging the conformal group to a suitable supergroup. In this chapter, we will substitute the ordinary conformal group by a quantum group and then we will show that there are adequate deformations of the Poincaré group and the homogenenous spaces (Minkowski and conformal) that preserve the relation (5.1), suitably adapted to the quantum setting. The spaces and groups are substituted by the deformations of the algebras of their polynomial functions, and the arrows are reversed.

One can go still further and apply the same philosophy to the supergroups and homogeneous superspaces that we obtained in Chapter 4. We will be able to perform the quantization in the chiral case. There is no conceptual problem in defining a similar deformation for the Minkowski and conformal superspaces: the problem is a practical one. The commutation relations among the generators of the quantum superalgebra become too involved to be computed directly, so one would have to resort to a different approach.

Quantum groups, as introduced by Drinfeld and Jimbo [43, 89], are quantum deformations of the universal enveloping (Hopf) algebra of the Lie algebra of a complex or real Lie group. They worked out the case of semisimple Lie algebras, extending the construction to the infinite dimensional Kac-Moody algebras. Fadeev, Reshetikhin and Takhtajan [46] gave examples of quantum groups through the formalism of the *R-matrix*. Later on, Manin, in Ref. [109] developed an alternative and more suggestive method to construct the deformation of the Hopf function algebra of a classical group in the special case of $\mathrm{SL}_n(\mathbb{C})$: this is the approach we are going to follow here. We want to remark that in all the cases, as in the more general framework of non commutative geometry, the idea of point and space has to be abandoned. What we will deform indeed is the algebra of global sections or functions over the corresponding space. Then, when we speak about a non commutative space or a quantum space, it should be clear that we refer to such deformation. The same is true when talking about deformations of superspaces, though functions here have to be interpreted as the global sections of the structural sheaf, as usual in supergeometry.

We want now to explain briefly the philosophy of the approach. When we think about the functions on the ordinary space $\mathbb{C}^n$ (or $\mathbb{R}^n$), we assume to have n independent variables $x_1, \dots, x_n$ *commuting* with each other, in other words:

$$x_i x_j = x_j x_i, \qquad \forall i, j = 1, \dots, n$$

Manin's idea is to impose modified commutation relations among the x_i's, depending on a certain parameter q:

$$x_i x_j = q x_j x_i, \qquad i < j. \tag{5.2}$$

This is one of the simplest non commutative relations among the spatial coordinates one can think of. The (non commutative) ring of algebraic functions generated by the x_i's and subject to the relations (5.2) is called the *complex quantum space* of dimension n.

In Manin's approach, the quantum deformations of the group $\mathrm{SL}_n(\mathbb{C})$ are constructed directly from the very simple non commutative relations among the spatial coordinates described above, together with the requirement that the quantum $\mathrm{SL}_n(\mathbb{C})$ acts on the quantum space. The same construction is mimicked for the super counterpart $\mathrm{SL}(m|n)$.

In Section 5.1 we are going to describe very briefly what quantum groups are in general and then we will give a more detailed description of the Manin construction that we have just sketched here. The rest of the chapter is dedicated to describe the quantization of the Minkowsi and conformal (super) spaces in this framework.

5.1 Quantum groups

Quantum groups represent a generalization of the concept of Lie or algebraic group. Many authors define a quantum group simply as a Hopf algebra (see Section 1.11), though the most interesting examples of quantum groups are Hopf algebras which are either, a deformation of the universal enveloping algebra naturally associated to a Lie group, or a deformation of the coordinate ring of an algebraic group.

Let us recall few important facts from Section 1.11. If G is a complex or real algebraic group with Lie algebra $\mathfrak{g}$, both, its universal enveloping algebra $\mathcal{U}(\mathfrak{g})$ and its coordinate ring $k[G]$ (k the real or complex field) have a natural Hopf algebra structure and these structures are in duality with each other. Schematically we have:

- *Hopf structure of the universal enveloping algebra* $\mathcal{U}(\mathfrak{g})$.
  The coproduct Δ, counit ϵ and antipode S are

$$\Delta(X) = X \otimes 1 + 1 \otimes X \,,$$
$$\epsilon(X) = 0, \qquad S(X) = -X, \qquad X \in \mathfrak{g} \,.$$

- *Hopf structure of the coordinate algebra* $k[G]$.
  The coproduct is defined as

$$\Delta(f) = \sum f^{(1)} \otimes f^{(2)} \quad \text{with} \quad \sum f^{(1)}(x) f^{(2)}(y) = f(xy),$$
$$x, y \in G, \; f \in k[G] \,,$$

  while the counit and antipode are

$$\epsilon(f)(x) = f(1_G), \qquad S(f)(x) = f(x^{-1}) \,.$$

- *Duality between* $\mathcal{U}(\mathfrak{g})$ *and* $k[G]$.
  We have a pairing:

$$\begin{array}{ccc} \mathcal{U}(\mathfrak{g}) \times k[G] & \longrightarrow & k \\ (u, f) & \longrightarrow & \langle u, f \rangle := u(f) \,. \end{array}$$

Here $u \in \mathcal{U}(\mathfrak{g})$ is understood as a left invariant differential operator. This works in the following way. Whenever $u \in \mathfrak{g}$ we have that

$$u(f) = \frac{\mathrm{d}}{\mathrm{d}t} f(\mathrm{e}^{tu}) \Big|_{t=0} \,.$$

More in general, if $u = X_1 \cdots X_r \in \mathcal{U}(\mathfrak{g})$, $X_i \in \mathfrak{g}$, then

$$u(f) = \frac{\partial}{\partial t_1} \cdots \frac{\partial}{\partial t_r} f(\mathrm{e}^{t_1 X_1} \cdots \mathrm{e}^{t_r X_r}) \Big|_{t_1=0,\ldots,t_r=0} \,.$$

The duality between $\mathcal{U}(\mathfrak{g})$ and $k[G]$ is a non degenerate pairing of Hopf algebras. This means that it satisfies the properties

$$\begin{array}{ll} \langle uv, x \rangle = \langle u \otimes v, \Delta(x) \rangle, & \langle u, xy \rangle = \langle \Delta(u), x \otimes y \rangle, \\ \langle 1, x \rangle = \epsilon(x), & \langle u, 1 \rangle = \epsilon(u) \\ \langle S(u), x \rangle = \langle u, S(x) \rangle, & \text{for} \quad u, v \in \mathcal{U}(\mathfrak{g}), \quad x, y \in k[G] \,. \end{array}$$

As customary we use the same letters Δ, ϵ, S to denote the comultiplication, counit and antipode in both Hopf algebras.

Notice that the algebra structure in the universal enveloping algebra carries the information about the group multiplication, while in the coordinate algebra it is the coproduct which contains the same information.

Let us now consider two examples of particular interest to us: the general linear and the special linear groups.

Example 5.1.1. We start with the algebraic group $\mathrm{GL}_n(k)$. Its polynomial algebra is given by

$$k[\mathrm{GL}_n] = \mathbb{C}[g_{ij}, d]/(d \cdot \det(g_{ij}) - 1), \qquad i, j = 1, \ldots, n.$$

The polynomial $\det(g_{ij})$ is the determinant, which is obtained by taking the determinant of the matrix formed by the generators

$$\det(g_{ij}) = \det \begin{pmatrix} g_{11} & g_{12} & \cdots & g_{1n} \\ g_{21} & g_{22} & \cdots & g_{2n} \\ \vdots & \vdots & \cdots & \vdots \\ g_{n1} & g_{n2} & \cdots & g_{nn} \end{pmatrix} = \sum_{\sigma} (-1)^{-\ell(\sigma)} g_{\sigma(1)1} g_{\sigma(2)2} \cdots g_{\sigma(n)n} \,. \tag{5.3}$$

σ runs over all the permutations of the first n integers. The generator d satisfies the relation $d \cdot \det(g_{ij}) = 1$ and behaves, for all purposes, as the inverse of the determinant. d encodes the information that matrices in $\mathrm{GL}_n(k)$ have non zero determinant. It is customary to denote d also with $\det(g_{ij})^{-1}$.

Explicitly, the coproduct, the counit and the antipode on $k[\mathrm{GL}_n]$ are

$$\Delta(g_{ij}) = \sum_k g_{ik} \otimes g_{kj}, \quad \epsilon(g_{ij}) = \delta_{ij}, \quad S(g_{ij}) = (-1)^{j-i} \det(g_{ij})^{-1} A^c_{ji} \,, \tag{5.4}$$

where A^c_{ji} is the determinant of the matrix obtained by suppressing the row j and the column i in the matrix that appears in (5.3). We put an index c (as for 'classical') to distinguish it from the similar quantum determinant that we shall introduce later.

Formally, we can write the coproduct as a matrix multiplication

$$\Delta \begin{pmatrix} g_{11} & \cdots & g_{1n} \\ \vdots & \cdots & \vdots \\ g_{n1} & \cdots & g_{nn} \end{pmatrix} = \begin{pmatrix} g_{11} & \cdots & g_{1n} \\ \vdots & \cdots & \vdots \\ g_{n1} & \cdots & g_{nn} \end{pmatrix} \otimes \begin{pmatrix} g_{11} & \cdots & g_{1n} \\ \vdots & \cdots & \vdots \\ g_{n1} & \cdots & g_{nn} \end{pmatrix},$$

and the counit can be expressed as

$$\epsilon \begin{pmatrix} g_{11} & \cdots & g_{1n} \\ \vdots & \ldots & \vdots \\ g_{n1} & \cdots & g_{nn} \end{pmatrix} = \begin{pmatrix} 1 & \cdots & 0 \\ \vdots & \ldots & \vdots \\ 0 & \cdots & 1 \end{pmatrix}.$$

These formulas are nothing but the first two equalities in (5.4) and one can also express the antipode in the same way. These conventions are found often in the literature on quantum groups.

We can define along the same lines the coordinate algebra of the special linear group:

$$k[\mathrm{SL}_n] = k[g_{ij}]/(\det(g_{ij}) - 1), \qquad i, j = 1, \ldots, n.$$

Notice that in this case we do not need the extra generator d, since we impose the condition $\det(g_{ij}) = 1$ and hence the determinant is automatically an invertible element. The coproduct, counit and antipode are defined in the same way as for $k[\mathrm{GL}_n]$. ■

We now want to give the basic definitions of the theory of quantum groups. We will only consider Manin's approach, that is, we will consider deformations of the coordinate ring of an algebraic group. For a comprehensive treatment and the alternative point of view with enveloping algebras the reader can consult, for example, the works of Chari and Pressley [28] or Kassel [93] and the references in the bibliography therein.

As before, let k be our ground field ($k = \mathbb{R}$ or $\mathbb{C}$) and let $k_q := k[q, q^{-1}]$, where q is an indeterminate[2].

Definition 5.1.2. Let A_q be an algebra over the ring k_q. We say that A_q is a *quantum deformation* of a given algebra A over the field k, if

$$A_q/(q-1) \cong A$$

as algebras. ■

Some authors use a more restrictive definition of quantum deformation, but we prefer to keep it this general.

Definition 5.1.3. Let A be the coordinate algebra of an algebraic group, with the Hopf algebra structure defined above. A *quantum group* A_q is a Hopf algebra such that

$$A_q/(q-1) \cong A$$

as Hopf algebras. ■

[2]Some authors prefer q to take values in k; most of what we say holds also in that setting, see Ref. [28] for more details.

In particular, if we regard only the algebra structure, a quantum group is a quantum deformation of the coordinate algebra of an algebraic group.

Remark 5.1.4. We are going to take Manin's approach (see [109]) to quantum groups, namely we will examine quantum deformations of the coordinate rings of complex algebraic groups, restricting our attention to the general linear and special linear groups, which are the most relevant examples for our purpose. Because of the duality described at the beginning of this section, it is only a matter of choice whether we decide to use the enveloping algebra or the coordinate ring associated with an algebraic group. The approach through the universal enveloping algebra is more general, in the sense that in Manin's approach one does strictly need an algebraic group and the algebra of its polynomial functions. In most of the interesting examples, we have both the universal enveloping algebra and the polynomial algebra available to operate our deformation. ■

Definition 5.1.5. Let X be an affine variety and $k[X]$ its coordinate ring. We say that the algebra $k_q[X]$ is a *quantum deformation* of the affine variety X if it is a quantum deformation of $k[X]$. Sometimes we shall also denote $k_q[X]$ with X_q for brevity.

Let X be a projective variety, together with an embedding into some projective space, and let $k[X]$ be the ($\mathbb{Z}$-homogeneous) coordinate ring of X with respect to such embedding. We say that $k_q[X]$ is a *quantum deformation* of the variety X, if it is a quantum deformation respecting the $\mathbb{Z}$-grading of the $\mathbb{Z}$-homogeneous ring $k[X]$. In both cases we refer to $k_q[X]$ (or more simply to X_q) as *quantum variety.* ■

Remark 5.1.6. As one realizes, our definition of quantum deformation of an affine or a projective variety is very general. While in the case of quantum groups we are requiring that the deformed coordinate algebra (or the deformed universal enveloping algebra) still maintains a Hopf structure, in the case of varieties one has more freedom, since there is no Hopf structure. However, we have in mind varieties that support the action of a group, and then we will be only interested in quantum deformations admitting an action of the corresponding quantum group. In the language of Hopf algebras, this means that there is a *coaction* (see Definition 1.11.8, which can be easily translated to the non super setting) of the quantum group (i.e. the Hopf algebra) on the quantum variety. This introduces a severe constraint on the possibilities. In particular, given the deformation of the quantum space that we are going to present, a few natural requirements in terms of the group action guarantee the uniqueness of the quantum deformation ■

Given the importance of the coaction of a quantum group on a quantum variety in this framework, we want to state it here, rephrasing Definition 1.11.8 in this more restricted setting.

Definition 5.1.7. We say that a quantum group G_q *acts* on the left on the quantum variety X_q, if we have a coaction

$$\lambda : X_q \longrightarrow G_q \otimes X_q$$

of the Hopf algebra G_q on the algebra X_q. In other words, we have the commutative diagrams

$$\begin{array}{ccc} G_q \otimes X_q & \xrightarrow{\Delta\otimes\mathrm{id}} & G_q \otimes G_q \otimes X_q \\ \lambda\uparrow & & \uparrow \mathrm{id}\otimes\lambda \\ X_q & \xrightarrow{\lambda} & G_q \otimes X_q \end{array} \qquad \begin{array}{ccc} X_q & \xrightarrow{\mathrm{id}} & X_q \\ \lambda\downarrow & & \downarrow\cong \\ G_q \otimes X_q & \xrightarrow{\epsilon\otimes\mathrm{id}} & k \otimes X_q \end{array}$$

When X_q is a quantum projective variety, we also ask that the action respects the $\mathbb{Z}$-grading.

In a similar fashion we can define right actions of G_q on X_q.

When the quantum variety X_q admits a coaction of a quantum group G_q we say that X_q is a *quantum homogeneous space.* ■

Observation 5.1.8. Notice that our definition of action involves just the coalgebra structure of G_q, hence our definition holds replacing G_q with a coalgebra or a bialgebra. ■

Let us now go back to Example 5.1.1 and examine the general linear group and its natural action using this new terminology that we have introduced.

The group $\mathrm{GL}_n(k)$ acts on the affine space k^n on the left and on the right. Let $\{e_i\}_{i=1}^n$ be the canonical basis of k^n and let (g_{ij}) be the matrix of the group element g in this basis.[3] Then the actions are:

$$\begin{array}{ccc} \mathrm{GL}_n(k) \times k^n & \longrightarrow & k^n \\ (g_{ij}), e_j & \longrightarrow & \sum_i g_{ij} e_i, \end{array} \qquad \begin{array}{ccc} k^n \times \mathrm{GL}_n(k) & \longrightarrow & k^n \\ e_i, (g_{ij}) & \longrightarrow & \sum_j e_j g_{ij}\,. \end{array}$$

Let now $k[\mathrm{GL}_n]$ be the coordinate ring associated with the algebraic group $\mathrm{GL}_n(k)$ and let $k[x_1, \ldots, x_n]$ be the coordinate ring of the affine space k^n.

[3] We are using for the entries of the group element the same symbol as for the algebra generators, that is, g_{ij}, the context making clear the meaning.

The actions described above give rise, by duality, to the coactions

$$\begin{array}{ccc} k[x_1,\dots,x_n] & \longrightarrow & k[\mathrm{GL}_n]\otimes k[x_1,\dots,x_n] \\ x_j & \longrightarrow & \sum_i g_{ji}\otimes x_i, \\ k[x_1,\dots,x_n] & \longrightarrow & k[x_1,\dots,x_n]\otimes k[\mathrm{GL}_n] \\ x_j & \longrightarrow & \sum_i x_i\otimes g_{ij}\,. \end{array}$$

As it happened for the coproduct, the coactions can be formally written in terms of matrices as

$$\begin{pmatrix} x_1 \\ \vdots \\ x_n \end{pmatrix} \longrightarrow \begin{pmatrix} g_{11} & \cdots & g_{1n} \\ \vdots & \dots & \vdots \\ g_{n1} & \cdots & g_{nn} \end{pmatrix} \otimes \begin{pmatrix} x_1 \\ \vdots \\ x_n \end{pmatrix},$$

$$\begin{pmatrix} x_1 \\ \vdots \\ x_n \end{pmatrix} \longrightarrow \begin{pmatrix} x_1 & \dots & x_n \end{pmatrix} \otimes \begin{pmatrix} g_{11} & \cdots & g_{1n} \\ \vdots & \dots & \vdots \\ g_{n1} & \cdots & g_{nn} \end{pmatrix}.$$

The same action and coaction hold replacing $\mathrm{GL}_n(k)$ with the special linear group $\mathrm{SL}_n(k)$.

We go now to the generalization of this picture, defining the quantum general and special linear groups together with their natural actions on *quantum spaces.*

Definition 5.1.9. A *quantum space* k_q^n is the non commutative algebra

$$k_q^n := k_q\langle x_1,\dots,x_n\rangle/(x_ix_j - q^{-1}x_ix_j), \qquad i<j\,,$$

where $k_q\langle x_1,\dots,x_n\rangle$ is the free algebra over the ring $k_q = k[q,q^{-1}]$ with generators $x_1,\dots,x_n$. ■

Notice that when q is specialized to 1, that is, when we take the quotient modulo the two sided ideal $(q-1)$, we obtain the polynomial algebra in n indeterminates, which is the coordinate ring of the affine space k^n. Hence the quantum space is a quantum deformation of the affine space k^n.

We can now construct a quantum deformation $k_q[\mathrm{GL}_n]$ of the general linear group $\mathrm{GL}_n(k)$ by imposing the existence of its natural coactions on the quantum space k_q^n, whose classical equivalents (i.e. for $q=1$) are described above. Let us start with an intermediate step: the *Manin quantum matrix bialgebra.*

Theorem 5.1.10. *Let $k_q\langle a_{ij}\rangle$ be the free algebra in the indeterminates a_{ij}, $i, j = 1, \dots n$. We consider the following algebra morphisms:*

$$\begin{array}{ccc} k_q\langle x_1, \dots, x_n\rangle & \xrightarrow{\lambda} & k_q\langle a_{ij}\rangle \otimes k_q\langle x_1, \dots, x_n\rangle \\ x_i & \longrightarrow & \sum_j a_{ij} \otimes x_j \,, \end{array}$$

$$\begin{array}{ccc} k_q\langle x_1, \dots, x_n\rangle & \xrightarrow{\rho} & k_q\langle x_1, \dots, x_n\rangle \otimes k_q\langle a_{ij}\rangle \\ x_i & \longrightarrow & \sum_j x_j \otimes a_{ji} \,. \end{array}$$

Then, these morphisms factor through the quantum space k_q^n, i.e. the indeterminates x_i satisfy

$$x_i x_j = q^{-1} x_j x_i, \qquad 1 \le i < j \le n$$

if and only if the indeterminates a_{ij} satisfy the Manin relations:

$$\begin{aligned} a_{ij}a_{kj} &= q^{-1}a_{kj}a_{ij},\ i<k, \quad a_{ij}a_{kl} = a_{kl}a_{ij},\ i<k, j>l, \ \text{ or } i>k, j<l, \\ a_{ij}a_{il} &= q^{-1}a_{il}a_{ij},\ j<l, \quad a_{ij}a_{kl} - a_{kl}a_{ij} = (q^{-1}-q)a_{ik}a_{jl},\ i<k, j<l\,. \end{aligned}$$

Proof. The proof is a direct calculation. The strategy is to consider first the subset of Manin relations among the elements

$$a_{ik},\ a_{il},\ a_{jk},\ a_{jl}$$

where i, j, k, l are fixed. This amounts to six different commutation relations. One then proves that the problem is reduced to show the equivalence among these relations and the relations of the quantum space

$$x_i x_j = q^{-1} x_j x_i, \quad x_i x_j = q^{-1} x_j x_i, \quad 1 \le i < j \le n, 1 \le k < l \le n.$$

It is useful to write these elements in matrix form, that is,

$$\begin{pmatrix} a_{ik} & a_{il} \\ a_{jk} & a_{jl} \end{pmatrix}. \tag{5.5}$$

The elements commute differently depending on their respective position. In fact, notice that the elements in the same column (or row) q-commute that is, $a_{ik}a_{jk} = q^r a_{jk}a_{ik}$, for $r = \pm 1$. The two elements on the principal diagonal have a long commutation relation and the elements on the other diagonal simply commute. ∎

We are prompted to give the following key definition.

Definition 5.1.11. The algebra of *quantum matrices* is the algebra $k_q[M_n]$ defined as

$$k_q[M_n] := k_q\langle a_{ij}\rangle / I_M ,$$

where a_{ij} are free indeterminates, with $1 \leq i, j \leq n$, and I_M is the (two-sided) ideal generated by the Manin relations described in Theorem 5.1.10. ∎

Notice that $k_q[M_n]/(q-1) \cong k[a_{ij}]$, in other words, when we specialize the parameter q to 1 we obtain the coordinate algebra of the affine space of the matrices $n \times n$. Then, $k_q[M_n]$ is a quantum deformation of such algebra.

As a Corollary of Theorem 5.1.10 we have the following proposition:

Proposition 5.1.12. *The quantum matrix algebra $k_q[M_n]$ is a bialgebra with comultiplication and counit given by:*

$$\Delta(a_{ij}) = \sum_k a_{ik} \otimes a_{kj}, \qquad \epsilon(a_{ij}) = \delta_{ij} .$$

∎

The definitions of the comultiplication Δ and the counit ϵ are sometimes summarized with the following notation:

$$\Delta \begin{pmatrix} a_{11} & \dots & a_{1n} \\ \vdots & & \vdots \\ a_{n1} & \dots & a_{nn} \end{pmatrix} = \begin{pmatrix} a_{11} & \dots & a_{1n} \\ \vdots & & \vdots \\ a_{n1} & \dots & a_{nn} \end{pmatrix} \otimes \begin{pmatrix} a_{11} & \dots & a_{1n} \\ \vdots & & \vdots \\ a_{n1} & \dots & a_{nn} \end{pmatrix}$$

$$\epsilon \begin{pmatrix} a_{11} & \dots & a_{1n} \\ \vdots & & \vdots \\ a_{n1} & \dots & a_{nn} \end{pmatrix} = \begin{pmatrix} 1 & \dots & 0 \\ \vdots & & \vdots \\ 0 & \dots & 1 \end{pmatrix} .$$

Notice that the expressions for Δ and ϵ on the generators a_{ij} are unchanged by the quantum deformation, that is, they are formally the same as in the classical matrix bialgebra, when given on the generators a_{ij} (compare with Example 5.1.1). However, we have to extend Δ and ϵ to the whole bialgebra $k_q[M_n]$ through the algebra products, and these are indeed different from their classical counterparts, as the Manin relations clearly show.

We have now the following proposition, whose proof is a direct check making use of Theorem 5.1.10.

Proposition 5.1.13. *We have natural left and right coactions of the bialgebra of quantum matrices on the quantum space k_q^n:*

$$\begin{array}{cccccc} k_q^n & \xrightarrow{\lambda} & k_q[M_n] \otimes k_q^n & k_q^n & \xrightarrow{\rho} & k_q^n \otimes k_q[M_n] \\ x_i & \longrightarrow & \sum_j a_{ij} \otimes x_j , & x_i & \longrightarrow & \sum_j x_j \otimes a_{ji} . \end{array}$$

■

As we discussed in Example 5.1.1, the coordinate algebra of the general linear group is obtained from the matrix bialgebra by adding an indeterminate (we called it d), which plays the role of the inverse of the determinant function. We now provide the equivalent notion in the quantum setting, namely we introduce the *quantum determinant*. The most natural way to do this is through the definition of the *quantum exterior algebra*.

Definition 5.1.14. We define the *quantum exterior algebra in n variables* as

$$\wedge_q^n := k_q\langle \chi_1, \dots, \chi_n \rangle / (\chi_i \chi_j + q \chi_j \chi_i ,\, \chi_i^2), \qquad i < j .$$

$\wedge_q^n$ is a quantum deformation of the ordinary exterior algebra in n-variables $\wedge(k^n)$. ■

Classically, we have an action of the general linear group $\mathrm{GL}_n(k)$ on the ordinary (i.e. non quantum) exterior algebra $\wedge(k^n)$:

$$\begin{array}{ccc} \rho : \mathrm{GL}_n(k) \times \wedge(k^n) & \longrightarrow & \wedge(k^n) \\ (g_{ij}), e_j & \longrightarrow & \sum_i g_{ij} e_i \end{array}$$

where $\{e_i\}$ is the standard basis of k^n and the action is given just on the generators e_i of the exterior algebra. By duality, this provides a coaction

$$\begin{array}{ccc} \wedge(k^{n*}) & \xrightarrow{\rho^*} & k[\mathrm{GL}_n] \otimes \wedge(k^{n*}) \\ e_j^* & \longrightarrow & \sum_i g_{ji} \otimes e_i^* . \end{array}$$

As one can readily check, we have that

$$\rho^*(e_1^* \wedge \cdots \wedge e_n^*) = \det(g_{ij}) \otimes e_1^* \wedge \cdots \wedge e_n^* ,$$

where $\det(g_{ij})$ denotes the determinant function in the variables g_{ij}'s.

We now proceed and examine the quantum setting. The proofs of the statements that follow are essentially tedious checks that can be found, for example, in the monograph [119]. We just state them here without proofs.

Proposition 5.1.15. *There is a well defined coaction of the matrix bialgebra on the quantum exterior algebra:*

$$\begin{aligned} \wedge_q^n &\xrightarrow{\ r\ } k_q[M_n] \otimes \wedge_q^n \\ \chi_i &\longrightarrow \textstyle\sum_j a_{ij} \otimes \chi_j \,. \end{aligned}$$

■

Definition 5.1.16. The *quantum determinant*, denoted as $\det_q(a_{ij})$ (or $\det_q$ for short), is the element of $k_q[M_n]$ defined by the equation

$$r(\chi_1 \cdots \chi_n) = \det_q(a_{ij}) \otimes \chi_1 \cdots \chi_n \,.$$

■

Proposition 5.1.17. *The quantum determinant is a central element in* $k_q[M_n]$. *Its explicit expression reads as*

$$\det_q(a_{ij}) = \sum_\sigma (-q)^{-\ell(\sigma)} a_{1\sigma(1)} \dots a_{n\sigma(n)} = \sum_\sigma (-q)^{-\ell(\sigma)} a_{\sigma(1)1} \dots a_{\sigma(n)n} \,,$$

where σ runs through all the permutations of the first n integers.

Furthermore, $\det_q$ *is a group-like element, that is,*

$$\Delta(\det_q) = \det_q \otimes \det_q \,.$$

■

Definition 5.1.18. The *quantum general linear group* is the algebra

$$k_q[\mathrm{GL}_n] = k_q[M_n][D]/(D \cdot \det_q - 1, \det_q \cdot D - 1) \,.$$

We shall denote the indeterminate D also with $\det_q^{-1}$.

The *quantum special linear group* is the algebra

$$k_q[\mathrm{SL}_n] = k_q[M_n]/(\det_q - 1) \,.$$

■

The fact that $\det_q$ is central means that left ideals are also right ideals and then two sided, so the quotient ring is well defined.

Most immediately one sees that the algebras $k_q[\mathrm{GL}_n]$ and $k_q[\mathrm{SL}_n]$ are quantum deformations of the general linear group $\mathrm{GL}_n(k)$ and the special linear group $\mathrm{SL}_n(k)$, respectively.

The next proposition, that we give without proof, justifies the term *quantum group* that we have used for $k_q[\mathrm{GL}_n]$ and $k_q[\mathrm{SL}_n]$.

Proposition 5.1.19. *Both $k_q[\mathrm{GL}_n]$ and $k_q[\mathrm{SL}_n]$ are Hopf algebras, in fact, they inherit the bialgebra structure from $k_q[M_n]$ and they admit an antipode given, respectively, by*

$$S_q(a_{ij}) = (-q)^{j-i}\mathrm{det}_q^{-1} A_{ji}, \qquad S_q(a_{ij}) = (-q)^{j-i} A_{ji}\,,$$

where A_{ji} is the quantum determinant in the quantum matrix bialgebra $k_q[M_{n-1}]$ generated by the indeterminates

$$a_{kl}, \qquad 1 \leq k, l \leq n, \quad k \neq j, \quad l \neq i\,.$$

■

In general, we shall denote with $D^{j_1 \dots j_r}_{i_1 \dots i_r}$ the quantum determinant in the quantum matrix bialgebra $k_q[M_{n-r}]$ generated by the indeterminates a_{rs} with $r \in \{i_1 \dots i_r\}$ and $s \in \{j_1 \dots j_r\}$, so

$$A_{ij} = D^{1 \dots \hat{j} \dots n}_{1 \dots \hat{i} \dots n}\,.$$

The notation '$\hat{i}$' means, as customary, that the index i is absent. Such determinants will play a key role in what follows.

Definition 5.1.20. Let R be an algebra and let $\{a_{ij},\ i, j = 1, \dots, n\}$ be a subset of R such that the elements a_{ij} satisfy the Manin relations. We then say that (a_{ij}) is a *quantum matrix*. So, for example, in $k_q[M_n]$, with arbitrary n, $\{a_{11}, a_{12}, a_{21}, a_{22}\}$ is a quantum matrix and we often write these elements arranged in a matrix form:

$$\begin{pmatrix} a_{11} & a_{12} \\ a_{21} & a_{22} \end{pmatrix}.$$

■

We end this section with two important observations.

Observation 5.1.21. The coactions in 5.1.13 are still well defined on both $k_q[\mathrm{GL}_n]$ and $k_q[\mathrm{SL}_n]$, in other words the quantum space k_q^n is a quantum homogeneous space for both the quantum general linear and quantum special linear groups. This follows from a straightforward calculation. ■

Observation 5.1.22. The algebra $k_q[\mathrm{GL}_n]$ has an additional property that often is required in the definition of deformation. As a k_q-module, it is isomorphic to $k_q \otimes k[\mathrm{GL}_n]$. This means that $k_q[\mathrm{GL}_n]$ is a free module over k_q. In the Appendix E we give a basis of it (in fact it is given for the more general case of the quantum supergroup $k_q\mathrm{GL}(m|n)$). In particular, the algebra $k_q[\mathrm{GL}_n]$ *torsion free*, that is,

$$r \cdot m = 0, \quad r \in k_q,\ m \in k_q[\mathrm{GL}_n] \qquad \Rightarrow \qquad r = 0 \quad \text{or} \quad m = 0\,.$$

All the examples of quantum algebras that we have considered here, namely, the quantum space k_q^n, the quantum matrices $k_q[M_n]$ and quantum groups $k_q[\mathrm{GL}_n]$ and $k_q[\mathrm{SL}_n]$, have the same property with respect to their undeformed counterparts. Hence, they are all torsion free k_q-modules. The same is true for the supersymmetric generalizations that we will study in Section 5.4.

This property is shown by exhibiting a particular basis, although we will not need to write it explicitly. The reader can consult the proof of this fact, for example in Chapter 7 of Ref. [28]. It will be used at some crucial point in our proofs.

■

5.2 The quantum complex conformal and Minkowski spaces

Let us briefly recall the classical (i.e. unquantized) picture that we described previously, in Chapter 2.

The complex conformal space is identified with the Grassmannian variety $G(2,4)$ of two dimensional subspaces inside a four dimensional complex vector space. Such variety comes naturally equipped with the action of the special linear group, which is identified with the complex conformal group. The complex Minkowski space is then realized inside the conformal space as an open set, called the *big cell*, and admits a natural action of the Poincaré group times dilations, which sits as a subgroup inside the conformal group.

Let us examine this picture more in detail. There is a natural transitive action of the special linear group $\mathrm{SL}_4(\mathbb{C})$ on the set of subspaces of dimension 2 inside $\mathbb{C}^4$. Let us fix the canonical basis in $\mathbb{C}^4$, $\{e_1, e_2, e_3, e_4\}$. Let I_0 be the stabilizer subgroup of the subspace $\langle e_1, e_2\rangle \in G(2,4)$,

$$I_0 = \left\{ \begin{pmatrix} L & M \\ 0 & R \end{pmatrix} \in \mathrm{SL}_4(\mathbb{C}) \right\},$$

with L, M, R being 2×2 matrices, and $\det L \cdot \det R = 1$. The Grassmannian variety $G(2,4)$ is the homogeneous space

$$G(2,4) = \mathrm{SL}_4(\mathbb{C})/I_0 .$$

A generic element W in $G(2,4)$ is represented by a 4×2 matrix of rank 2:

$$G(2,4) = \left\{ \begin{pmatrix} a_1 & b_1 \\ a_2 & b_2 \\ a_3 & b_3 \\ a_4 & b_4 \end{pmatrix} \right\} \Big/ \mathrm{GL}_2(\mathbb{C}) .$$

We have to quotient by the right action of $\mathrm{GL}_2(\mathbb{C})$ in order to take into account the change of basis in the description of the subspace. We call the *big cell* the open set in $G(2,4)$ consisting of those subspaces W whose matrix is such that $a_1b_2 - a_2b_1$ invertible:

$$\mathrm{M} = \left\{ \begin{pmatrix} 1 & 0 \\ 0 & 1 \\ m_{11} & m_{12} \\ m_{21} & m_{22} \end{pmatrix} \right\} \subset G(2,4). \tag{5.6}$$

In Chapter 2 Section 2.5 M was also denoted as U_{12} and later on as $\mathbf{M}^c$.

Notice that the right action of $\mathrm{GL}_2(\mathbb{C})$ always allows us to choose a unique representative for a generic element $W \in G(2,4)$ in the form (5.6). Since the entries m_{ij} can be totally arbitrary, $\mathrm{M} \approx \mathbb{C}^4$. M is the complex Minkowski space.

The subgroup of $\mathrm{SL}_4(\mathbb{C})$ that leaves the big cell invariant consists of all the matrices of the form

$$F_0^c = \left\{ \begin{pmatrix} x & 0 \\ tx & y \end{pmatrix} \;\middle|\; \det x \cdot \det y = 1 \right\}. \tag{5.7}$$

The symbols x, y and t are 2×2 matrices with complex coefficients (see Chapter 2 for more details). The group F_0^c is called the complex Poincaré group times dilations. It has a well defined action on M that reads as follows

$$\begin{pmatrix} x & 0 \\ tx & y \end{pmatrix}, \begin{pmatrix} I_2 \\ m \end{pmatrix} \longmapsto \begin{pmatrix} I_2 \\ ymx^{-1} + t \end{pmatrix} \in \mathrm{M}. \tag{5.8}$$

and I_2 denotes the identity matrix of order 2.

The Poincaré group (plus dilations) F_0^c is a semidirect product

$$(\mathrm{SL}_2(\mathbb{C}) \times \mathrm{SL}_2(\mathbb{C}) \times \mathbb{C}^\times) \ltimes \mathbb{C}^4.$$

The complexified Lorentz group is then $\mathrm{SL}_2(\mathbb{C}) \times \mathrm{SL}_2(\mathbb{C})$, represented here inside F_0^c, by the matrices of the form

$$\begin{pmatrix} x & 0 \\ 0 & y \end{pmatrix}, \qquad \det x = \det y = 1,$$

and it acts on the Minkowski space by the tensor product representation

$$\text{fundamental of } \mathrm{SL}_2(\mathbb{C}) \otimes \text{contragradient of } \mathrm{SL}_2(\mathbb{C}).$$

The extra factor $\mathbb{C}$ represents the dilations, since it multiplies the coordinates of Minkowski space by a common factor. This representation is the one used in the *twistor* approach to Minkowski space (see for example Ref. [124]). In fact, in the basis of the Pauli matrices,

$$\sigma_0 = \begin{pmatrix} 1 & 0 \\ 0 & 1 \end{pmatrix}, \quad \sigma_1 = \begin{pmatrix} 0 & 1 \\ 1 & 0 \end{pmatrix}, \quad \sigma_2 = \begin{pmatrix} 0 & -\mathrm{i} \\ \mathrm{i} & 0 \end{pmatrix}, \quad \sigma_3 = \begin{pmatrix} 1 & 0 \\ 0 & -1 \end{pmatrix},$$

an arbitrary matrix m in M can be written as

$$m = \begin{pmatrix} m_{11} & m_{12} \\ m_{21} & m_{22} \end{pmatrix} = x^0\sigma_0 + x^1\sigma_1 + x^2\sigma_2 + x^3\sigma_3 = \begin{pmatrix} x^0 + x^3 & x^1 - \mathrm{i}x^2 \\ x^1 + \mathrm{i}x^2 & x^0 - x^3 \end{pmatrix}.$$

Then

$$\det m = (x^0)^2 - (x^1)^2 - (x^2)^2 - (x^3)^2,$$

which is invariant under the action of the complexified Lorentz group $\mathrm{SL}_2(\mathbb{C}) \times \mathrm{SL}_2(\mathbb{C})$, so that (x^0, x^1, x^2, x^3) can be recognized as the ordinary coordinates of Minkowski space.

We now want to associate to the Poincaré group, the special linear group and the Minkowski space, the algebra of their polynomial functions in order to be able to proceed to their quantum deformations.

The coordinate algebras of the groups $\mathrm{SL}_4(\mathbb{C})$ and F_0^c, are immediately given by

$$\mathbb{C}[\mathrm{SL}_4] = \mathbb{C}[g_{ij}]/(\det(g) - 1), \qquad i, j = 1, \ldots, 4,$$

$$\mathbb{C}[F_0^c] = \mathbb{C}[g_{ij}]/(\det(g) - 1, g_{rs}), \qquad i, j = 1, \ldots, 4, \; r = 1, 2, \; s = 3, 4.$$

The last one can also be expressed as

$$\mathbb{C}[F_0^c] \cong \mathbb{C}[x_{ij}, y_{kl}, t_{kj}]/(\det(x) \cdot \det(y) - 1), \qquad i, j = 1, 2, \; k, l = 3, 4.$$

Here det denotes as usual the determinant function and the indeterminates x_{ij}, y_{kl} and t_{kj} are defined in (5.7). Both of these algebras carry a natural Hopf algebra structure, since they are the algebras of polynomial functions on groups (see Chapter 1, Section 1.11). Explicitly, such structure on $\mathbb{C}[\mathrm{SL}_4]$ is

$$\Delta(g_{ij}) = \sum_k g_{ik} \otimes g_{kj}, \qquad \epsilon(g_{ij}) = \delta_{ij}, \qquad S(g_{ij}) = (-1)^{j-i} A^c_{ji},$$

where A^c_{ji} denotes the determinant of the matrix of indeterminates (g_{kl}) obtained by suppressing the jth row and ith column. As usual, Δ, ϵ and S denote the comultiplication, the counit and the antipode respectively.

The Hopf algebra structure on $\mathbb{C}[F_0^c]$ is the same as in $\mathbb{C}[\mathrm{SL}_4]$ when expressed in terms of the indeterminates g_{ij}. In fact, $\mathbb{C}[F_0^c]$ is the quotient of $\mathbb{C}[\mathrm{SL}_4]$ by the Hopf ideal (g_{rs}), $r=1,2$, $s=3,4$. In terms of x, y and t we can write

$$\Delta\begin{pmatrix} x & 0 \\ tx & y \end{pmatrix} = \begin{pmatrix} x & 0 \\ tx & y \end{pmatrix} \otimes \begin{pmatrix} x & 0 \\ tx & y \end{pmatrix}, \qquad \epsilon\begin{pmatrix} x & 0 \\ tx & y \end{pmatrix} = \begin{pmatrix} I & 0 \\ 0 & I \end{pmatrix},$$

$$\Delta(x) = x \otimes x, \quad \Delta(y) = y \otimes y, \quad \Delta(t) = t \otimes 1 + yS(x) \otimes t,$$

where

$$S(x) = S\begin{pmatrix} x_{11} & x_{12} \\ x_{21} & x_{22} \end{pmatrix} = \begin{pmatrix} x_{22}\det(x)^{-1} & -x_{12}\det(x)^{-1} \\ -x_{21}\det(x)^{-1} & x_{11}\det(x)^{-1} \end{pmatrix}$$

denotes the antipode for the Hopf algebra $\mathbb{C}[x_{11}, x_{12}, x_{21}, x_{22}][\det(x)^{-1}]$, corresponding to the algebraic group $\mathrm{GL}_2(\mathbb{C})$.

The Grassmannian $G(2,4)$, being a projective variety, does not have a canonical algebra of polynomial functions associated with it. However, in Section 2.6 we studied the *Plücker embedding*, a projective embedding of $G(2,4)$ in $\mathbb{P}^5$. In Theorem 2.6.2 we showed that the Plücker embedding establishes an isomorphism of algebraic varieties between $G(2,4)$ and the *Klein quadric* in $\mathbb{P}^5$, defined by the homogeneous equation

$$\lambda_{12}\lambda_{34} - \lambda_{13}\lambda_{24} + \lambda_{14}\lambda_{23} = 0. \tag{5.9}$$

The homogeneous coordinates λ_{ij}, $1 \leq i < j \leq 4$ in $\mathbb{P}^5$ are called the Plücker coordinates.

We can then associate to $G(2,4)$, together with its projective embedding, the homogeneous (i.e. $\mathbb{Z}$-graded) algebra

$$\mathbb{C}[G(2,4)] = \mathbb{C}[\lambda_{ij}]/(\lambda_{12}\lambda_{34} - \lambda_{13}\lambda_{24} + \lambda_{14}\lambda_{23}), \qquad 1 \leq i < j \leq 4. \tag{5.10}$$

The algebra $\mathbb{C}[G(2,4)]$ can be effectively retrieved as a subalgebra of $\mathbb{C}[\mathrm{SL}_4(\mathbb{C})]$. Let us write, as usual, the generators of $\mathbb{C}[\mathrm{SL}_4(\mathbb{C})]$ in matrix form

$$\begin{pmatrix} g_{11} & g_{12} & g_{13} & g_{14} \\ g_{21} & g_{22} & g_{23} & g_{24} \\ g_{31} & g_{32} & g_{33} & g_{34} \\ g_{41} & g_{42} & g_{43} & g_{44} \end{pmatrix}.$$

Then, the elements formed by taking the 2×2 determinants of the first two columns

$$d_{ij} = g_{i1}g_{j2} - g_{j1}g_{i2}, \qquad 1 \leq i < j \leq 4,$$

satisfy the equation of the Klein quadric (5.9). Moreover, as a consequence of the First and Second Fundamental Theorems of invariant theory that we discussed in Section 2.9, (5.9) is the only relation that these minors satisfy. Therefore, the subalgebra of $\mathbb{C}[\mathrm{SL}_4(\mathbb{C})]$ generated by d_{ij}

$$\mathbb{C}[d_{ij}] \subset \mathbb{C}[\mathrm{SL}_4(\mathbb{C})], \qquad 1 \leq i < j \leq 4\,,$$

is isomorphic to (5.10). From now on, we will use this isomorphism to identify

$$\mathbb{C}[G(2,4)] \cong \mathbb{C}[d_{ij}] \subset \mathbb{C}[\mathrm{SL}_4(\mathbb{C})], \qquad 1 \leq i < j \leq 4\,.$$

This identification will allow us to induce a deformation of $\mathbb{C}[G(2,4)]$ given a deformation of $\mathbb{C}[\mathrm{SL}_4(\mathbb{C})]$.

The Minkowski space M is the big cell inside $G(2,4)$ corresponding to those subspaces with invertible coordinate λ_{12}. Hence the coordinate algebra of the Minkowski space $\mathbb{C}[\mathrm{M}]$ is the *projective localization* at λ_{12} of the $\mathbb{Z}$-graded algebra $\mathbb{C}[G(2,4)]$ (see Section 1.6). Let us recall the construction. One considers first the standard localization at λ_{12} of the algebra $\mathbb{C}[G(2,4)]$ introducing the extra generator λ_{12}^{-1} and the condition $\lambda_{12}\lambda_{12}^{-1} = 1$. We denote it as

$$\mathbb{C}[G(2,4), \lambda_{12}^{-1}] = \mathbb{C}[G(2,4)][\lambda_{12}^{-1}]/(\lambda_{12}\lambda_{12}^{-1} - 1)\,. \tag{5.11}$$

If we assign degree -1 to the generator λ_{12}^{-1}, then $\mathbb{C}[G(2,4), \lambda_{12}^{-1}]$ is a $\mathbb{Z}$-graded algebra. The projective localization of $\mathbb{C}[G(2,4)]$ at λ_{12} is the subalgebra of the elements with zero $\mathbb{Z}$-degree; such algebra is generated by the elements

$$\lambda_{ij}\lambda_{12}^{-1}, \qquad i,j = 1,\dots,4, \quad (i,j) \neq (1,2)\,.$$

The next proposition summarizes our discussion and clarifies the structure of $\mathbb{C}[\mathrm{M}]$.

Proposition 5.2.1. *The coordinate algebra $\mathbb{C}[\mathrm{M}]$ of the Minkowski space is the subalgebra of $\mathbb{Z}$-degree zero elements in $\mathbb{C}[G(2,4)][\lambda_{12}^{-1}]$, hence it is the homogeneous ring inside $\mathbb{C}[G(2,4)][\lambda_{12}^{-1}]$ generated by the elements*

$$\lambda_{13}\lambda_{12}^{-1}, \quad \lambda_{14}\lambda_{12}^{-1}, \quad \lambda_{23}\lambda_{12}^{-1}, \quad \lambda_{24}\lambda_{12}^{-1}\,. \tag{5.12}$$

We have an isomorphism of $\mathbb{C}[\mathrm{M}]$ with the polynomial subring $\mathbb{C}[t_{ij}]$ of $\mathbb{C}[F_0^c]$:

$$\phi : \mathbb{C}[\mathrm{M}] \longrightarrow \mathbb{C}[t_{ij}]$$

$$\phi(\lambda_{13}\lambda_{12}^{-1}) = t_{32},\ \phi(\lambda_{14}\lambda_{12}^{-1}) = t_{42},\ \phi(\lambda_{23}\lambda_{12}^{-1}) = -t_{31},\ \phi(\lambda_{24}\lambda_{12}^{-1}) = -t_{41}\,.$$

Proof. (Sketch). The first claim is a simple check. Notice that we do not have the generator $\lambda_{34}\lambda_{12}^{-1}$ since it is expressible in terms of the others using the Plücker relation:

$$\lambda_{34}\lambda_{12}^{-1} = (\lambda_{13}\lambda_{12}^{-1})(\lambda_{24}\lambda_{12}^{-1}) - (\lambda_{14}\lambda_{12}^{-1})(\lambda_{23}\lambda_{12}^{-1}).$$

For the second claim, one needs to show that the generators in (5.12) do not satisfy any relation. But it is clear that any relation among the generators (5.12) would immediately imply a relation among λ_{ij} for $(i,j) \neq (3,4)$ in $\mathbb{C}[G(2,4)]$. This would be different from the Plücker relation, which is not possible. ∎

We now want to give a quantization of the complex conformal and Minkowski spaces. The algebras we start from are the algebras of polynomial functions over the Minkowski and conformal spaces. We shall provide a quantum deformation of such algebras and of the algebras of the Poincaré and conformal groups, in such way that there are respective coactions deforming the classical ones.

We notice that, so far, all the coordinate rings that are relevant for us are realized in terms of the coordinate ring of $\mathrm{SL}_4(\mathbb{C})$. $\mathbb{C}[G(2,4)]$ is a subring, $\mathbb{C}[F_0^c]$ is a quotient and $\mathbb{C}[\mathrm{M}]$ is a localization of a subring of $\mathrm{SL}_4(\mathbb{C})$. So our starting point will be a quantum deformation of such coordinate ring.

Let $\mathbb{C}_q = \mathbb{C}[q, q^{-1}]$, where q is an indeterminate and let the notation be as in the previous section.

Definition 5.2.2. We define *complex quantum conformal group* as the Hopf algebra

$$\mathbb{C}_q[\mathrm{SL}_4] = \mathbb{C}_q[M_4]/(\det_q - 1),$$

were $\mathbb{C}_q[M_4]$ denotes the algebra of 4×4 quantum matrices (see Definition 5.1.11). The comultiplication, counit and antipode are given by

$$\Delta(g_{ij}) = \sum_k g_{ik} \otimes g_{kj}, \qquad \epsilon(g_{ij}) = \delta_{ij}, \qquad S(g_{ij}) = (-q)^{j-i} A_{ji},$$

where A_{ji} is the quantum determinant of the quantum matrix obtained by removing the row j and column i from the quantum matrix (g_{ij}). ∎

We now proceed to give the definition of quantum conformal space and quantum Minkowski space in analogy with what we have described in the ordinary setting (see Proposition 5.2.1).

Definition 5.2.3. We define *quantum Grassmannian* $\mathbb{C}_q[G(2,4)]$ as the subalgebra of $\mathbb{C}_q[\mathrm{SL}_4]$ generated by the quantum determinants

$$D_{ij}^{12} = a_{i1}a_{j2} - q^{-1}a_{j1}a_{i2}, \quad 1 \leq i < j \leq 4\,.$$

To ease the notation we shall write D_{ij} in place of D_{ij}^{12}, when there is no danger of confusion. ■

We want to give a presentation of this ring in terms of generators and relations. This will allow us to work with it more easily and to see more clearly its properties. The procedure was introduced in Ref. [54]. The problem was taken up again in Refs. [55,56] for general Grassmannians and flag manifolds. Here we will report only the special case of $\mathbb{C}_q[G(2,4)]$.

The first thing that we need to do is to establish the commutation relations among the quantum determinants D_{ij} inside the quantum group. This is not an easy task: one can compute the commutator of two such determinants in terms of the generators g_{ij} of the quantum group, but then the result has to be reexpressed again in terms of quantum determinants if we want a presentation of the ring $\mathbb{C}_q[G(2,4)]$ in analogy with the classical setting. There is a priori no guarantee that given two quantum determinants the expression:

$$D_{ij}D_{kl} - q^r D_{kl}D_{ij}$$

(for a suitable r) is again expressible in terms of products of other quantum determinants. Furthermore, it is not just enough to give this expression in terms of quantum determinants, because one also wishes to do so in such a way to obtain a basis for $\mathbb{C}_q[G(2,4)]$ which reduces to a known basis in the classical setting (where we have commutativity, so one can freely reorder the determinants). In other words, one must be able to give an order on the quantum determinant indices in such way that the indices appearing in the reordering of two quantum determinants are strictly lower than the ones of the original quantum determinants. While in the commutative setting we have more freedom in choosing such order, this is no longer the case in the quantum setting. In fact extra care must be exerted, as we are going to see. One can appreciate this phenomenon already in the choice of basis for the quantum matrix bialgebra (see Refs. [93,109]).

In Ref. [57] it is proven the very remarkable fact that such commutation rules exists among *generic* quantum minors (i.e. determinants of any size and respective position) inside $\mathbb{C}_q[\mathrm{SL}_n]$ and, furthermore, in Ref. [64] such quantum minors are given a geometric interpretation in terms of quantum Schubert varieties and cells.

The commutation relations are in general very involved. In Refs. [56,57] a recursive algorithm to work them out is given, while in Ref. [78] such relations are given in a closed form using the formalism of the R-matrix.

Since the calculations are of a very technical nature, we will not attempt to reproduce here the strategies used in the references above to perform them. The reader that is interested in knowing the details can consult them directly. However, since we only deal with quantum determinants of order two, one can also compute the commutation relations directly, with long calculations, without the knowledge of the inner workings of the general theory.

To complete the presentation of the quantum Grassmannian, we will find a relation reducing to the Plücker relation when $q = 1$.

The next proposition tells us the explicit form of all these relations and thus allows us to provide an explicit presentation of the ring of the quantum Grassmannian.

Proposition 5.2.4. *Let us consider the free algebra* $\mathbb{C}_q\langle\lambda_{ij}\rangle$, $1 \leq i < j \leq 4$ *and the lexicographic ordering on the set of pairs* (i, j)*:*

$$(1,2) < (1,3) < (1,4) < (2,3) < (2,4) < (3,4)\,.$$

Let $I_{G(2,4)}$ *be the two sided ideal generated by the relations*

$$\lambda_{i_1 i_2}\lambda_{j_1 j_2} = q^{-1}\lambda_{j_1 j_2}\lambda_{i_1 i_2}\,, \qquad (i_1, i_2) < (j_1, j_2)$$

with i_1, i_2, j_1, j_2 *not all distinct and*

$$\lambda_{12}\lambda_{34} = q^{-2}\lambda_{34}\lambda_{12}\,, \qquad \lambda_{14}\lambda_{23} = \lambda_{23}\lambda_{14}\,,$$

$$\lambda_{13}\lambda_{24} = q^{-2}\lambda_{24}\lambda_{13} - (q^{-1} - q)\lambda_{12}\lambda_{34}\,,$$

$$\lambda_{12}\lambda_{34} - q^{-1}\lambda_{13}\lambda_{24} + q^{-2}\lambda_{14}\lambda_{23} = 0\,. \qquad (p)$$

Then, there is an isomorphism

$$\begin{array}{ccc} \mathbb{C}_q[G(2,4)] & \longrightarrow & \mathbb{C}_q\langle\lambda_{ij}\rangle/I_{G(2,4)} \\ D_{ij} & \longrightarrow & \lambda_{ij}\,, \end{array} \qquad 1 \leq i < j \leq 4\,.$$

We will refer to the relation (p) as the *quantum Plücker relation*, since for $q = 1$ it reduces to the usual Plücker relation for the embedding $G(2,4) \subset \mathbb{P}^5(\mathbb{C})$.

Proof. (Sketch, we invite the reader to consult Refs. [54, 56, 57, 64] as well as the classical Refs. [71, 72]). Consider the map

$$\begin{array}{ccc} \mathbb{C}_q\langle\lambda_{ij}\rangle/I_{G(2,4)} & \longrightarrow & \mathbb{C}_q[G(2,4)] \\ \\ \lambda_{ij} & \longrightarrow & D_{ij}\,. \end{array}$$

This map is well defined since the relations in $I_{G(2,4)}$ are satisfied by the D_{ij}'s in $\mathbb{C}_q[G(2,4)]$. This follows from a long but direct calculation.

From the Definition 5.2.3, it follows that the ring $\mathbb{C}_q[G(2,4)]$ is generated, as $\mathbb{C}_q$-module, by the monomials

$$D_{I_1}\cdots D_{I_r}, \qquad I_i=(i_1,i_2), \qquad 1\le i_1<i_2\le 4\,,$$

although they are not all independent.

The first observation is that the commutation relations always allow us to write a generic monomial $D_{I_1}\cdots D_{I_r}$ as a linear combination of the lexicographically ordered monomials, that is monomials

$$D_{L_1}\cdots D_{L_r}, \qquad \text{with} \quad L_1\le\cdots\le L_r\,,$$

where '$\le$' is the lexicographic ordering.

The second observation is the following. In the classical case, with the classical determinants d_J (see Section 2.9, where the determinants are denoted as y_J), the Plücker relation allowed us to express any lexicographically ordered monomial in terms of the *standard monomials*, that is monomials $d_{J_1}\cdots d_{J_r}$ were the multiindices $J_1,\dots,J_r$ formed a *Young tableau* (see Section 2.9). Since the quantum Plücker relation has the same form as the classical one except for the coefficients, which are nevertheless invertible, following the same formal procedure we have that the standard monomials are a set of generators for $\mathbb{C}_q[G(2,4)]$.

We now show that the standard monomials are independent. Assume that there exists an arbitrary relation among the standard monomials

$$\sum_{J_k} R_{J_1,\dots,J_r}\,D_{J_1}\cdots D_{J_r}=0, \qquad R_{J_1,\dots,J_r}\in\mathbb{C}_q\,.$$

Since the algebra $\mathbb{C}_q[\mathrm{SL}_n]$ is torsion free (see Observation 5.1.22) we can divide the all the coefficients $R_{J_1,\dots,J_r}$ by the power of $(q-1)$ that is its maximum common divisor. Then we set $q=1$. This will give a relation among the given monomials in the commutative case, which is not possible since we know that they are linearly independent.

Hence there are no other relations in $\mathbb{C}_q[G(2,4)]$ besides the Plücker relation. The morphism is injective and in turn, it is an isomorphism. ∎

Since we have given a basis, the algebra $\mathbb{C}_q[G(2,4)]]$ is a free module (hence torsion free) over $\mathbb{C}_q$.

Proposition 5.2.5. *$\mathbb{C}_q[G(2,4)]$ is a $\mathbb{C}_q[\mathrm{SL}_4]$-comodule. The coaction map is the comultiplication Δ in $\mathbb{C}_q[\mathrm{SL}_4]$, restricted to $\mathbb{C}_q[G(2,4)]$, and we denote it as $\tilde{\Delta} := \Delta\Big|_{\mathbb{C}_q[G(2,4)]}$*

$$\begin{array}{ccc} \mathbb{C}_q[G(2,4)] & \xrightarrow{\tilde{\Delta}} & \mathbb{C}_q[\mathrm{SL}_4] \otimes \mathbb{C}_q[G(2,4)] \\ D^{12}_{i_1 i_2} & \longrightarrow & \sum_{1 \le j_1 < j_2 \le 4} D^{j_1 j_2}_{i_1 i_2} \otimes D^{12}_{j_1 j_2} \,. \end{array}$$

Proof. The proof amounts to a direct, though long, calculation. The crucial fact here is that the second factor in the tensor product belongs indeed to $\mathbb{C}_q[G(2,4)]$. ■

The quantum Grassmannian is then a quantum homogeneous space for the quantum group $\mathbb{C}_q[\mathrm{SL}_4]$. We will refer to $\mathbb{C}_q[G(2,4)]$ as the *quantum conformal space*, while we call $\mathbb{C}_q[\mathrm{SL}_4]$ the *quantum conformal group.*

We now tackle the definition of the quantum Minkowski space. We would like to imitate the classical definition by constructing a *projective localization* of $\mathbb{C}_q[G(2,4)]$ at D_{12}. However, D_{12} is not a central element and such localizations are not well defined unless the *Ore condition* is verified. We shall proceed avoiding such difficulty with a direct definition, however for completeness we briefly discuss the Ore condition in the next observation.

Observation 5.2.6. Let R be a (non commutative) ring with unit and no zero divisors. Let $S \subset R$ be a subset of R that is closed under multiplication. The Ore condition states that

$$\text{for } (r,s) \in R \times S, \text{ there exists } (r',s') \in R \times S \text{ such that } rs' = sr' \,.$$

The localization is then defined as the pairs $(r,s) \in R \times S$ modulo the equivalence relation

$$(r_1,s_1) \sim (r_2,s_2) \text{ if } r_1 s_2' = s_1 r_2' \text{ with } (r_2',s_2') \text{ such that } r_2 s_2' = s_2 r_2' \,.$$

The existence of (r_2',s_2') is assured by the Ore condition. One can check that the relation defined above is indeed an equivalence relation.

On the set of equivalence classes, the product

$$[(r_1,s_1)][(r_2,s_2)] = [(r_1 r_2', s_1' s_2)], \qquad \text{where} \qquad r_2 s_1' = s_1 r_2' \,,$$

is well defined. As in the commutative case, one can write the equivalence classes as fractions

$$r\,\frac{1}{s} := [(r,s)] \,,$$

which is perhaps more intuitive. ■

We now proceed to give our definition of localization. Then, in Remark 5.2.9, we shall make contact with the previous observation.

Definition 5.2.7. We define the *localization* of $\mathbb{C}_q[\mathrm{SL}_4]$ at D_{12} as the ring

$$\mathbb{C}_q[\mathrm{SL}_4, {D_{12}}^{-1}] = \mathbb{C}_q[\mathrm{SL}_4]\langle {D_{12}}^{-1}\rangle / (D_{12}{D_{12}}^{-1} - 1,\, {D_{12}}^{-1}D_{12} - 1)\,. \tag{5.13}$$

■

Notice that in this definition, ${D_{12}}^{-1}$ is an indeterminate satisfying the relations $D_{12}{D_{12}}^{-1} = 1$, ${D_{12}}^{-1}D_{12} = 1$, hence behaving in $\mathbb{C}_q[\mathrm{SL}_4, {D_{12}}^{-1}]$ as the inverse of the quantum determinant D_{12}. This justifies the chosen terminology.

Proposition 5.2.8. *Let the notation be as above. Then*

$$\mathbb{C}_q[\mathrm{SL}_4, D_{12}^{-1}] = \mathbb{C}_q[\mathrm{SL}_4]\langle {D_{12}}^{-1}\rangle / (I_{{D_{12}}^{-1}})$$

where $I_{{D_{12}}^{-1}}$ is the ideal generated by the relations:

$$D_{12}{D_{12}}^{-1} - 1, \quad {D_{12}}^{-1}D_{12} - 1,$$

$${D_{12}}^{-1}a_{ij} = a_{ij}{D_{12}}^{-1}, \quad {D_{12}}^{-1}a_{kj} = qa_{kj}{D_{12}}^{-1}, \quad {D_{12}}^{-1}a_{il} = qa_{il}{D_{12}}^{-1},$$

$${D_{12}}^{-1}a_{kl} = a_{kl}{D_{12}}^{-1} + (q^{-1} - q)\big(D_{12}^{-1}D_{1k}a_{2l}D_{12}^{-1} - q^{-1}D_{12}^{-1}D_{2k}a_{1l}D_{12}^{-1}\big),$$

$${D_{12}}^{-1}D_{1k} = qD_{1k}{D_{12}}^{-1}, \qquad {D_{12}}^{-1}D_{2k} = qD_{2k}{D_{12}}^{-1},$$

with $i, j = 1, 2$ and $k, l = 3, 4$.

Proof. (Sketch). The fact the element ${D_{12}}^{-1}$ verifies the given commutation relations comes from a tedious calculation, which is based on the equivalent commutation relations for D_{12}. For example, one computes directly that

$$D_{12}a_{31} = q^{-1}a_{31}D_{12}\,.$$

Then multiplying both sides by ${D_{12}}^{-1}$ on the left and on the right and using the relations defining the ideal, one obtains a special case of the second commutation relation.

To prove that these are all of the relations in $\mathbb{C}_q[\mathrm{SL}_4, D_{12}^{-1}]$ one uses the fact that $\mathbb{C}_q[\mathrm{SL}_4]$ is a free module (Observation 5.1.22). In Appendix E we have given a basis of it (see Refs. [109, 110]) in terms of ordered monomials. Then, using a similar reasoning to the one in the proof of Proposition 5.2.4, such monomials multiplied by D_{12}^{-r} give also a basis for $\mathbb{C}_q[\mathrm{SL}_4, D_{12}^{-1}]$. ■

Remark 5.2.9. In the Observation 5.2.6 we have introduced the Ore condition. One can verify with some tedious calculations that the set $S = \{D_{12}^n\}$ indeed satisfies the Ore condition, hence Definition 5.2.7 leads to a well defined quotient ring. However, we prefer to use Definition 5.2.7, since it is more explicit and more suitable for the calculations. ■

We are ready for the main definition of this section.

Definition 5.2.10. We define the *quantum Minkowski space* $\mathbb{C}_q[\mathrm{M}]$ as the *projective localization* of $\mathbb{C}_q[G(2,4)]$ at D_{12} (denoted as $\mathbb{C}_q[G(2,4)]\langle {D_{12}}^{-1}\rangle_{\text{loc}}$), that is, the subring of $\mathbb{C}_q[\mathrm{SL}_4, {D_{12}}^{-1}]$ generated by the elements

$$D_{13}{D_{12}}^{-1},\; D_{14}{D_{12}}^{-1},\; D_{23}{D_{12}}^{-1},\; D_{24}{D_{12}}^{-1},\; D_{34}{D_{12}}^{-1}.$$

Notice that the last generator can be omitted since the Plücker relation yields:

$$D_{34}{D_{12}}^{-1} = D_{13}{D_{12}}^{-1}D_{24}{D_{12}}^{-1} - q^{-1}D_{14}{D_{12}}^{-1}D_{23}{D_{12}}^{-1}.$$

Hence,

$$\mathbb{C}_q[\mathrm{M}] = \mathbb{C}_q\langle D_{13}{D_{12}}^{-1}, D_{14}{D_{12}}^{-1}, D_{23}{D_{12}}^{-1}, D_{24}{D_{12}}^{-1}\rangle \subset \mathbb{C}_q[\mathrm{SL}_4, {D_{12}}^{-1}].$$

■

The next proposition provides a characterization of $\mathbb{C}_q[\mathrm{M}]$ in terms of generators and relations. The proof is a direct calculation, based on the relations of Proposition 5.2.8.

Proposition 5.2.11. *The quantum Minkowski space $\mathbb{C}_q[\mathrm{M}]$ is isomorphic to a matrix bialgebra:*

$$\begin{aligned} \mathbb{C}_q[\mathrm{M}] &\longrightarrow \mathbb{C}_q[M_2] \\ D_{13}{D_{12}}^{-1} &\longrightarrow a_{11} \\ D_{14}{D_{12}}^{-1} &\longrightarrow a_{21} \\ D_{23}{D_{12}}^{-1} &\longrightarrow a_{12} \\ D_{24}{D_{12}}^{-1} &\longrightarrow a_{22}\,, \end{aligned}$$

where the a_{ij} are subject to the Manin relations. ■

From now on we will denote the generators of $\mathbb{C}_q[\mathrm{M}]$ as m_{ij}:

$$\begin{pmatrix} m_{11} & m_{12} \\ m_{21} & m_{22} \end{pmatrix} = \begin{pmatrix} D_{13}{D_{12}}^{-1} & D_{23}{D_{12}}^{-1} \\ D_{14}{D_{12}}^{-1} & D_{24}{D_{12}}^{-1} \end{pmatrix}.$$

(m_{ij}) is then a quantum matrix.

We now want to define the quantum Poincaré group. We start with an heuristic argument.

Let us consider the classical equality of matrices in $\mathrm{SL}_4(\mathbb{C})$ which holds whenever $a_{11}a_{22} - a_{12}a_{21}$ is invertible:

$$\begin{pmatrix} x & 0 \\ tx & y \end{pmatrix} \begin{pmatrix} I_2 & s \\ 0 & I_2 \end{pmatrix} = \begin{pmatrix} a_{11} & \dots & a_{14} \\ \vdots & & \vdots \\ a_{41} & \dots & a_{44} \end{pmatrix},$$

were x, y, t and s are 2×2-matrices.

Now, assume that (a_{ij}) is a quantum matrix. Then, taking into account the commutation relations, after a small calculation we get the equalities

$$x = (x_{ij}) = \begin{pmatrix} a_{11} & a_{12} \\ a_{21} & a_{22} \end{pmatrix},$$

$$t = (t_{kj}) = \begin{pmatrix} -q^{-1} D_{23} {D_{12}}^{-1} & D_{13} {D_{12}}^{-1} \\ -q^{-1} D_{24} {D_{12}}^{-1} & D_{14} {D_{12}}^{-1} \end{pmatrix},$$

$$s = (s_{il}) = \begin{pmatrix} -q {D_{12}}^{-1} D_{12}^{23} & -q {D_{12}}^{-1} D_{12}^{24} \\ {D_{12}}^{-1} D_{12}^{13} & {D_{12}}^{-1} D_{12}^{14} \end{pmatrix},$$

$$y = (y_{kl}) = \begin{pmatrix} D_{123}^{123} {D_{12}}^{-1} & D_{124}^{123} {D_{12}}^{-1} \\ D_{123}^{124} {D_{12}}^{-1} & D_{124}^{124} {D_{12}}^{-1} \end{pmatrix}, \tag{5.14}$$

where $1 \leq i, j \leq 2$, $3 \leq k, l \leq 4$.

We recall $D_{i_1 \dots i_r}^{j_1 \dots j_r}$ is the quantum determinant obtained by taking the rows $i_1, \dots, i_r$ and columns $j_1, \dots, j_r$ (we may omit the column index, when we take the first r columns).

Notice that the entries of the matrix t correspond to the generators of the quantum Minkowski space, up to a multiplicative constant. For the moment, we write the elements of t, x, y, s in matrix form, for convenience.

This heuristic reasoning leads to the following definition, in analogy with the classical setting.

Definition 5.2.12. We define *quantum Poincaré group times dilations* $\mathbb{C}_q[F_0^c]$ as the subalgebra of $\mathbb{C}_q[\mathrm{SL}_4]$ generated by the elements in x, y

and t described in (5.14). This is a quantum deformation of $\mathbb{C}[F_0^c]$, the coordinate algebra of the complex Poincaré group times dilations.

We define *quantum translations* as the subalgebra $\mathbb{C}_q[t]$ in $\mathbb{C}_q[\mathrm{SL}_4]$ generated by the elements t_{kj}. This is a quantum deformation of the coordinate algebra of the translation group.

The deformations considered here are algebra deformations: no statements are made at this point about their Hopf algebra structure. ∎

The next proposition establishes the relation between the quantum Minkowski and the quantum Poincaré group, justifying the quantum group denomination that we have used in the previous definition.

Proposition 5.2.13.

(1) *The quantum Minkowski space $\mathbb{C}_q[\mathrm{M}]$ is isomorphic to the translation algebra. The isomorphism is given by*

$$\begin{array}{rcl} \mathbb{C}_q[t] & \longrightarrow & \mathbb{C}_q[\mathrm{M}] \\ t_{32} = D_{13}{D_{12}}^{-1} & \longrightarrow & m_{11} \\ t_{42} = D_{14}{D_{12}}^{-1} & \longrightarrow & m_{21} \\ t_{31} = -q^{-1}D_{23}{D_{12}}^{-1} & \longrightarrow & m_{12} \\ t_{41} = -q^{-1}D_{24}{D_{12}}^{-1} & \longrightarrow & m_{22}\,. \end{array}$$

Hence, $\mathbb{C}_q[\mathrm{M}]$ becomes identified with the subring $\mathbb{C}_q[t]$ in $\mathbb{C}_q[F_0^c]$. In particular, both $\mathbb{C}_q[t]$ and $\mathbb{C}_q[\mathrm{M}]$ are isomorphic to a quantum matrix bialgebra.

(2) *The quantum Poincaré group $\mathbb{C}_q[F_0^c]$ is isomorphic to*

$$\mathbb{C}_q[F_0^c] \cong \mathbb{C}_q[\mathrm{SL}_n]/(a_{13}, a_{14}, a_{23}, a_{24})\,,$$

and as a quotient of $\mathbb{C}_q[\mathrm{SL}_n]$, it inherits a natural quantum group structure from it.

The isomorphism is given by:

$$\begin{array}{ccc} \mathbb{C}_q[F_0^c] & \longrightarrow & \mathbb{C}_q[\mathrm{SL}_n]/(a_{13}, a_{14}, a_{23}, a_{24}) \\ x & \longrightarrow & x \\ y & \longrightarrow & \begin{pmatrix} a_{33} & a_{34} \\ a_{43} & a_{44} \end{pmatrix} \\ t & \longrightarrow & \begin{pmatrix} a_{32} & a_{31} \\ a_{42} & a_{41} \end{pmatrix}. \end{array}$$

(3) The quantum Minkowski space admits a natural coaction of the quantum Poincaré group. In the identification $\mathbb{C}_q[\mathrm{M}] \cong \mathbb{C}_q[t]$ such coaction corresponds to the restriction to $\mathbb{C}_q[t] \cong \mathbb{C}_q[\mathrm{M}]$ of the comultiplication Δ in $\mathbb{C}_q[F_0^c]$

$$\Delta|_{\mathbb{C}_q[M]} : \mathbb{C}_q[\mathrm{M}] \longrightarrow \mathbb{C}_q[F_0^c] \otimes \mathbb{C}_q[\mathrm{M}]$$

where

$$\Delta|_{\mathbb{C}_q[M]} \begin{pmatrix} m_{12} & m_{11} \\ m_{22} & m_{21} \end{pmatrix} = \begin{pmatrix} t_{31} & t_{32} \\ t_{41} & t_{42} \end{pmatrix} \otimes \begin{pmatrix} 1 & 0 \\ 0 & 1 \end{pmatrix}$$

$$+ \begin{pmatrix} y_{33} & y_{34} \\ y_{43} & y_{44} \end{pmatrix} S \begin{pmatrix} x_{11} & x_{12} \\ x_{21} & x_{22} \end{pmatrix} \otimes \begin{pmatrix} m_{12} & m_{11} \\ m_{22} & m_{21} \end{pmatrix},$$

and

$$S \begin{pmatrix} x_{11} & x_{12} \\ x_{21} & x_{22} \end{pmatrix} = \begin{pmatrix} x_{22} {D_{12}}^{-1} & -q x_{12} {D_{12}}^{-1} \\ -q^{-1} x_{21} {D_{12}}^{-1} & x_{11} {D_{12}}^{-1} \end{pmatrix}.$$

Hence, $\mathbb{C}_q[\mathrm{M}]$ is a quantum homogeneous variety.

Proof. As the first two statements, the proof consists of easy but tedious calculations. As for the point 3, notice that Δ in $\mathbb{C}_q[F_0^c]$ gives:

$$\Delta \begin{pmatrix} x & 0 \\ tx & y \end{pmatrix} = \begin{pmatrix} x & 0 \\ tx & y \end{pmatrix} \otimes \begin{pmatrix} x & 0 \\ tx & y \end{pmatrix}, \qquad \epsilon \begin{pmatrix} x & 0 \\ tx & y \end{pmatrix} = \begin{pmatrix} I_2 & 0 \\ 0 & I_2 \end{pmatrix}.$$

Hence

$$\Delta(tx) = tx \otimes x + y \otimes tx .$$

By multiplying this relation by $S(x) \otimes S(x)$ on both sides and using few properties of Hopf algebras (see Chapter 1), we obtain

$$\Delta(t) = t \otimes 1 + yS(x) \otimes t .$$

Now, recalling that

$$t = \begin{pmatrix} t_{31} & t_{32} \\ t_{41} & t_{42} \end{pmatrix} \cong \begin{pmatrix} m_{12} & m_{11} \\ m_{22} & m_{21} \end{pmatrix},$$

we obtain the result. ■

5.3 The real quantum Minkowski space

We now want to provide a real form of the quantum Minkowski space. In the quantum setting, a real form of a complex quantum group G_q is a real form of the Hopf algebra, as defined in Section 1.12. We recall here Definition 1.12.1.

Definition 5.3.1. Let G_q be a quantum group. We call the pair $(G_q, *)$ a *real form* of G_q if $* : G_q \to G_q$ is a $*$-structure on the Hopf algebra G_q, in other words, $* : G_q \longrightarrow G_q$ is an antilinear map satisfying the properties:

- $*$ is involutive, i.e. $(a^*)^* = a$.
- $*$ is antimultiplicative $(ab)^* = b^* a^*$.
- $*$ is a morphism of real coalgebras, $\Delta(a^*) = \Delta(a)^*$.
- If S is the antipode of G_q, then $S(x)^* = S^{-1}(x^*)$ for all $x \in G_q$.

■

If A is just an algebra, then an antilinear map $* : A \to A$ satisfying conditions 1 and 2 is also called a $*$-structure on A. In particular, if A_q is a deformation of the coordinate algebra of the variety, a $*$-structure on A_q is also called a real form of the quantum variety.

We now define real forms of quantum homogeneous varieties.

Definition 5.3.2. Let X_q be complex quantum homogeneous variety for the quantum group G_q, with coaction

$$\lambda : X_q \longrightarrow G_q \otimes X_q \, .$$

Let $(X_q, *_X)$ be a real form of X_q (as an algebra) and let $(G_q, *_G)$ be a real form of G_q.

We say that $(X_q, *_X)$ is a *real form* of X_q as a quantum homogeneous variety for $(G_q, *_G)$ if

$$(*_G \otimes *_X) \circ \lambda = \lambda \circ *_X \, .$$

In other words the real forms of X_q and G_q are compatible with each other with respect to the coaction. ■

In Chapter 1 we discussed the relation between the $*$-structures and the real forms in the classical setting (for commutative algebras). We saw that there is a one-to-one correspondence between the $*$-structures on a given algebra and the real forms of the corresponding geometric object. It is in this sense that we call the pairs $(G_q, *_G)$ and $(X_q, *_X)$ real forms of G_q and X_q, respectively.

We now want to define a real form of the quantum Minkowski space as a quantum homogeneous space for the quantum Poincaré group.

We start with the definition, on the quantum Minkowski space $\mathbb{C}_q[\mathrm{M}]$, of an antilinear involution. We give it on the generators:

$$\begin{array}{ccc} \mathbb{C}_q[\mathrm{M}] & \xrightarrow{*_{M,q}} & \mathbb{C}_q[\mathrm{M}] \\ \begin{pmatrix} m_{11} & m_{12} \\ m_{21} & m_{22} \end{pmatrix} & \longrightarrow & \begin{pmatrix} m_{22} & m_{12} \\ m_{21} & m_{11} \end{pmatrix} . \end{array}$$

As one can check with a simple calculation, $*_{M,q}$ is well defined, i.e., the commutation relations are respected.

According to Proposition 5.2.13, we can identify the quantum Minkowski space with the quantum translation subalgebra inside $\mathbb{C}_q[F_0^c]$. In terms of the generators t_{ij} of the quantum translation subalgebra,

$$\begin{pmatrix} t_{31} & t_{32} \\ t_{41} & t_{42} \end{pmatrix} = \begin{pmatrix} m_{12} & m_{11} \\ m_{22} & m_{21} \end{pmatrix} ,$$

the $*$-structure $*_{M,q}$ takes the more familiar form

$$\begin{pmatrix} t_{31} & t_{32} \\ t_{41} & t_{42} \end{pmatrix} \longrightarrow \begin{pmatrix} t_{31} & t_{41} \\ t_{32} & t_{42} \end{pmatrix} .$$

We will use the coordinates t_{ij}'s in place of the coordinates m_{ij}'s.

Remark 5.3.3. Let us stress that, while $\begin{pmatrix} m_{11} & m_{12} \\ m_{21} & m_{22} \end{pmatrix}$ is a quantum matrix, $\begin{pmatrix} t_{31} & t_{32} \\ t_{41} & t_{42} \end{pmatrix}$ is not. It becomes a quantum matrix by exchanging the first column with the second one. One must exert special care with this distinction when computing commutation relations and quantum determinants. ■

We now define a star structure on the quantum Poincaré group times dilations.

Definition 5.3.4. We define the antilinear involution on the Poincaré quantum group:

$$\begin{array}{ccc} \mathbb{C}_q[F_0^c] & \xrightarrow{*_{F,q}} & \mathbb{C}_q[F_0^c] \\ x_{ij} & \longrightarrow & S(y_{ji}) \\ y_{ij} & \longrightarrow & S(x_{ji}) \\ t_{ij} & \longrightarrow & t_{ji} \, , \end{array}$$

where $S(x_{ij})$ denotes the antipode for the quantum matrix x_{ij}:

$$S\begin{pmatrix} x_{11} & x_{12} \\ x_{21} & x_{22} \end{pmatrix} = \begin{pmatrix} x_{22}D_{12}^{-1} & -qx_{12}D_{12}^{-1} \\ -q^{-1}x_{21}D_{12}^{-1} & x_{11}D_{12}^{-1} \end{pmatrix},$$

and similarly for $S(y_{ij})$. ∎

As one can readily check, this map is well defined and gives a $*$-structure on $\mathbb{C}_q[F_0^c]$, hence $(\mathbb{C}_q[F_0^c], *_{F,q})$ is a real form of the quantum Poincaré group in the sense of Definition 5.3.1. The next proposition states that the two involutions $*_{M,q}$ and $*_{F,q}$ are compatible with each other.

Proposition 5.3.5. *Consider the coaction λ of $\mathbb{C}_q[F_0^c]$ on the quantum Minkowski space (see 5.2.13):*

$$\begin{array}{ccc} \mathbb{C}_q[\mathrm{M}] & \longrightarrow & \mathbb{C}_q[F_0^c] \otimes \mathbb{C}_q[\mathrm{M}] \\ (t) & \longrightarrow & (y)S(x) \otimes (t) + (t) \otimes 1 \,. \end{array}$$

Then

$$(*_{F,q} \times *_{M,q}) \circ \lambda = \lambda \circ *_{M,q} \,.$$

Proof. Direct computation. ∎

The previous proposition implies that $(\mathbb{C}_q[\mathrm{M}], *_{M,q})$ is a real form of the quantum Minkowski space as an homogeneous space for the real form of the Poincaré quantum group $(\mathbb{C}_q[F_0^c], *_{F,q})$.

When $q = 1$, both involutions become multiplicative. Then, the set of fixed points of these involutions are real algebras, that is, they are real forms in the standard, commutative sense. It is not difficult to show that the real forms defined in this way coincide with the real forms of the Minkowski space and the Poincaré group times translations studied in (2.11).

We now turn to examine how the Minkowski quadratic form (see Chapter 2, (2.61)) is suitably quantized in this setting.

It is convenient to perform the same change of coordinates that we used in the classical case, involving the Pauli matrices (see Section 2.7):

$$t = \begin{pmatrix} t_{31} & t_{32} \\ t_{41} & t_{42} \end{pmatrix} = x^\mu \sigma_\mu = \begin{pmatrix} x^0 + x^3 & x^1 - \mathrm{i}x^2 \\ x^1 + ix^2 & x^0 - x^3 \end{pmatrix},$$

so

$$x^0 = \frac{1}{2}(t_{31} + t_{42}), \qquad x^1 = \frac{1}{2}(t_{32} + t_{41}),$$

$$x^2 = \frac{1}{2\mathrm{i}}(t_{41} - t_{32}), \qquad x^3 = \frac{1}{2}(t_{31} - t_{42}) \,.$$

The coordinates $x^0, \dots, x^3$ are left unchanged by the involution. Indeed, when $q = 1$, these coordinates generate, over the reals, the real Minkowski space.

Let us now compute the quantum determinant in the new coordinates (recall here Remark 5.3.3). We have

$$\begin{aligned} Q_q := \det_q \begin{pmatrix} t_{32} & t_{31} \\ t_{42} & t_{41} \end{pmatrix} &= t_{32}t_{41} - q^{-1}t_{31}t_{42} \\ &= -q^{-1}(x^0)^2 + q^{-1}(x^3)^2 + (x^1)^2 + (x^2)^2 \,. \end{aligned} \tag{5.15}$$

In the classical setting, this quadratic form is preserved by the action of the Lorentz group. In the quantum case, we can easily define the *quantum Lorentz group* as

$$\mathbb{C}_q[\mathbf{L}^c] = \mathbb{C}_q[F_0^c]/(t, \det_q(x) - 1, \det_q(y) - 1)\,.$$

$\mathbb{C}_q[\mathbf{L}^c]$ is obtained from the Poincaré quantum group times dilations $\mathbb{C}_q[F_0^c]$, by taking the quotient by the translations and setting the determinants of the quantum matrices (x) and (y) equal to 1. It is clear that this gives a quantum deformation of the classical Lorentz group. As we will see in the next proposition, the quantum determinant (5.15) is coinvariant under the coaction of the quantum Lorentz group on the quantum Minkowski space. Hence, we call Q_q the *quantum Minkowski quadratic form.*

The proof of the following proposition consists of a tedious but direct calculation.

Proposition 5.3.6.

(1) The quantum Lorentz group acts on the quantum Minkowski space as

$$\begin{aligned} \mathbb{C}_q[\mathrm{M}] &\xrightarrow{\lambda_L} \mathbb{C}_q[\mathbf{L}^c] \otimes \mathbb{C}_q[\mathrm{M}] \\ (t) &\longrightarrow (y)S(x) \otimes (t)\,. \end{aligned}$$

(2) The map:

$$\begin{aligned} \mathbb{C}_q[\mathbf{L}^c] &\xrightarrow{*_{\mathbf{L},q}} \mathbb{C}_q[\mathbf{L}^c] \\ x_{ij} &\longrightarrow S(y_{ji}) \\ y_{ij} &\longrightarrow S(x_{ji})\,, \end{aligned}$$

is a $*$-structure of $\mathbb{C}_q[\mathbf{L}^c]$. We call the real form $(\mathbb{C}_q[\mathbf{L}^c], *_{\mathbf{L},q})$ the real quantum Lorentz group. Furthermore, the real quantum Minkowski *space $(\mathbb{C}_q[\mathrm{M}], *_{M,q})$ is a quantum homogeneous space for the real quantum Lorentz group. In other words, we have*

$$(*_{L,q} \times *_{M,q}) \circ \lambda_L = \lambda_L \circ *_{M,q} .$$

(3) *The quantum Minkowski quadratic form is a coinvariant for the action of the quantum Lorentz group on the Minkowski quantum space, that is,*

$$\lambda_L(Q_q) = 1 \otimes Q_q .$$

■

5.4 Quantum supergroups

A quantum supergroup is a *Hopf superalgebra* which is the quantum deformation of either, the universal enveloping superalgebra of a Lie supergroup, or the coordinate superalgebra of an algebraic supergroup.[4] In this section we proceed in strict analogy with Section 5.1. First, we define the bialgebra of supermatrices and then we proceed to the definition of the general linear and special linear quantum supergroups, via the notion of *quantum berezinian* which replaces in this analogy the quantum determinant.

Let k be our ground field ($k = \mathbb{R}$ or $\mathbb{C}$ in our applications) and let $k_q = k[q, q^{-1}]$.

Definition 5.4.1. Let $k_q\langle a_{ij}\rangle$ denote the free algebra over k_q generated by the variables a_{ij}, $i, j = 1, \ldots, m+n$. The parity of the generator a_{ij}, $\pi(a_{ij})$ is defined in terms of the parity of the indices i and j.We have

$$\pi(a_{ij}) = p(i) + p(j) ,$$

with $p(i) = 0$ for $i = 1, \ldots, m$ and $p(i) = 1$ for $i = m+1, \ldots, m+n$.

Following Manin [110], we define the algebra of *quantum supermatrices* as

$$k_q[M(m|n)] =_{\text{def}} k_q\langle a_{ij}\rangle / I_M ,$$

[4]For the definitions and a short summary of properties of Hopf algebras and superalgebras see Chapter 1 Section 1.11.

where I_M is the two sided, homogeneous ideal generated by the relations:

$$a_{ij}a_{il} = (-1)^{\pi(a_{ij})\pi(a_{il})} q^{(-1)^{p(i)+1}} a_{il}a_{ij}, \quad j < l,$$

$$a_{ij}a_{kj} = (-1)^{\pi(a_{ij})\pi(a_{kj})} q^{(-1)^{p(j)+1}} a_{kj}a_{ij}, \quad i < k,$$

$$a_{ij}a_{kl} = (-1)^{\pi(a_{ij})\pi(a_{kl})} a_{kl}a_{ij}, \quad i < k, j > l \quad or \quad i > k, j < l,$$

$$\begin{aligned} a_{ij}a_{kl} &- (-1)^{\pi(a_{ij})\pi(a_{kl})} a_{kl}a_{ij} \\ &= (-1)^{\pi(a_{ij})\pi(a_{kj})} (q^{-1} - q) a_{kj}a_{il} \quad i < k, j < l\,. \end{aligned}$$

■

Notice that I_M is a $\mathbb{Z}_2$-homogeneous ideal, that is, it is generated by polynomials homogeneous in the $\mathbb{Z}_2$-grading. Hence the quotient $k_q[M(m|n)] = k_q\langle a_{ij}\rangle / I_M$ has a well defined superalgebra structure induced by the multiplication in the tensor algebra $k_q\langle a_{ij}\rangle$ (see Chapter 1 for such multiplication). It is then immediate to see that $k_q[M(m|n)]$ is a deformation of the polynomial superalgebra in $m^2 + n^2$ even and $2mn$ odd variables (compare with Section 1.1 and with Examples 1.1.6, 1.4.5, 1.8.3).

We now want to show that $k_q[M(m|n)]$ is a bialgebra, for which we need to provide natural comultiplication and counit. We first need to introduce some notation that we shall use consistently in what follows.

Let us organize the indeterminates generating $k_q[M(m|n)]$ in matrix form. This is often a convenient notational device in quantum group theory; however, one has always to remember that these are matrices of inderminates and have no geometric significance.

$$\begin{pmatrix} a_{11} & \dots & a_{1m+n} \\ \vdots & & \vdots \\ a_{m+n,1} & \dots & a_{m+n,m+n} \end{pmatrix}.$$

The commutation relations in Definition 5.4.1 give immediately that the upper diagonal, $m \times m$ block is a quantum matrix (Definition 5.1.20), that is, the indeterminates:

$$\begin{pmatrix} a_{11} & \dots & a_{1m} \\ \vdots & & \vdots \\ a_{m,1} & \dots & a_{m,m} \end{pmatrix}$$

satisfy the classical Manin relations as in Theorem 5.1.10.

The lower diagonal, $n \times n$ block,

$$\begin{pmatrix} a_{m+1,m+1} & \cdots & a_{m+1,m+n} \\ \vdots & & \vdots \\ a_{m+n,m+1} & \cdots & a_{m+n,m+n} \end{pmatrix},$$

on the other hand, is a quantum matrix where q is replaced with q^{-1}, that is the entries satisfy the Manin relations with q^{-1} in place of q.

This is a crucial property. In fact, it allows us to define quantum determinants for both quantum matrices, the upper diagonal block and the lower diagonal block, respectively,

$$D_1 =_{\text{def}} \sum_{\sigma \in S_m} (-q)^{-\ell(\sigma)} a_{1\sigma(1)} \dots a_{m\sigma(m)},$$

$$D_2 =_{\text{def}} \sum_{\sigma \in S_n} (-q)^{\ell(\sigma)} a_{m+1,m+\sigma(1)} \dots a_{m+n,m+\sigma(n)} \,.$$

We will then be able to make use of the properties of quantum determinants in this new context. Notice that in the expression of D_1 we have $(-q)^{-\ell(\sigma)}$, while in the expression of D_2 we have $(-q)^{\ell(\sigma)}$, according to the commutation relations. This happens because in the lower diagonal blocks the indeterminates obey the Manin relations with q replaced by q^{-1}. D_1 is central in the superalgebra generated by the elements a_{ij}, $i, j = 1, \dots, m$ in $k_q[M(m|n)]$, while D_2 is central in the superalgebra generated by the elements a_{ij}, $i, j = m+1, \dots, m+n$ in $k_q[M(m|n)]$.

Proposition 5.4.2. *$k_q[M(m|n)]$ is a super bialgebra with comultiplication and counit:*

$$\Delta(a_{ij}) = \sum a_{ik} \otimes a_{kj}, \qquad \epsilon(a_{ij}) = \delta_{ij} \,,$$

or equivalently, in the notation that we have introduced:

$$\Delta\begin{pmatrix} a_{11} & \dots & a_{1m+n} \\ \vdots & & \vdots \\ a_{m+n,1} & \dots & a_{m+n,m+n} \end{pmatrix}$$

$$= \begin{pmatrix} a_{11} & \dots & a_{1m+n} \\ \vdots & & \vdots \\ a_{m+n,1} & \dots & a_{m+n,m+n} \end{pmatrix} \otimes \begin{pmatrix} a_{11} & \dots & a_{1m+n} \\ \vdots & & \vdots \\ a_{m+n,1} & \dots & a_{m+n,m+n} \end{pmatrix},$$

$$\epsilon\begin{pmatrix} a_{11} & \dots & a_{1m+n} \\ \vdots & & \vdots \\ a_{m+n,1} & \dots & a_{m+n,m+n} \end{pmatrix} = I_{m+n}\,,$$

where I_{m+n} *is the identity matrix of order* $m+n$.

Proof. The proof consists of tedious, but straightforwards checks. ■

We are ready to define the general linear supergroup, which will be most interesting for us.

Definition 5.4.3. We define *quantum general linear supergroup*

$$k_q[\mathrm{GL}(m|n)] =_{\mathrm{def}} k_q[M(m|n)][{D_1}^{-1}, {D_2}^{-1}]$$

where ${D_1}^{-1}$, ${D_2}^{-1}$ are even indeterminates such that

$$D_1 D_1^{-1} = 1 = {D_1}^{-1} D_1, \qquad D_2 {D_2}^{-1} = 1 = {D_2}^{-1} D_2.$$

■

In order to fully describe $k_q[\mathrm{GL}(m|n)]$ it is necessary to express the commutation relations of D_1 and D_2 with the generators. These relations are not difficult to work out. We describe them in the Appendix E, since they are not relevant for our treatment.

$k_q[\mathrm{GL}(m|n)]$ is a Hopf superalgebra, where the comultiplication and counit are inherited by the ones in $k_q[M(m|n)]$. The antipode S is however more involved and it is detailed in Refs. [125] and [163]. Again, we have provided a brief description of S in Appendix E.

$k_q[\mathrm{GL}(m|n)]$ is a deformation of the coordinate ring of the general linear supergroup of Definition 1.8.4.

As we have seen in the ordinary setting, there are natural coactions of both the quantum super matrices and the quantum general linear supergroup on the *quantum superspace*.

Definition 5.4.4. We define the *quantum superspace* $k_q^{m|n}$ as the non commutative superalgebra

$$k_q^{m|n} := k_q\langle x_1, \dots, x_m, \xi_1, \dots, \xi_n\rangle / (x_i x_j - q^{-1} x_j x_i,\ \xi_k \xi_l + q^{-1} \xi_l \xi_k,\ \xi_r^2),$$

with $1 \le i < j \le m$ and $1 \le k < l \le n$, $r = 1, \dots, n$. ■

As it happens for its non super counterpart, k_q^n (see 5.1.9), we have most immediately that $k_q^{m|n}$ is a deformation of the superspace $k^{m|n}$ (refer to Chapter 1), that is, $k_q^{m|n}/(q-1)$ is the polynomial superalgebra in m even and n odd variables that corresponds to the polynomial functions on the affine superspace $k^{m|n}$.

Proposition 5.4.5. *There are well defined coactions of both, the quantum supermatrices and the quantum general linear supergroup on the quantum superspace*

$$\begin{array}{ccccccc} k_q^{m|n} & \xrightarrow{\lambda} & k_q[M(m|n)] \otimes k_q^{m|n} & \qquad & k_q^{m|n} & \xrightarrow{\lambda'} & k_q[\mathrm{GL}(m|n)] \otimes k_q^{m|n} \\ x_i & \longrightarrow & \sum_j a_{ij} \otimes x_j, & & x_i & \longrightarrow & \sum_j a_{ij} \otimes x_j \,, \end{array}$$

where to ease the notation we have denoted ξ_k with x_{m+k}.

Proof. The proof amounts to a direct, but tedious check that these maps are well defined with respect to the commutation relations. ■

We end this section with the definition of the *quantum Berezinian*, which takes the place of the quantum determinant in the quantum supergroup setting. It is important to observe that such a notion is not required to define the quantum general linear supergroup. In fact, in the classical setting a super matrix is invertible if and only if it has invertible diagonal blocks. For this reason, in the definition of $k_q[\mathrm{GL}(m|n)]$, we have required the invertibility of the two quantum determinants D_1 and D_2, which are the quantum analogues of the determinants of the diagonal blocks.

On the other hand, the quantum Berezinian is needed if we want to define the *quantum special linear supergroup*. Notice that the classical Berezinian is defined only for those matrices which are invertible: this is an important difference with the determinant, which is defined for all the square matrices. This makes the quantum correspondent much harder to deal with. In our definitions we shall follow Ref. [163] with one word of warning: in that reference the indeterminate q is replaced by its inverse q^{-1}.

Definition 5.4.6. Let the notation be as above. We define the *quantum Berezinian* as the element in $k_q[\mathrm{GL}(m|n)]$

$$\mathrm{Ber}_q := \det_q(a_{ij})\det_{q^{-1}}(S(a_{kl}))$$

where $\det_q(a_{ij})$ is the quantum determinant of the quantum matrix a_{ij}, $1 \leq i,j \leq m$, while $\det_{q^{-1}}(S(a_{kl}))$ is the quantum determinant with q replaced with q^{-1} of the quantum matrix $S(a_{kl})$ with $1 \leq k,l \leq n$ (the image under the antipode of the lower diagonal block). ■

Example 5.4.7. For $k_q[\mathrm{GL}(1|1)]$ we have:

$$\mathrm{Ber}_q = x_{22}^{-1}(x_{11} - \xi_{12}x_{22}^{-1}\xi_{21}).$$

■

In general, Ber_q is central and group-like, that is, it commutes will all of the elements in $k_q[\mathrm{GL}(m|n)]$ and

$$\Delta(\mathrm{Ber}_q) = \mathrm{Ber}_q \otimes \mathrm{Ber}_q.$$

These are highly non trivial facts, basically impossible to get through direct calculation and can be found in [163], where the authors derive them using the R-matrix formalism.

We can now define the quantum special linear supergroup.

Definition 5.4.8. We define *quantum special linear supergroup* the superalgebra

$$k_q[\mathrm{SL}(m|n)] := k_q[\mathrm{GL}(m|n)]/(\mathrm{Ber}_q - 1).$$

Remark 5.4.9. As in the non super case, the fact that Ber_q is central guarantees that the two sided ideal $(\mathrm{Ber}_q - 1)$ is not too big. Otherwise, it may result that the quotient algebra is trivial. ■

5.5 The quantum chiral conformal superspace

In this section we want to construct a quantum deformation of the chiral conformal superspace described in Sections 4.7 and 4.8, the Grassmannian supervariety $\mathrm{Gr}^{\mathrm{ch}} = G(2|0, 4|1)$ of $2|0$ planes inside the superspace $\mathbb{C}^{4|1}$. This was the first supersymmetric extension that we found of the conformal space $G(2,4)$ and although it does not have a real form, it plays a role in supersymmetric theories (see Section 4.14). The content of this section is a summary of the results in Refs. [26, 27].

We will work out this deformation in the same spirit as we did for the non super case in Section 5.2.

Let our ground field be $\mathbb{C}$. We start by recalling the superalgebra associated with $\mathrm{Gr}^{\mathrm{ch}}$. The topological space is $G(2,4)$ the Grassmannian of 2-planes in $\mathbb{C}^4$. As we have seen in Chapter 4, we can view $\mathrm{Gr}^{\mathrm{ch}}$ both as an analytic supermanifold and as an algebraic supervariety. We are now interested in the algebraic point of view, since we are after the quantum deformation, which takes place starting from the function superalgebra. Since $\mathrm{Gr}^{\mathrm{ch}}$ is not an affine supervariety, there is no intrinsic coordinate superalgebra associated with it, however, once we provide a projective embedding of $\mathrm{Gr}^{\mathrm{ch}}$ into some projective superspace, we will obtain a coordinate superalgebra associated uniquely to such embedding. In general the Grassmannian supervariety does not admit always a projective embedding, however in our special case it does: we have the Plücker embedding (see Chapter 4, Section 4.8)

$$\mathrm{Gr}^{\mathrm{ch}} \longrightarrow \mathbb{P}^{6|4}\,.$$

With respect to such embedding, the homogeneous superring associated with $\mathrm{Gr}^{\mathrm{ch}}$ was given in terms of generators and relations as

$$\mathbb{C}[\mathrm{Gr}^{\mathrm{ch}}] = \mathbb{C}[y_{ij}, \eta_k, a]/\mathcal{I}_P, \qquad 1 \le i < j \le 4,\, 1 \le k \le 4\,,$$

with y_{ij} and a even coordinates, η_k odd coordinates and $\mathcal{I}_P$ being the ideal of the super Plücker relations:

$$\begin{aligned}
&y_{12}y_{34} - y_{13}y_{24} + y_{14}y_{23} = 0, && \text{(standard Plücker relation)},\\
&y_{ij}\eta_k - y_{ik}\eta_j + y_{jk}\eta_i = 0, && 1 \le i < j < k \le 4,\\
&\eta_i\eta_j = a y_{ij}, && 1 \le i < j \le 4,\\
&\eta_i a = 0, && 1 \le i \le 4 = 0\,.
\end{aligned}$$

As in the non super case, we can regard $\mathbb{C}[\mathrm{Gr}^{\mathrm{ch}}]$ as a subring of the supergroup ring

$$\mathbb{C}[\mathrm{SL}(4|1)] = \mathbb{C}[g_{ij}]/(\mathrm{Ber} - 1), \qquad i,j = 1,\dots,5\,,$$

(see Section 1.8). As usual, the supergroup generators are written in matrix form

$$\left(\begin{array}{cccc|c}
g_{11} & g_{12} & g_{13} & g_{14} & g_{15}\\
g_{21} & g_{22} & g_{23} & g_{24} & g_{25}\\
g_{31} & g_{32} & g_{33} & g_{34} & g_{35}\\
g_{41} & g_{42} & g_{43} & g_{44} & g_{45}\\
\hline
g_{51} & g_{52} & g_{53} & g_{54} & g_{55}
\end{array}\right), \tag{5.16}$$

where we have highlighted the diagonal (even) and off-diagonal (odd) blocks. Ber stands for the Berezinian (the super version of the determinant) of the above supermatrix of indeterminates (see Definition 1.8.6).

In fact, in Section 4.9, we showed that the homogeneous superring $\mathbb{C}[\mathrm{Gr}^{\mathrm{ch}}]$ is generated inside the superalgebra of supermatrices $\mathbb{C}[a_{ij}]$ by the elements

$$\begin{aligned} d_{ij} &= a_{i1}a_{j2} - a_{i2}a_{j1}, \quad 1 \leq i < j \leq 5, \\ d_{55} &= a_{51}a_{52} . \end{aligned}$$

This is the content of the Second Fundamental Theorem of Invariant theory, adapted to our special setting; it is a profound result in invariant theory and we have stated it, together with a sketch of its proof, in Proposition 4.9.5. Let $\mathbb{C}[d_{ij}, d_{55}]$ be the superalgebra generated by the elements d_{ij}, d_{55} in $\mathbb{C}[\mathrm{SL}(4|1)]$ (by an abuse of notation we use the same symbols d_{ij}, d_{55} to denote both the elements in $\mathbb{C}[a_{ij}]$ and in its quotient $\mathbb{C}[\mathrm{SL}(4|1)]$). Since $\mathbb{C}[d_{ij}, d_{55}]$ has intersection zero with the ideal $(\mathrm{Ber}_q - 1)$, we have that

$$\mathbb{C}[\mathrm{Gr}^{\mathrm{ch}}] \cong \mathbb{C}[d_{ij}, d_{55}] \subset \mathbb{C}[\mathrm{SL}(4|1)].$$

The elements d_{ij}, $1 \leq i < j \leq 5$, are the determinants of the 2×2 minors of the first two columns of the matrix (5.16). Notice that the elements d_{i5} are odd, while d_{ij} with $1 \leq i < j \leq 4$ and d_{55} are even.

The explicit morphism ϕ mapping $\mathbb{C}[\mathrm{Gr}^{\mathrm{ch}}]$ isomorphically onto the subalgebra described above is just

$$\begin{array}{ccc} \mathbb{C}[\mathrm{Gr}^{\mathrm{ch}}] & \xrightarrow{\ \phi\ } & \mathbb{C}[\mathrm{SL}(4|1)] \\ y_{ij} & \longrightarrow & d_{ij} \\ \eta_i & \longrightarrow & d_{i5} \\ a & \longrightarrow & d_{55}, \end{array} \qquad 1 \leq i < j \leq 4 .$$

The Grassmannian $\mathrm{Gr}^{\mathrm{ch}}$ admits an action of the conformal supergroup $\mathrm{SL}(4|1)$. In fact, as we saw in Section 4.7, $\mathrm{Gr}^{\mathrm{ch}}$ is realized as the quotient of the supergroup $\mathrm{SL}(4|1)$ by a suitable parabolic subgroup (the one corresponding to the superalgebra $\mathfrak{f}_2$ in (4.32)).

The chiral Minkowski superspace is then realized as the big cell inside $\mathrm{Gr}^{\mathrm{ch}}$. Its coordinate superalgebra is the projective localization of $\mathbb{C}[\mathrm{Gr}^{\mathrm{ch}}]$ in d_{12} (see Section 4.10).

Notice that, as for the ordinary setting, all of the supergeometric objects that we want to quantize are obtained directly from the supergroup $\mathrm{SL}(4|1)$

in the sense that their coordinate superalgebras are either quotients of the coordinate superalgebra $\mathbb{C}[\mathrm{SL}(4|1)]$ of global sections on $\mathrm{SL}(4|1)$, or subalgebras of it. Since we already have detailed its quantization in Section 5.4, we can then proceed to give the key definitions.

Definition 5.5.1. The *quantum super Grassmannian* of $2|0$ planes in $4|1$ dimensional superspace is the non commutative superalgebra $\mathbb{C}_q[\mathrm{Gr}^{\mathrm{ch}}]$ generated by the following quantum super determinants in $\mathbb{C}_q[\mathrm{SL}(4|1)]$:

$$\begin{aligned} D_{ij} &= g_{i1}g_{j2} - q^{-1}g_{i2}g_{j1}, && 1 \leq i < j \leq 4, \\ D_{i5} &= g_{i1}g_{52} - q^{-1}g_{i2}g_{51}, && 1 \leq i \leq 4, \\ D_{55} &= g_{51}g_{52}\,. \end{aligned}$$

■

For clarity we write explicitly all the generators:

$$\begin{aligned} &D_{12}, \quad D_{13}, \quad D_{14}, \quad D_{23}, \quad D_{24}, \quad D_{34}, \quad D_{55}, && \text{(even)} \\ &D_{15}, \quad D_{25}, \quad D_{35}, \quad D_{45}, && \text{(odd)}. \end{aligned} \tag{5.17}$$

Notice that D_{55} is an even nilpotent element. The parity is easily given by the rule:

$$|D_{ij}| = (-1)^{|i|+|j|},$$

where, for a given index k, we have $|k| = 0$ for $k = 1, \dots, 4$ and $|k| = 1$ for $k = 5$.

When q is specialized to 1, that is, when we take the quotient by the two-sided ideal $(q-1)$, we observe that $\mathbb{C}_q[\mathrm{Gr}^{\mathrm{ch}}]/(q-1)$ is the coordinate superalgebra of the Grassmannian $G(2|0, 4|1)$ with respect to the super Plücker embedding. Hence, $\mathbb{C}_q[\mathrm{Gr}^{\mathrm{ch}}]$ is a quantum deformation of the Grassmannian supervariety in the sense of Definition 5.1.2.

We now want to find a presentation of $\mathbb{C}_q[\mathrm{Gr}^{\mathrm{ch}}]$ in terms of generators and relations. In order to do so, we need first to work out the commutation relations among the quantum minors D_{ij} and then, as in the classical setting, there will be additional relations among the generators. These are called the *quantum super Plücker relations.*

The commutation relations can be checked directly with tedious calculations. They are as follows:

- If i, j, k, l are *not* all distinct and D_{ij}, D_{kl} are not both odd, we have
$$D_{ij}D_{kl} = q^{-1}D_{kl}D_{ij}, \quad (i,j) < (k,l), \quad 1 \leq i,j,k,l \leq 5, \quad i < j,\ k < l\,, \tag{5.18}$$
where '$<$' refers to the lexicographic ordering.

- If i, j, k, l are all distinct and D_{ij}, D_{kl} are not both odd, we have

$$D_{ij}D_{kl} = q^{-2}D_{kl}D_{ij}, \qquad 1 \le i < j < k < l \le 5,$$

$$D_{ij}D_{kl} = q^{-2}D_{kl}D_{ij} - (q^{-1} - q)D_{ik}D_{jl}, \quad 1 \le i < k < j < l \le 5,$$

$$D_{ij}D_{kl} = D_{kl}D_{ij}, \qquad 1 \le i < k < l < j \le 5. \tag{5.19}$$

As for the remaining relations relative to the pairs:

$$(D_{ij}, D_{55}), \qquad (D_{i5}, D_{j5}), \qquad (D_{i5}, D_{55}),$$

after some computations one gets (for $1 \le i < j \le 4$)

$$D_{ij}D_{55} = q^{-2}D_{55}D_{ij},$$

$$D_{i5}D_{j5} = -q^{-1}D_{j5}D_{i5} - (q^{-1} - q)D_{ij}D_{55} = -qD_{j5}D_{i5},$$

$$D_{i5}D_{55} = D_{55}D_{i5} = 0. \tag{5.20}$$

This concludes the discussion of the commutation relations. As for the quantum super Plücker relations, by direct calculation we have

$$D_{12}D_{34} - q^{-1}D_{13}D_{24} + q^{-2}D_{14}D_{23} = 0, \tag{5.21}$$

$$D_{ij}D_{k5} - q^{-1}D_{ik}D_{j5} + q^{-2}D_{i5}D_{jk} = 0, \qquad 1 \le i < j < k \le 4, \tag{5.22}$$

$$D_{i5}D_{j5} = qD_{ij}D_{55}, \qquad 1 \le i < j \le 4. \tag{5.23}$$

We are ready for the explicit presentation of $\mathbb{C}_q[\mathrm{Gr}^{\mathrm{ch}}]$ (again see Refs. [26, 27] for more details).

Proposition 5.5.2.

(1) The quantum Grassmannian superring is given in terms of generators and relations as

$$\mathbb{C}_q[\mathrm{Gr}^{\mathrm{ch}}] = \mathbb{C}_q\langle X_{ij}\rangle / I_{Gr}, \quad \textit{for} \quad 1 \le i < j \le 5 \quad \textit{or} \quad i = j = 5,$$

where I_{Gr} is the two-sided ideal generated by the commutations relations (5.18), (5.19), (5.20) and the quantum super Plücker relations (5.21), (5.22), (5.23), replacing D_{ij} by the indeterminates X_{ij}.

(2) *The quantum Grassmannian superring is the free ring over $\mathbb{C}_q$ generated by the monomials in the quantum determinants*

$$D_{i_1 j_1} \cdots D_{i_r j_r} \,,$$

where $(i_1, j_1), \ldots, (i_r, j_r)$ form the rows of a super standard tableau (for the definition see Sections 2.9, 4.7, 4.9).

Proof. (Sketch). The proof is the same as the proof of the corresponding classical result that one can find in [72] and that we have recalled in the proof of Proposition 5.2.4. We briefly sketch it, since its importance in our construction.

By definition, the generic monomials in the quantum determinants D_{ij} generate the ring $\mathbb{C}_q[\mathrm{Gr}^{\mathrm{ch}}]$ as $\mathbb{C}_q$-module. Using the commutation relations, we can write any monomial as a lexicographically ordered monomial. Then, using the quantum Plücker relations and the straightening algorithm [71], we can rewrite any lexicographically ordered monomial as a linear combination of super standard monomials, i.e. monomials in the quantum determinants whose indices form a super standard tableau (see Section 4.9).

Notice that there are two obstacles in applying the straightening algorithm to the quantum setting, but they are both easily overcome. The first is the presence of the coefficients q in the Plücker relations. This is not a problem, since q is invertible. The second is the non commutativity: when we are commuting two quantum determinants one may argue that the pairs of the forbidden kind, that is (i, j) and (k, l) with $i < k < l < j$ may arise. However a closer look to the commutation relations shows that this is never the case.

So both (1) and (2) will be done if we can show that the standard monomials in the D_{ij}'s are linearly independent. Assume there is a relation R among such monomials. Clearly $R \equiv 0 \bmod (q-1)$ since there are no relations among the standard classical monomials, hence $R = (q-1)R'$. Since the superalgebra $\mathbb{C}_q[\mathrm{SL}(m|n)]$ containing $\mathbb{C}_q[\mathrm{Gr}^{\mathrm{ch}}]$ is torsion free, we can divide by $(q-1)^r$ so that $R' \not\equiv 0 \bmod (q-1)$. This will yield a non trivial relation among the classical super monomials, hence the relation R we start with cannot exist (see Proposition 4.9.5 in Chapter 4). ■

The quantum Grassmannian that we have constructed is a deformation of the coordinate superalgebra of the supervariety $G(2|0, 4|1)$ and it admits a coaction of the quantum supergroup $\mathbb{C}_q[\mathrm{SL}(4|1)]$.

Proposition 5.5.3. $\mathbb{C}_q[\mathrm{Gr}^{\mathrm{ch}}]$ *is a quantum homogeneous superspace for the quantum supergroup* $\mathbb{C}_q[\mathrm{SL}(4|1)]$, *i.e., there is a coaction of* $\mathbb{C}_q[\mathrm{SL}(4|1)]$ *on* $\mathbb{C}_q[\mathrm{Gr}^{\mathrm{ch}}]$ *given via the restriction of the comultiplication on* $\mathbb{C}_q[\mathrm{SL}(4|1)]$ *as*

$$\Delta|_{\mathbb{C}_q[\mathrm{Gr}^{\mathrm{ch}}]} : \mathbb{C}_q[\mathrm{Gr}^{\mathrm{ch}}] \longrightarrow \mathbb{C}_q[\mathrm{SL}(4|1)] \otimes \mathbb{C}_q[\mathrm{Gr}^{\mathrm{ch}}].$$

Proof. We just have to check that the restriction is well defined, that is,

$$\Delta|_{\mathbb{C}_q[\mathrm{Gr}^{\mathrm{ch}}]}(\mathbb{C}_q[\mathrm{Gr}^{\mathrm{ch}}]) \subset \mathbb{C}_q[\mathrm{SL}(4|1)] \otimes \mathbb{C}_q[\mathrm{Gr}^{\mathrm{ch}}].$$

Then, the coaction property

$$\begin{array}{ccc} \mathbb{C}_q[\mathrm{Gr}^{\mathrm{ch}}] & \xrightarrow{\Delta|_{\mathbb{C}_q[\mathrm{Gr}^{\mathrm{ch}}]}} & \mathbb{C}_q[\mathrm{SL}(4|1)] \otimes \mathbb{C}_q[\mathrm{Gr}^{\mathrm{ch}}] \\ {\scriptstyle \Delta|_{\mathbb{C}_q[\mathrm{Gr}^{\mathrm{ch}}]}}\downarrow & & \downarrow{\scriptstyle \Delta\otimes\mathrm{id}} \\ \mathbb{C}_q[\mathrm{SL}(4|1)] \otimes \mathbb{C}_q[\mathrm{Gr}^{\mathrm{ch}}] & \xrightarrow[\mathrm{id}\otimes\Delta|_{\mathbb{C}_q[\mathrm{Gr}^{\mathrm{ch}}]}]{} & \mathbb{C}_q[\mathrm{SL}(4|1)] \otimes \mathbb{C}_q[\mathrm{SL}(4|1)] \otimes \mathbb{C}_q[\mathrm{Gr}^{\mathrm{ch}}] \end{array}$$

is guaranteed by the coassociativity of Δ.

So we need to verify that

$$\Delta(D_{ij}),\ \Delta(D_{i5}),\ \Delta(D_{55}) \in \mathbb{C}_q[\mathrm{SL}(4|1)] \otimes \mathbb{C}_q[\mathrm{Gr}^{\mathrm{ch}}].$$

Let us denote the generic 2×2 quantum minor (even or odd) as

$$D_{ij}^{kl} = g_{ik}g_{jl} - q^{-1}g_{il}g_{jk}, \tag{5.24}$$

so, in the previous notation $D_{ij} = D_{ij}^{12}$. A direct check shows that

$$\Delta(D_{ij}^{12}) = \sum_{1\le k<l\le 4} D_{ij}^{kl} \otimes D_{kl}^{12}. \tag{5.25}$$

This proves immediately that

$$\Delta(D_{ij}),\ \Delta(D_{i5}) \in \mathbb{C}_q[\mathrm{SL}(4|1)] \otimes \mathbb{C}_q[\mathrm{Gr}^{\mathrm{ch}}], \quad \text{for } 1 \le i < j \le 4.$$

The statement for $\Delta(D_{55})$ is a straightforward (long) check. ■

5.6 The quantum chiral Minkowski superspace

We now turn to the quantum deformation of the big cell inside $\mathrm{Gr}^{\mathrm{ch}}$: it will be our model for the *quantum chiral Minkowski superspace*.[5]

[5] The content of this section is again a summary of results appearing in [26, 27] where we send the reader for further details.

We start by discussing the classical setting, in order to write the coordinate superalgebras associated with the Poincaré supergroup and the Minkowski superspace. For all of the notation refer to Chapter 4, where we have described the classical super setting. We shall then proceed to obtain a quantum deformation of them.

Let $\mathbb{C}[\mathrm{SL}(4|1)]$ be the coordinate superalgebra of the algebraic supergroup $\mathrm{SL}(4|1)$, whose generators we write in supermatrix form (5.16)

$$\left(\begin{array}{cccc|c} g_{11} & g_{12} & g_{13} & g_{14} & g_{15} \\ g_{21} & g_{22} & g_{23} & g_{24} & g_{25} \\ g_{31} & g_{32} & g_{33} & g_{34} & g_{35} \\ g_{41} & g_{42} & g_{43} & g_{44} & g_{45} \\ \hline g_{51} & g_{52} & g_{53} & g_{54} & g_{55} \end{array}\right) .$$

We now consider the algebra

$$\mathbb{C}[F_l^{\mathrm{ch}}] := \mathbb{C}[\mathrm{SL}(4|1)]/\mathcal{I} ,$$

where $\mathcal{I}$ is the (two-sided) ideal generated by

$$g_{1j}, g_{2j}, \quad \text{for} \quad j = 3, 4 \quad \text{and} \quad g_{15}, g_{25} .$$

Remark 5.6.1. In functor of points notation, the supergroup F_l^{ch} is given as

$$F_l^{\mathrm{ch}}(T) = \left\{ \begin{pmatrix} a_{11} & a_{12} & 0 & 0 & 0 \\ a_{21} & a_{22} & 0 & 0 & 0 \\ a_{31} & a_{32} & a_{33} & a_{34} & \alpha_{35} \\ a_{41} & a_{42} & a_{43} & a_{44} & \alpha_{45} \\ \alpha_{51} & \alpha_{52} & \alpha_{53} & \alpha_{54} & a_{55} \end{pmatrix} \right\} \subset \mathrm{SL}(4|1)(T) , \tag{5.26}$$

where T is a complex affine supervariety, a_{ij}, α_{kl} even and odd, respectively, elements of $\mathbb{C}[T]$, the superalgebra of global sections on T (see Chapter 1, Section 1.7). We shall not use the machinery of the functor of points in this section, however it is important to keep the geometric point of view in mind when dealing with the quantum deformations. ■

$\mathbb{C}[F_l^{\mathrm{ch}}]$ is a Hopf superalgebra with comultiplication, counit and antipode naturally inherited from $\mathbb{C}[\mathrm{SL}(4|1)]$:

$$\Delta(g_{ij}) = \sum_k g_{ik} \otimes g_{kj} , \qquad \epsilon(g_{ij}) = \delta_{ij} ,$$

and for the antipode, see Appendix E.

Δ, ϵ, S are well defined on $\mathbb{C}[F_l^{\mathrm{ch}}]$. This comes immediately from two equivalent observations. In fact, $\mathcal{I}$ is a *Hopf ideal* of $\mathbb{C}[\mathrm{SL}(4|1)]$, that is, $\mathcal{I}$

is a superalgebra ideal, a coalgebra coideal and $S(\mathcal{I}) \subset \mathcal{I}$. This makes the quotient $\mathbb{C}[F_l^{\mathrm{ch}}] = \mathbb{C}[\mathrm{SL}(4|1)]/\mathcal{I}$ a Hopf superalgebra (see also Section 1.11). Equivalently, in the functor of points approach, we see immediately that F_l^{ch} is a subgroup of $\mathrm{SL}(4|1)$ hence the functor of points of F_l^{ch} is group valued (see Remark 5.6.1), so the comultiplication, counit and antipode inherited from $\mathrm{SL}(4|1)$ are well defined (see also Section 1.8). Notice furtherly that F_l^{ch} is isomorphic to F_u^{ch}, the upper parabolic subgroup.

Let us now make the following two different changes of variables in $\mathbb{C}[F_l^{\mathrm{ch}}]$ that shall turn out particularly useful for later considerations:

$$\begin{pmatrix} g_{11} & g_{12} & 0 & 0 & 0 \\ g_{21} & g_{22} & 0 & 0 & 0 \\ g_{31} & g_{32} & g_{33} & g_{34} & g_{35} \\ g_{41} & g_{42} & g_{43} & g_{44} & g_{45} \\ g_{51} & g_{52} & g_{53} & g_{54} & g_{55} \end{pmatrix} = \begin{pmatrix} x & 0 & 0 \\ tx & y & y\eta \\ \tilde{\tau}x & d\xi & d \end{pmatrix} = \begin{pmatrix} x & 0 & 0 \\ tx & y & y\eta \\ d\tau & d\xi & d \end{pmatrix}, \tag{5.27}$$

where x, y and t are 2×2 matrices, η a 2×1 matrix, τ and ξ 1×2 matrices and d a 1×1 matrix. We remind the reader that such equalities are not to be understood on matrices, but on the generators of the corresponding superalgebras, using the notation that we have introduced following the Definition 5.4.1.

Notice that the only difference between the two sets of variables is that we replace τ with $\tilde{\tau}$ and we have

$$d\tau = \tilde{\tau}x \,. \tag{5.28}$$

The next proposition tells us that both of these are sets of generators for $\mathbb{C}[F_l^{\mathrm{ch}}]$.

Proposition 5.6.2. *Let the notation be as above. The Hopf superalgebra $\mathbb{C}[F_l^{\mathrm{ch}}]$ is generated by each of the following sets of variables:*

- *x, y, t, $\tilde{\tau}$, ξ, η and d,*
- *x, y, t, τ, ξ, η and d,*

defined in terms of the generators g_{ij} *as*

$$x = \begin{pmatrix} g_{11} & g_{12} \\ g_{21} & g_{22} \end{pmatrix}, \qquad y = \begin{pmatrix} g_{33} & g_{34} \\ g_{43} & g_{44} \end{pmatrix},$$

$$t = \begin{pmatrix} -d_{23}d_{12}^{-1} & d_{13}d_{12}^{-1} \\ -d_{24}d_{12}^{-1} & d_{14}d_{12}^{-1} \end{pmatrix}, \qquad d = g_{55},$$

$$\tilde{\tau} = (-d_{25}d_{12}^{-1}, d_{15}d_{12}^{-1}), \qquad \tau = (g_{55}^{-1}g_{51}, g_{55}^{-1}g_{52}),$$

$$\eta = \begin{pmatrix} {d_{34}^{34}}^{-1} g_{35} \\ {d_{34}^{34}}^{-1} g_{45} \end{pmatrix}, \qquad \xi = \left(g_{55}^{-1}g_{53}, g_{55}^{-1}g_{54}\right), \tag{5.29}$$

where, for $1 \leq i < j \leq 4$ *we have defined*

$d_{ij} = g_{i1}g_{j2} - g_{j1}g_{i2}, \qquad d_{i5} = g_{i1}g_{52} - g_{51}g_{i2}, \qquad d_{34}^{34} = g_{33}g_{44} - g_{34}g_{43}\,.$
(Convention: we omit the superindex 12 *in* d_{ij} *whenever there is no danger of confusion.)*

Proof. One must show that any generator g_{ij} of $\mathbb{C}[F_l^{\text{ch}}]$ is expressible in terms of x, y, t, $\tilde{\tau}$, ξ, η and d (or equivalently in terms of x, y, t, τ, ξ, η and d). The only generators to be checked are g_{31}, g_{32}, g_{41}, g_{42}, g_{51}, g_{52}, g_{53}, g_{54}, g_{35}, g_{45}. For the odd generators, g_{51}, g_{52}, g_{53}, g_{54}, g_{35} and g_{45} the calculation is immediate. As for the others we have

$$tx = \begin{pmatrix} g_{31} & g_{32} \\ g_{41} & g_{42} \end{pmatrix}$$

hence by right multiplication by x^{-1} we are done. ∎

We can consider the ideal (ξ) in $\mathbb{C}[F_l^{\text{ch}}]$. It is not difficult to see that it is a Hopf ideal. The quotient is the Hopf algebra

$$\mathbb{C}[F_l^{\text{ch}}]/(\xi) = \mathbb{C}[F^c]\,,$$

where F^c is the complex Poincaré group times dilations (4.61).

The chiral Minkowski superspace M^{ch} sits inside Gr^{ch} as its big cell (see Section 4.10). Its functor of points, described in Chapter 4, (4.51), is given, for T a supervariety, by

$$T \longrightarrow \mathrm{M}^{\text{ch}}(T) = \text{span}_{\mathcal{O}(T)} \left\{ \begin{pmatrix} 1 \\ 0 \\ t_{31} \\ t_{41} \\ \theta_{51} \end{pmatrix}, \begin{pmatrix} 0 \\ 1 \\ t_{32} \\ t_{42} \\ \theta_{52} \end{pmatrix} \right\} \subset \mathcal{O}(T)^{4|1}\,. \tag{5.30}$$

Hence, one can readily see that it is an affine superspace of dimension $4|2$. As it happens for the ordinary (i.e. non super) setting, its coordinate superalgebra is obtained as the projective localization of the coordinate superalgebra $\mathbb{C}[\mathrm{Gr}^{\mathrm{ch}}]$ of the chiral Grassmannian supervariety at the even element d_{12}. In fact, the invertibility of such determinant singles out, in the functor of points notation, the subspaces in $\mathrm{Gr}^{\mathrm{ch}}(T)$ belonging to $\mathrm{M}^{\mathrm{ch}}(T)$ and allows us to write such elements in the unique form (5.30), which was discussed in detail in Sections 4.7, 4.10.

The procedure of the projective localization in the supergeometric context offers no difficulty, since we are localizing at an even element, hence the procedure is carried along exactly in the same way as for the ordinary setting: we take the $\mathbb{Z}$-degree zero elements in the ordinary localization $\mathbb{C}[\mathrm{Gr}^{\mathrm{ch}}][d_{12}^{-1}]$ (see Sections 5.2 and 1.6), where this construction is briefly reviewed). We can then state the following proposition.

Proposition 5.6.3. *The coordinate superalgebra* $\mathbb{C}[\mathrm{M}^{\mathrm{ch}}]$ *of the chiral Minkowski superspace is the projective localization of* $\mathbb{C}[\mathrm{Gr}^{\mathrm{ch}}]$ *with respect to* d_{12}*. Moreover, it is the* $\mathbb{Z}$*-homogeneous subring of* $\mathbb{C}[\mathrm{Gr}^{\mathrm{ch}}][d_{12}^{-1}]$ *generated by the elements:*

$$d_{13}d_{12}^{-1}, \quad d_{23}d_{12}^{-1}, \quad d_{14}d_{12}^{-1}, \quad d_{24}d_{12}^{-1}, \quad d_{15}d_{12}^{-1}, \quad d_{25}d_{12}^{-1}\,.$$

Furthermore $\mathbb{C}[\mathrm{M}^{\mathrm{ch}}] = \mathbb{C}[m_{ij}, \mu_j]$, $i, j = 1, 2$. *In other words* $\mathbb{C}[\mathrm{M}^{\mathrm{ch}}]$ *is a polynomial superalgebra in 4 even and 2 odd inderminates.*

Proof. (Sketch). Our previous discussion establishes the realization of $\mathbb{C}[\mathrm{M}^{\mathrm{ch}}]$ as the projective localization of the $\mathbb{C}[\mathrm{Gr}^{\mathrm{ch}}]$ superring at the element d_{12}. Hence $\mathbb{C}[\mathrm{M}^{\mathrm{ch}}]$ is generated by:

$$d_{13}d_{12}^{-1}, \quad d_{23}d_{12}^{-1}, \quad d_{14}d_{12}^{-1}, \quad d_{24}d_{12}^{-1}, \quad d_{34}d_{12}^{-1},$$

$$d_{15}d_{12}^{-1}, \quad d_{25}d_{12}^{-1}, \quad d_{35}d_{12}^{-1}, \quad d_{45}d_{12}^{-1}, \quad d_{55}d_{12}^{-1}.$$

However the Plücker relations allow us to express $d_{34}d_{12}^{-1}$, $d_{35}d_{12}^{-1}$, $d_{45}d_{12}^{-1}$, $d_{55}d_{12}^{-1}$ in terms of the other generators. ∎

We now want to realize the superalgebra representing the chiral Minkowski superspace as a subalgebra of $\mathbb{C}[F_l^{\mathrm{ch}}]$, in perfect analogy to what we did in Section 5.2 for the ordinary setting.

Proposition 5.6.4. *The subalgebra* $\mathbb{C}[t, \tilde{\tau}]$ *of* $\mathbb{C}[F_l^{\mathrm{ch}}]$ *generated by the elements in* $(t, \tilde{\tau})$ *coincides with* $\mathbb{C}[\mathrm{M}^{\mathrm{ch}}]$*. There is a well defined natural coaction* $\tilde{\Delta}$ *of* $\mathbb{C}[F_l^{\mathrm{ch}}]$ *on* $\mathbb{C}[t, \tilde{\tau}]$ *induced by the comultiplication in* $\mathbb{C}[F_l^{\mathrm{ch}}]$*, which* *corresponds to the natural action of the supergroup* F_l^{ch} *on* M^{ch}*. Explicitly it takes the form*

$$\tilde{\Delta} : \mathbb{C}[\mathrm{M}^{\text{ch}}] \xrightarrow{\tilde{\Delta}} \mathbb{C}[F_l^{\text{ch}}] \otimes \mathbb{C}[\mathrm{M}^{\text{ch}}],$$

with

$$\tilde{\Delta}(t) = t \otimes 1 + yS(x) \otimes t + y\eta S(x) \otimes \tilde{\tau},$$

$$\tilde{\Delta}(\tilde{\tau}) = (d \otimes 1)(\tau \otimes 1 + \xi \otimes t + 1 \otimes \tilde{\tau})(S(x) \otimes 1)\,,$$

(compare this expression with 5.2.13 (3), of which it is a generalization).

Proof. Let us write the comultiplication in $\mathbb{C}[F_l^{\text{ch}}]$ using the generators x, y, t, $\tilde{\tau}$, ξ, η and d.

$$\Delta \begin{pmatrix} x & 0 & 0 \\ tx & y & y\eta \\ \tilde{\tau}x & d\xi & d \end{pmatrix} = \begin{pmatrix} x & 0 & 0 \\ tx & y & y\eta \\ \tilde{\tau}x & d\xi & d \end{pmatrix} \otimes \begin{pmatrix} x & 0 & 0 \\ tx & y & y\eta \\ \tilde{\tau}x & d\xi & d \end{pmatrix}.$$

That is

$$\begin{aligned} \Delta(tx) &= tx \otimes x + y \otimes tx + y\eta \otimes \tilde{\tau}x\,, \\ \Delta(\tilde{\tau}x) &= \tilde{\tau}x \otimes x + d\xi \otimes tx + d \otimes \tilde{\tau}x\,, \end{aligned} \tag{5.31}$$

from which the first statement is clear.

Now let us recall the formula in Section 4.10 giving the action of F_l^{ch} on M^{ch} in the functor of points notation:

$$\begin{array}{ccc} F_l^{\text{ch}}(T) \times \mathrm{M}^{\text{ch}}(T) & \longrightarrow & \mathrm{M}^{\text{ch}}(T) \\ \left(\begin{pmatrix} x & 0 & 0 \\ tx & y & y\eta \\ d\tau & d\xi & d \end{pmatrix}, \begin{pmatrix} I \\ A \\ \alpha \end{pmatrix} \right) & \longrightarrow & \begin{pmatrix} I \\ y(A + \eta\alpha)x^{-1} + t \\ d(\alpha + \tau + \xi A)x^{-1} \end{pmatrix}, \end{array} \tag{5.32}$$

where in this case x, y, η, d, ξ, A, α are matrices with elements in $\mathcal{O}(T)$ for a supervariety T. A comparison between (5.31) and (5.32) gives immediately our second statement, taking into account the duality between the coordinate superalgebra and the functor of points approach. ∎

We now turn to the quantum setting. In order to keep our notation minimal, we use the same letters as in the classical case to denote the generators of the quantum big cell and the quantum supergroups. We shall, however, use capital letters to denote the quantum determinants, to be consistent with our previous notation.

Let $\mathbb{C}_q[F_l^{\mathrm{ch}}]$ be the superalgebra:

$$\mathbb{C}_q[F_l^{\mathrm{ch}}] := \mathbb{C}_q[\mathrm{SL}(4|1)]/\mathcal{I}_q\,,$$

where $\mathcal{I}_q$ is the (two-sided) ideal in $\mathbb{C}_q[\mathrm{SL}(4|1)]$ generated by

$$g_{1j}, g_{2j}, \quad \text{for} \quad j = 3, 4 \quad \text{and} \quad g_{15}, g_{25}\,. \tag{5.33}$$

As one can readily check $\mathcal{I}_q$ is an Hopf ideal and, consequently, $\mathbb{C}_q[F_l^{\mathrm{ch}}]$ is a Hopf superalgebra with comultiplication and counit naturally inherited from $\mathbb{C}_q[\mathrm{SL}(4|1)]$.

As in the classical case, it is convenient to change coordinates as in formula (5.27). This time, however, one has to pay attention to the order in which we take the variables. The new coordinates (that is, generators) for $\mathbb{C}_q[F_l^{\mathrm{ch}}]$ are, explicitly:

$$x = \begin{pmatrix} g_{11} & g_{12} \\ g_{21} & g_{22} \end{pmatrix}, \qquad t = \begin{pmatrix} -q^{-1}D_{23}D_{12}^{-1} & D_{13}D_{12}^{-1} \\ -q^{-1}D_{24}D_{12}^{-1} & D_{14}D_{12}^{-1} \end{pmatrix},$$

$$y = \begin{pmatrix} g_{33} & g_{34} \\ g_{43} & g_{44} \end{pmatrix}, \qquad d = g_{55},$$

$$\tilde{\tau} = \begin{pmatrix} -q^{-1}D_{25}D_{12}^{-1} & D_{15}D_{12}^{-1} \end{pmatrix}, \qquad \xi = \begin{pmatrix} g_{55}^{-1}\gamma_{53} & g_{55}^{-1}\gamma_{54} \end{pmatrix},$$

$$\eta = y^{-1}\begin{pmatrix} \gamma_{35} \\ \gamma_{45} \end{pmatrix} = (D_{34}^{34})^{-1}\begin{pmatrix} g_{44} & -q^{-1}g_{34} \\ -qg_{43} & g_{33} \end{pmatrix} = \begin{pmatrix} -q^{-1}{D_{34}^{34}}^{-1}D_{34}^{45} \\ {D_{34}^{34}}^{-1}D_{34}^{35} \end{pmatrix}.$$

Proposition 5.6.5. *The Hopf superalgebra $\mathbb{C}_q[F_l^{\mathrm{ch}}]$ is equivalently generated by the elements x, y, t, d, η, ξ and $\tilde{\tau}$.*

Proof. The proof amounts to a tedious calculation similar to the one in Section 5.2 for the ordinary quantum setting and to the one in Proposition 5.6.2 for the classical super setting. ∎

Obviously, $\mathbb{C}_q[F_l^{\mathrm{ch}}]$ is a deformation of $\mathbb{C}[F_l^{\mathrm{ch}}]$. As in the classical case, taking the quotient by the Hopf ideal (ξ), one obtains a superalgebra that is a deformation of the algebra of the Poincaré supergroup times dilations F^c (see (4.61)):

$$\left(\mathbb{C}_q[F_l^{\mathrm{ch}}]\,/\,(\xi)\right)/(q-1) \cong \mathbb{C}[F^c].$$

One can then call this quotient the *quantum Poincaré supergroup times dilations* and denote it as $\mathbb{C}_q[F^c]$. This algebra is then generated by the images in the quotient of x, y, t, d, η and $\tilde{\tau}$. Expressing the generators in a matrix form, one has

$$\begin{pmatrix} x & 0 & 0 \\ tx & y & y\eta \\ \tilde{\tau}x & 0 & d \end{pmatrix}.$$

We now want to give a presentation of $\mathbb{C}_q[F^c]$. We have that

$$\mathbb{C}_q[F^c] = \mathbb{C}_q[F_l^{\mathrm{ch}}]\,/\,(\xi) = \mathbb{C}_q\langle x, y, t, d, \eta, \tilde{\tau}\rangle / I_{F^c,q}\,,$$

where $I_{F^c,q}$ is the ideal generated by the relations that we are going to describe schematically downbelow.

- The indeterminates x and y behave respectively as quantum (even) matrices, that is, their entries are subject to the Manin relations. In other words, we have for x

$$x_{11}x_{12} = q^{-1}x_{12}x_{11}, \quad x_{11}x_{21} = q^{-1}x_{21}x_{11}, \quad x_{21}x_{22} = q^{-1}x_{22}x_{21},$$

$$x_{12}x_{22} = q^{-1}x_{22}x_{12}, \quad x_{12}x_{21} = x_{21}x_{12},$$

$$x_{11}x_{22} - x_{22}x_{11} = (q^{-1} - q)x_{12}x_{21}\,.$$

- One has similar relations for y, with q is replaced with q^{-1}:

$$y_{11}y_{12} = qy_{12}y_{11}, \quad y_{11}y_{21} = qy_{21}y_{11}, \quad y_{21}y_{22} = qy_{22}y_{21},$$

$$y_{12}y_{22} = qy_{22}y_{12}, \quad y_{12}y_{21} = y_{21}y_{12},$$

$$y_{11}y_{22} - y_{22}y_{11} = (q - q^{-1})y_{12}y_{21}\,.$$

- Moreover, the entries in x and y commute with each other:

$$x_{ij}y_{kl} = y_{kl}x_{ij}, \qquad \forall i, j, k, l\,.$$

- The entries in x and t, $\tilde{\tau}$ commute in the following way. Let $i = 1, 2$ and $j = 3, 4$. Then

$$x_{1i}t_{j1} = q^{-1}t_{j1}x_{1i}, \qquad x_{2i}t_{j1} = t_{j1}x_{2i},$$

$$x_{1i}\tilde{\tau}_{51} = q^{-1}\tilde{\tau}_{51}x_{1i}, \qquad x_{2i}\tilde{\tau}_{51} = \tilde{\tau}_{51}x_{2i},$$

$$x_{1i}\tilde{\tau}_{52} = \tilde{\tau}_{52}x_{1i}, \qquad x_{2i}\tilde{\tau}_{52} = q^{-1}\tilde{\tau}_{52}x_{2i}\,.$$

- x commutes with η and d.
- y, t and $\tilde{\tau}$ satisfy similar relations to the ones satified by x, t and τ, respectively, with the rows exchanged with the columns.
- y and η commute following the rules of quantum super matrices, very much the same calculation and relations expressed in Proposition 5.6.7.
- y and d commute.
- The commutation relations among t and $\tilde{\tau}$ are expressed in Proposition 5.6.7.
- t and η commute.
- t, τ and d satisfy the following relations

$$t_{ij}d = dt_{ij} - (q^{-1} - q)\eta_{i5}\tilde{\tau}_{5i},$$
$$\tilde{\tau}_{5j}d = d\tilde{\tau}_{5j}\,.$$

- $\tilde{\tau}$ and η commute with each other.
- Finally,

$$\eta_{j5}d = q^{-1}d\eta_{j5}\,.$$

We have provided a full description of the ideal $I_{F^c,q}$, hence a presentation of the complex quantum Poincaré supergroup $\mathbb{C}_q[F^c]$.

In analogy with the classical (non quantum) super setting, we give the following definition.

Definition 5.6.6. We define the *quantum chiral Minkowski space* $\mathbb{C}_q[\mathrm{M}^{\mathrm{ch}}]$ as the subring of $\mathbb{C}_q[F_l^{\mathrm{ch}}]$ generated by t and $\tilde{\tau}$. ■

Notice that the superalgebra $\mathbb{C}_q[\mathrm{M}^{\mathrm{ch}}]$ is generated inside $\mathbb{C}_q[F_l^{\mathrm{ch}}]$ by the elements:

$$t = \begin{pmatrix} -q^{-1}D_{23}D_{12}^{-1} & D_{13}D_{12}^{-1} \\ -q^{-1}D_{24}D_{12}^{-1} & D_{14}D_{12}^{-1} \end{pmatrix} \qquad \tilde{\tau} = \begin{pmatrix} -q^{-1}D_{25}D_{12}^{-1} & D_{15}D_{12}^{-1} \end{pmatrix}.$$

Comparing with the ordinary counterpart (see Proposition 5.6.3) we see that $\mathbb{C}_q[\mathrm{M}^{\mathrm{ch}}]$ can be interpreted as a quantum projective super localization, in analogy with what we have done for the ordinary quantum Minkowski space (see Definition 5.2.10). It is indeed surprising that, as in the classical case (see Proposition 5.2.11), $\mathbb{C}_q[\mathrm{M}^{\mathrm{ch}}]$ is a quantum super matrix bialgebra. This is a consequence of Proposition 5.6.7 below.

We have then that $\mathbb{C}_q[\mathrm{M}^{\mathrm{ch}}]$ defined above is a quantization of the big cell, that is, $\mathbb{C}_q[\mathrm{M}^{\mathrm{ch}}]/(q-1)$ is the coordinate ring of the big cell inside the supervariety $\mathrm{Gr}^{\mathrm{ch}}$.

We compute now the quantum commutation relations among the generators of the quantum big cell $\mathbb{C}_q[\mathrm{M}^{\mathrm{ch}}]$, which is our chiral Minkowski superspace, and then see that the quantum big cell admits a well defined coaction of the quantum supergroup $\mathbb{C}_q[F_l^{\mathrm{ch}}]$.

Proposition 5.6.7. *The quantum big cell superring* $\mathbb{C}_q[\mathrm{M}^{\mathrm{ch}}]$ *has the following presentation:*

$$\mathbb{C}_q[\mathrm{M}^{\mathrm{ch}}] = \mathbb{C}_q\langle t_{ij}, \tilde{\tau}_{5j}\rangle \,/\, I_U, \qquad 3 \le i \le 4,\ j = 1, 2\,,$$

where I_U *is the ideal generated by the relations:*

$$t_{i1}t_{i2} = q\,t_{i2}t_{i1}, \qquad t_{3j}t_{4j} = q^{-1}\,t_{4j}t_{3j}, \qquad 1 \le j \le 2, \quad 3 \le i \le 4\,,$$

$$t_{31}t_{42} = t_{42}t_{31}, \qquad t_{32}t_{41} = t_{41}t_{32} + (q^{-1} - q)t_{42}t_{31},$$

$$\tilde{\tau}_{51}\tilde{\tau}_{52} = -q^{-1}\tilde{\tau}_{52}\tilde{\tau}_{51}, \qquad t_{ij}\tilde{\tau}_{5j} = q^{-1}\tilde{\tau}_{5j}t_{ij}, \qquad 1 \le j \le 2\,,$$

$$t_{i1}\tilde{\tau}_{52} = \tilde{\tau}_{52}t_{i1}, \qquad t_{i2}\tilde{\tau}_{51} = \tilde{\tau}_{51}t_{i2} + (q^{-1} - q)t_{i1}\tilde{\tau}_{52}\,.$$

Proof. Direct calculation. ■

As in the classical setting, we have the following proposition.

Proposition 5.6.8. *The quantum chiral Minkowski superspace* $\mathbb{C}_q[\mathrm{M}^{\mathrm{ch}}]$ *admits a coaction of* $\mathbb{C}_q[F_l^{\mathrm{ch}}]$ *obtained by restricting suitably the comultiplication in* $\mathbb{C}_q[F_l^{\mathrm{ch}}]$. *In other words we have a well defined morphism:*

$$\tilde{\Delta} : \mathbb{C}_q[\mathrm{M}^{\mathrm{ch}}] \longrightarrow \mathbb{C}_q[F_l^{\mathrm{ch}}] \otimes \mathbb{C}_q[\mathrm{M}^{\mathrm{ch}}]\,,$$

satisfying the coaction properties and given explicitly by (see Proposition 5.6.4),

$$\tilde{\Delta}(t) = t \otimes 1 + yS(x) \otimes t + y\eta S(x) \otimes \tilde{\tau},$$

$$\tilde{\Delta}(\tilde{\tau}) = (d \otimes 1)(\tau \otimes 1 + \xi \otimes t + 1 \otimes \tilde{\tau})(S(x) \otimes 1)$$

by choosing as before generators x, y, t, d, τ, η, ξ *for* $\mathbb{C}_q[F_l^{\mathrm{ch}}]$ *and* t, $\tilde{\tau}$ *for* $\mathbb{C}_q[\mathrm{M}^{\mathrm{ch}}]$ *with* $d\tau = \tilde{\tau}x$.

Furthermore, this coaction gives a well defined coaction for the quantization $\mathbb{C}_q[F^c]$.

Proof. Direct calculation. ∎

5.7 References

Our main references for the treatment of the quantum general and special linear groups and quantum spaces are the foundational works by Manin Refs. [108, 109], while for the corresponding supergeometric objects the generalization was done by Manin in Ref. [110]. In these works, however, Manin gives no explicit construction for the antipode for the general and special linear supergroups, though its existence is proven. For the construction of the antipode and its main properties in the ordinary case see Ref. [119], while for its super counterpart see Refs. [125, 163], where the theory is developed with the equivalent approach of the R-matrix.

As for the quantization of the ordinary Grassmannian and the Minkowski space, together with the treatment of real forms, see Ref. [54]. The super counterpart, with its physical interpretation of the Grassmannian as chiral superspace, is treated in detail in Ref. [26].

Appendix A

Categories

In this appendix we give a brief summary of definitions and main properties concerning categories. For more details we invite the reader to consult one of the many available comprehensive treatments, like for example Ref. [99].

Let us start with the definition of category. A *category* $\mathcal{C}$ consists of a collection of objects, $\mathrm{Ob}(\mathcal{C})$, and sets of *morphisms* between objects, denoted, for any A and B in $\mathrm{Ob}(\mathcal{C})$, as $\mathrm{Hom}_{\mathcal{C}}(A, B)$.

We furtherly require that for all $A, B, C \in \mathcal{C}$, there exists an association

$$\mathrm{Hom}_{\mathcal{C}}(B, C) \times \mathrm{Hom}_{\mathcal{C}}(A, B) \longrightarrow \mathrm{Hom}_{\mathcal{C}}(A, C)$$

called the 'composition law' $(f, g) \to f \circ g$ which satisfies the properties:

(1) the law '$\circ$' is associative;
(2) for all $A, B \in \mathrm{Ob}(\mathcal{C})$, there exists $\mathrm{id}_A \in \mathrm{Hom}_{\mathcal{C}}(A, A)$ so that $f \circ \mathrm{id}_A = f$ for all $f \in \mathrm{Hom}_{\mathcal{C}}(A, B)$ and $\mathrm{id}_A \circ g = g$ for all $g \in \mathrm{Hom}_{\mathcal{C}}(B, A)$;
(3) $\mathrm{Hom}_{\mathcal{C}}(A, B)$ and $\mathrm{Hom}_{\mathcal{C}}(A', B')$ are disjoint unless $A = A'$, $B = B'$ in which case they are equal.

For example, the category of groups, denoted with (grps), consists of the collection of the objects, the groups, and the set of morphisms between any two objects consists of all group morphisms between the two given groups. The category of sets, denoted with (sets), has for objects the sets and for morphisms the functions between sets. We may consider the same set of objects with different morphisms. For example, we may consider another category of sets taking as morphisms the inclusions between sets.

Notice that, given two objects, the set of morphisms between them may be empty. For example, in the last example mentioned above, the set of morphisms between two sets A and B, with $A \not\subset B$, is empty.

A morphism $f \in \mathrm{Hom}_{\mathcal{C}}(A, B)$ is invertible if there exists another morphism $g \in \mathrm{Hom}_{\mathcal{C}}(B, A)$ such that $f \circ g$ and $g \circ f$ are the identities respectively in $\mathrm{Hom}_{\mathcal{C}}(B, B)$ and $\mathrm{Hom}_{\mathcal{C}}(A, A)$. If f is invertible, we say that f is an *isomorphism.*

Once the category is understood, it is conventional to write $A \in \mathcal{C}$ instead of $A \in \mathrm{Ob}(\mathcal{C})$ for objects. We may also suppress the '$\mathcal{C}$' from $\mathrm{Hom}_{\mathcal{C}}$ and just write Hom whenever there is no danger of confusion.

Essentially, a category is a collection of objects which share some basic structure, along with maps between objects which preserve that structure.

A category $\mathcal{C}'$ is a *subcategory* of the category $\mathcal{C}$ if $\mathrm{Ob}(\mathcal{C}') \subset \mathrm{Ob}(\mathcal{C})$ and if for all $A, B \in \mathcal{C}'$, $\mathrm{Hom}_{\mathcal{C}'}(A, B) \subset \mathrm{Hom}_{\mathcal{C}}(A, B)$ so that the composition law '$\circ$' on $\mathcal{C}'$ is induced by that on $\mathcal{C}$.

For example, the category (agrps) of abelian groups and group morphisms is a subcategory of the category of groups (grps).

We now turn to two of the most important concepts in category theory: *functors* and *natural transformations.*

Let $\mathcal{C}_1$ and $\mathcal{C}_2$ be two categories. A *covariant (resp. contravariant) functor* $F : \mathcal{C}_1 \longrightarrow \mathcal{C}_2$ consists of

(1) a map $F : \mathrm{Ob}(\mathcal{C}_1) \longrightarrow \mathrm{Ob}(\mathcal{C}_2)$ and
(2) a map $F_{A,B} : \mathrm{Hom}_{\mathcal{C}_1}(A, B) \longrightarrow \mathrm{Hom}_{\mathcal{C}_2}(F(A), F(B))$ (resp. $F_{A,B} : \mathrm{Hom}_{\mathcal{C}_1}(A, B) \longrightarrow \mathrm{Hom}_{\mathcal{C}_2}(F(B), F(A))$) so that

- $F(\mathrm{id}_A) = \mathrm{id}_{F(A)}$ and
- $F(f \circ g) = F(f) \circ F(g)$, (respectively $F(f \circ g) = F(g) \circ F(f)$),

for all $A, B \in \mathrm{Ob}(\mathcal{C}_1)$.

It is customary to denote $F_{A,B}$ simply as F whenever there is no danger of confusion.

When we say 'functor' we mean covariant functor. A contravariant functor $F : \mathcal{C}_1 \to \mathcal{C}_2$ is the same as a covariant functor from $\mathcal{C}_1^o \longrightarrow \mathcal{C}_2$ where $\mathcal{C}_1^o$ denotes the *opposite* category i. e. the category where all morphism arrows are reversed.

Let F_1, F_2 be two functors from $\mathcal{C}_1$ to $\mathcal{C}_2$. We say that there is a *natural transformation* of functors $\varphi : F_1 \longrightarrow F_2$ if for all $A \in \mathcal{C}_1$ there is a set of morphisms $\varphi_A : F_1(A) \longrightarrow F_2(A)$ so that for any $f \in \mathrm{Hom}_{\mathcal{C}_1}(A, B)$

$(B \in \mathcal{C}_1)$, the following diagram commutes:

$$\begin{array}{ccc} F_1(A) & \xrightarrow{\varphi_A} & F_2(A) \\ \downarrow{\scriptstyle F_1(f)} & & \downarrow{\scriptstyle F_2(f)} \\ F_1(B) & \xrightarrow{\varphi_B} & F_2(B). \end{array} \tag{A.1}$$

We say that the family of morphisms φ_A is *functorial* in A.

We say that two functors $F, G : \mathcal{C}_1 \longrightarrow \mathcal{C}_2$ are *isomorphic* $(F \cong G)$ if there exist two natural transformations $\phi : F \longrightarrow G$ and $\psi : G \longrightarrow F$ such that $\phi \circ \psi = \text{id}$ and $\psi \circ \phi = \text{id}$. In other words $F(A)$ is isomorphic to $G(A)$ via the two maps ψ_A and ϕ_A for all objects A in a functorial way i.e. in such a way the diagram (A.1) holds.

The functors from $\mathcal{C}_1$ to $\mathcal{C}_2$ for any two given categories together with their natural transformations form a category.

The notion of equivalence of categories is important since it allows to identify two categories which are apparently different.

We say that two categories $\mathcal{C}_1$ and $\mathcal{C}_2$ are *equivalent* if there exists two functors $F : \mathcal{C}_1 \longrightarrow \mathcal{C}_2$ and $G : \mathcal{C}_2 \longrightarrow \mathcal{C}_1$ such that $FG \cong \text{id}_{\mathcal{C}_2}$, $GF \cong \text{id}_{\mathcal{C}_1}$, where $\text{id}_{\mathcal{C}}$ denotes the identity functor of a given category, defined in the obvious way.

If F is a functor from the category $\mathcal{C}_1$ to the category $\mathcal{C}_2$ it induces, for any two objects A, $B \in \mathcal{C}_1$, a function:

$$F_{A,B} : \text{Hom}_{\mathcal{C}_1}(A, B) \longrightarrow \text{Hom}_{\mathcal{C}_2}(F(A), F(B)).$$

Let F be a functor. We say that F is *faithful* if $F_{A,B}$ is injective; we say F is *full* if $F_{A,B}$ is surjective and we say that F is *fully faithful* if $F_{A,B}$ is bijective.

Next we want to formally define what it means for a functor to be *representable*. Let us first define the representation functors.

Let $\mathcal{C}$ be a category and A a fixed object in $\mathcal{C}$. We define the two *representation functors* Hom^A, Hom_A as

$$\text{Hom}^A : \mathcal{C}^o \longrightarrow (\text{sets}), \qquad B \mapsto \text{Hom}_{\mathcal{C}}(B, A),$$

$$\text{Hom}_A : \mathcal{C} \longrightarrow (\text{sets}), \qquad B \mapsto \text{Hom}_{\mathcal{C}}(A, B),$$

where (sets) denotes the category of sets. On the arrow $f \in \text{Hom}(B, C)$ we have:

$$\text{Hom}^A(f)\phi = \phi \circ f, \quad \phi \in \text{Hom}^A(B) \quad \text{Hom}_A(f)\psi = f \circ \psi, \quad \psi \in \text{Hom}_A(B).$$

Let F be a functor from the category $\mathcal{C}$ to the category of sets. We say that F is *representable by* $X \in \mathcal{C}$ if $F \cong \mathrm{Hom}_X$ or $F \cong \mathrm{Hom}^X$.

We end our brief exposition of categories with the construction of the fibered product.

Given functors A, B, C, from a category $\mathcal{C}$ to the category of (sets), and given natural transformations $f : A \longrightarrow C$, $g : B \longrightarrow C$ the *fibered product* $A \times_C B$ is the universal object making the following diagram commute:

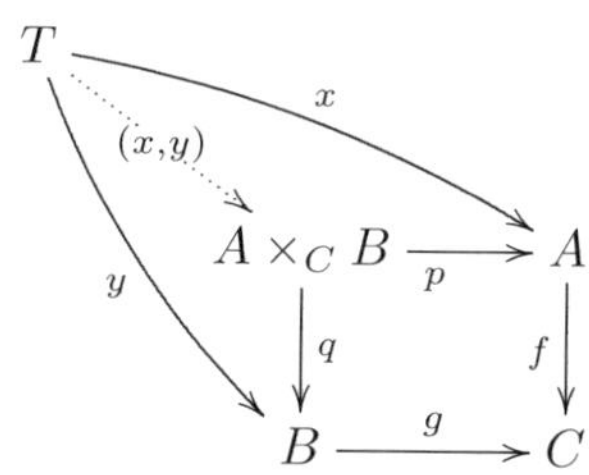

One can see that

$$(A \times_C B)(R) = A(R) \times_{C(R)} B(R) = \{(a,b) \in A(R) \times B(R) \mid f(a) = g(b)\}.$$

If g is injective, that is, if $g_R : B(R) \subset C(R)$, we have that $(A \times_C B)(R) = f^{-1}(B(R))$.

Appendix B

Representability Criterion

In this appendix we want to discuss a *representability criterion*, which allows us to single out among all of the functors from the category of supermanifolds (both, differentiable and complex analytic) to the category of sets those that are representable, i.e., those that are the functors of points of a supermanifold. The criterion we are going to describe holds also for the category of supervarieties, making the appropriate changes.

As usual, we denote with (smflds) both, the categories of real differentiable and complex analytic supermanifolds. Our treatment of representability works in the same way in the two categories. We use k to denote the real or the complex field, accordingly.

We start by restating, in our context, the definition of representable functor that we introduced very briefly in Appendix A.

Definition B.0.1. We say that a functor $F : (\text{smflds})^o \longrightarrow (\text{sets})$ is *representable* if it is naturally isomorphic to $\text{Hom}(\ ,X)$ for a suitable supermanifold X. In other words, F is representable if and only if F is naturally isomorphic to the functor of points of X. We also say that X is the supermanifold *representing* F. ■

In this appendix we shall make a notational distinction between a supermanifold X and its functor of points h_X.

If F is representable, the supermanifold X representing it and the natural isomorphism $F \cong h_X$ are unique. For the sake of clarity, we shall not make an explicit mention of such natural isomorphism in our formulas.

We now proceed to give a generalization of the notions of *open submanifold* and of *open cover*.

Definition B.0.2. Let U and F be two functors $(\text{smflds})^o \longrightarrow (\text{sets})$. The functor U is a *subfunctor* of F if $U(R) \subset F(R)$ for all $R \in$ (smflds). We express this by writing $U \subset F$.

We say that U is an *open subfunctor* of a functor F (not necessarily representable) if for all natural transformations $f : h_T \longrightarrow F$ with $T \in$ (smflds) we have $f^{-1}(U) = h_V$, where V is open in T. If U is also representable we say that U is an *open supermanifold subfunctor*. ■

Notice that the functor F is not assumed to be representable. In fact, the notion of open supermanifold subfunctor is instrumental to the proof of the representability of functors.

Definition B.0.3. Let $\mathcal{U}_\alpha$ be open subfunctors of F. We say that $\{\mathcal{U}_\alpha\}$ is an *open cover* of a functor $F : (\text{smflds})^o \longrightarrow (\text{sets})$ if for all supermanifolds T and natural transformations $f : h_T \longrightarrow F$, $f^{-1}(\mathcal{U}_\alpha) = h_{V_\alpha}$ and V_α cover T.

A functor $F : (\text{smflds})^o \longrightarrow (\text{sets})$ is said to be *local* if it has the sheaf property, that is, if when restricted to the open sets of a given supermanifold, it is a sheaf. ■

Notice that any functor $F : (\text{smflds})^o \longrightarrow (\text{sets})$, when restricted to the category of open subsupermanifolds of a given supermanifold, defines a presheaf.

We are ready to state a representability criterion which gives necessary and sufficient conditions for a functor from the (opposite) category of supermanifolds to the category of sets to be representable.

Theorem B.0.4. *Representability Criterion. Let F be a functor*

$$F : (\text{smflds})^o \longrightarrow (\text{sets}) .$$

The functor F is representable if and only if:

(1) F is local, i. e., it has the sheaf property;
(2) F is covered by open supermanifold functors.

Proof. For completeness we include a sketch of the proof. For more details see Section 9.4 of Ref. [25].

If F is representable, $F = h_X$ and one can check directly that it has the two properties listed above.

We now prove it in the other direction. Let $\{h_{X_\alpha}\}_{\alpha \in A}$ be the open supermanifold subfunctors that cover F, according to the hypothesis 2. The

X_α's are supermanifolds by hypothesis, so $X_\alpha = (|X_\alpha|, \mathcal{O}_{X_\alpha})$. We want to construct, starting from such X_α's, a supermanifold $X = (|X|, \mathcal{O}_X)$, such that $h_X = F$.

Let $h_{X_{\alpha\beta}} = h_{X_\alpha} \times_F h_{X_\beta}$. By the very definition of fiber product, we have that $h_{X_{\alpha\beta}}(T) = h_{X_\alpha}(T) \cap h_{X_\beta}(T)$. We have the commutative diagram:

$$\begin{array}{ccc} h_{X_{\alpha\beta}} = h_{X_\alpha} \times_F h_{X_\beta} & \xrightarrow{j_{\beta,\alpha}} & h_{X_\beta} \\ \Big\downarrow j_{\alpha,\beta} & & \Big\downarrow i_\beta \\ h_{X_\alpha} & \xrightarrow{i_\alpha} & F \end{array}$$

So $h_{X_{\alpha\beta}}$ is the functor of points of the open subsupermanifold which is the intersection of the two open supermanifolds X_α and X_β.

The construction of the supermanifold X starts by defining its underlying topological space $|X|$. We define

$$|X| =_{\text{def}} \coprod_\alpha |X_\alpha| / \sim ,$$

where $\sim$ is the following equivalence relation: $\forall x_\alpha \in |X_\alpha|, \ x_\beta \in |X_\beta|$

$$x_\alpha \sim x_\beta \iff \exists x_{\alpha\beta} \in |X_{\alpha\beta}| \text{ with } j_{\alpha,\beta}(x_{\alpha\beta}) = x_\alpha, \ j_{\beta,\alpha}(x_{\alpha\beta}) = x_\beta .$$

This is an equivalence relation. The topologies on the $|X_\alpha|$'s will form a base for the topology on $|X|$. The map $\pi_\alpha : |X_\alpha| \hookrightarrow |X|$ is an injective map into the topological space $|X|$.

We now need to define a sheaf of superalgebras $\mathcal{O}_X$ by gluing the sheaves $\mathcal{O}_{X_\alpha}$. Let U be open in $|X|$ and let $U_\alpha = \pi_\alpha^{-1}(U)$. We define

$$\mathcal{O}_X(U) =_{\text{def}} \left\{ (f_\alpha) \in \coprod_{\alpha \in I} \mathcal{O}_{X_\alpha}(U_\alpha) \ \middle| \ j^*_{\beta,\gamma}(f_\beta) = j^*_{\gamma,\beta}(f_\gamma), \forall \beta, \gamma \in I \right\} .$$

The condition $j^*_{\beta,\gamma}(f_\beta) = j^*_{\gamma,\beta}(f_\gamma)$ simply states that to be an element of $\mathcal{O}_X(U)$, the collection $\{f_\alpha\}$ must be such that f_β and f_γ agree on the intersection of X_β and X_γ for any β and γ.

One can check directly that $\mathcal{O}_X$ is a sheaf of superalgebras and that $h_X = F$. ∎

Appendix C

Lie Superalgebras and Lie Supergroups of Classical Type

In this appendix, we give a brief account on the classification of complex Lie superalgebras of classical type. Our main reference is the fundamental work by Kac [90].

Let k be our ground field, $\mathrm{char}(k) = 0$. We recall some definitions from Chapter 1. A super vector space $\mathfrak{g} = \mathfrak{g}_0 \oplus \mathfrak{g}_1$ is a Lie superalgebra if we have a bracket $[\ ,\ ] : \mathfrak{g} \times \mathfrak{g} \longrightarrow \mathfrak{g}$ satisfying the following properties:

(1) *Antisymmetry:* $[x, y] + (-1)^{p(x)\,p(y)}[y, x] = 0$
(2) *Jacobi identity:* $(-1)^{p(x)\,p(z)}[x, [y, z]] + (-1)^{p(y)p(x)}[y, [z, x]] + (-1)^{p(z)\,p(y)}\,[z, [x, y]] = 0.$

If $V = V_0 \oplus V_1$ is a super vector space then $\mathrm{End}(V)$, the endomorphisms of V, is a super vector space with the grading $\mathrm{End}(V) = \mathrm{End}(V)_0 \oplus \mathrm{End}(V)_1$, where $\mathrm{End}(V)_0$ are the morphisms which preserve the parity while $\mathrm{End}(V)_1$ are the morphisms which reverse the parity. If V is finite dimensional and we choose a basis of homogeneous elements for V (writing first the even ones), then $\mathrm{End}(V)_0$ is the set of all diagonal block matrices, while $\mathrm{End}(V)_1$ is the set of all off-diagonal block matrices. $\mathrm{End}(V)$ is a Lie superalgebra with bracket

$$[A, B] := AB - (-1)^{|A||B|}\,BA \quad \text{for all homogeneous } A, B \in \mathrm{End}(V).$$

The standard example is $V := k^{p|q} = k^p \oplus k^q$, with $V_0 := k^p$ and $V_1 := k^q$. In this case, we also write $\mathrm{End}\left(k^{p|q}\right) := \mathrm{End}(V)$ or $\mathfrak{gl}(p\,|q) := \mathrm{End}(V)$.

In $\mathrm{End}(V)$ we can define the *supertrace* as follows:

$$\mathrm{str}\begin{pmatrix} A & B \\ C & D \end{pmatrix} := \mathrm{tr}(A) - \mathrm{tr}(D).$$

Definition C.0.5. A non abelian Lie superalgebra $\mathfrak{g}$ is called a *classical Lie superalgebra* if it is simple, i.e. it has no nontrivial (homogeneous) ideals, and $\mathfrak{g}_1$ is completely reducible as a $\mathfrak{g}_0$–module. Furthermore, $\mathfrak{g}$ is said to be *basic* if, in addition, it admits a non-degenerate, invariant bilinear form. ∎

Examples C.0.6. (See Ref. [90])

(1) $\mathfrak{sl}(m|n)$. The superalgebra $\mathfrak{sl}(m|n)$ is the subset of $\mathfrak{gl}(m|n)$ consisting of all matrices with supertrace zero. This is a Lie subalgebra of $\mathfrak{gl}(m|n)$, with $\mathbb{Z}_2$-grading

$$\mathfrak{sl}(m|\,n)_0 \;=\; \mathfrak{sl}(m)\oplus\mathfrak{sl}(n)\oplus\mathfrak{gl}(1)\;, \qquad \mathfrak{sl}(m|\,n)_1 \;=\; (\mathbf{m}\otimes\overline{\mathbf{n}})\oplus(\overline{\mathbf{m}}\otimes\mathbf{n})\,,$$

where $\mathbf{r}$ is the defining representation of $\mathfrak{sl}(r)$ and $\overline{\mathbf{r}}$ is its dual, for any r. When $m\neq n$ this Lie superalgebra is a *classical* one.

(2) $\mathfrak{osp}(p\,|q)$. Let ϕ denote a nondegenerate consistent supersymmetric bilinear form in $V := k^{p|q}$. This means that V_0 and V_1 are mutually orthogonal and the restriction of ϕ to V_0 is symmetric and to V_1 is skewsymmetric (in particular, $q=2\,n$ is even). We define in $\mathfrak{gl}(p\,|q)$ the subalgebra $\mathfrak{osp}(p\,|q) := \mathfrak{osp}(p\,|q)_0 \oplus \mathfrak{osp}(p\,|q)_1$ by setting, for $s=0,1$,

$$\mathfrak{osp}(p\,|q)_s \;:=\; \Big\{\ell\in\mathfrak{gl}(p\,|q)\;\Big|\;\phi\big(\ell(x),y\big) = -(-1)^{s\,|x|}\,\phi\big(x,\ell(y)\big) \quad \forall\, x,y\in k^{p|q}\Big\}\,.$$

We call $\mathfrak{osp}(p\,|q)$ the *orthosymplectic* Lie superalgebra. Again, all the superalgebras $\mathfrak{osp}(p|\,q)$ are *classical* Lie superalgebras, actually Lie subalgebras of $\mathfrak{gl}(p\,|q)$. Note also that $\mathfrak{osp}(0|\,q)$ is the symplectic Lie algebra, while $\mathfrak{osp}(p|\,0)$ is the orthogonal Lie algebra.

We can also describe the explicit matrix form of the elements of $\mathfrak{osp}(p\,|q)$. First, note that, in a suitable basis, the bilinear form ϕ has matrix of the form

$$\phi \;=\; \begin{pmatrix} 0 & I_m & 0 & 0 & 0\\ I_m & 0 & 0 & 0 & 0\\ 0 & 0 & 1 & 0 & 0\\ 0 & 0 & 0 & 0 & I_n\\ 0 & 0 & 0 & -I_n & 0 \end{pmatrix}, \qquad \phi \;=\; \begin{pmatrix} 0 & I_m & 0 & 0\\ I_m & 0 & 0 & 0\\ 0 & 0 & 0 & I_n\\ 0 & 0 & -I_n & 0 \end{pmatrix},$$

according to whether $(p,q)=(2m+1,2n)$ or $(p,q)=(2m,2n)$. Then, in the block form given by the partition of rows and columns according to

$(p \mid q) = m+m+1+n+n$ for $p = 2m+1$ or to $(p+q) = m+m+n+n$ for $p = 2n$, the orthosymplectic Lie superalgebras $\mathfrak{osp}(p\,|q)$ read as follows:

$$\mathfrak{osp}(2m+1|2n) = \left\{ \begin{pmatrix} A & B & u & X & X_1 \\ C & -A^t & v & Y & Y_1 \\ -v^t & -u^t & 0 & z^t & z_1^t \\ Y_1^t & X_1^t & z_1 & D & E \\ -Y^t & -X^t & -z & F & -D \end{pmatrix} \quad \text{with} \quad \begin{array}{c} B = -B^T \\ C = -C^T \\ E = E^T \\ F = F^T \end{array} \right\},$$

$$\mathfrak{osp}(2\,m\,|2\,n) = \left\{ \begin{pmatrix} A & B & X & X_1 \\ C & -A^t & Y & Y_1 \\ Y_1^t & X_1^t & D & E \\ -Y^t & -X^t & F & -D \end{pmatrix} \quad \text{with} \quad \begin{array}{c} B = -B^T \\ C = -C^T \\ E = E^T \\ F = F^T \end{array} \right\}.$$

Moreover, if $m, n \geq 2$, then, in the previous notation, we have

$$\begin{aligned} \mathfrak{osp}(2m+1|2n)_0 &= \mathfrak{o}(2m+1) \oplus \mathfrak{sp}(2n), & \mathfrak{osp}(2m|2n)_0 &= \mathfrak{o}(2m) \oplus \mathfrak{sp}(2n) \\ \mathfrak{osp}(p\,|2n)_1 &= \mathbf{p} \otimes \mathbf{2n} \quad \forall\, p > 2, & \mathfrak{osp}(2|2n)_1 &= (\mathbf{2n})^{\oplus 2}. \end{aligned}$$

■

Definition C.0.7. We define the following Lie superalgebras:

(1) $A(m,n) := \mathfrak{sl}(m+1|\,n+1), \quad \forall \quad m \neq n;$
(2) $A(n,n) := \mathfrak{sl}(n+1|\,n+1)\big/ kI_{2n};$
(3) $B(m,n) := \mathfrak{osp}(2m+1|\,2n), \quad \forall \quad m \geq 0, \quad n \geq 1;$
(4) $C(n) := \mathfrak{osp}(2|\,2n-2), \quad \forall \quad n \geq 2\,;$
(5) $D(m,n) := \mathfrak{osp}(2m|\,2n), \quad \forall \quad m \geq 2, \quad n \geq 1;$
(6) $P(n) := \left\{ \begin{pmatrix} A & B \\ C & -A^t \end{pmatrix} \in \mathfrak{gl}(n+1|\,n+1) \;\middle|\; \begin{array}{c} A \in \mathfrak{sl}(n+1) \\ B^t = B\,,\; C^t = -C \end{array} \right\};$
(7) $Q(n) := \left\{ \begin{pmatrix} A & B \\ B & A \end{pmatrix} \in \mathfrak{gl}(n+1|\,n+1) \;\middle|\; B \in \mathfrak{sl}(n+1) \right\}\big/ kI_{2(n+1)}.$

■

The importance of these examples lies in the following Theorem [90]:

Theorem C.0.8. *Let $k = \mathbb{C}$. Then the classical Lie superalgebras over $\mathbb{C}$ are either isomorphic to a simple Lie algebra or to one of the following* *classical Lie superalgebras:*

$$\begin{array}{ll} A(m,n),\ m \geq n \geq\ 0,\ m+n>0, & B(m,n),\ m \geq 0,\ n \geq 1, \\ D(m,n),\ m \geq 2,\ n \geq 1, & C(n),\ n \geq 3, \\ P(n),\ n \geq 2, & Q(n),\ n \geq 2\,, \end{array}$$

given in Definition C.0.7 and additionally

$$D(2,1;a),\ a \in k \setminus \{0,-1\}, \qquad F(4), \qquad \textit{and} \qquad G(3)\,.$$

■

For the definition of the last three items, and for the proof of the Theorem, the reader can consult Ref. [90].

Appendix D

Super Harish-Chandra Pairs

The theory of Super Harish-Chandra Pairs (SHCP) provides an equivalent way to approach supergroups. We are going to state the main definitions and results only, sending the reader to the original work of Koszul [98] for more details (see also Chapter 7 in Ref. [25] and Ref. [24]).

For all the notation and terminology, refer to Chapter 1.

Definition D.0.9. Let $(G_0, \mathfrak{g})$ be, respectively, a group (a real Lie group, a complex analytic group or an affine algebraic group) and a super Lie algebra (real or complex). Assume that:

(1) $\mathfrak{g}_0 \simeq \mathrm{Lie}(G_0)$,
(2) G_0 acts on $\mathfrak{g}$ and this action, restricted to $\mathfrak{g}_0$, is the adjoint representation of G_0 on $\mathrm{Lie}(G_0)$. Moreover, the differential of such action is the Lie bracket. We shall denote such an action with Ad or as $g.X$, $g \in G_0$, $X \in \mathfrak{g}$.

Then $(G_0, \mathfrak{g})$ is called a *super Harish-Chandra pair (SHCP)*.

A *morphism* of SHCP's is simply a pair of morphisms $\psi = (\psi_0, \rho^\psi)$ preserving the SHCP structure, that is:

(1) $\psi_0 : G_0 \to H_0$ is a group morphism (in the Lie, analytic or algebraic category);
(2) $\rho^\psi : \mathfrak{g} \to \mathfrak{h}$ is a super Lie algebra morphism;
(3) ψ_0 and ρ^ψ are compatible in the sense that:

$$\rho^\psi_{|\mathfrak{g}_0} \simeq \mathrm{d}\psi_0, \qquad \mathrm{Ad}(\psi_0(g)) \circ \rho^\psi = \rho^\psi \circ \mathrm{Ad}(g) .$$

When G_0 is a Lie or an analytic group we say $(G_0, \mathfrak{g})$ is a *Lie or analytic SHCP*. When G_0 is an affine algebraic group we say it is an *algebraic SHCP*. ■

The interest of SHCPs is that the category of (Lie, analytic or algebraic) SHCPs, that we denote as (shcps), is equivalent to the category of supergroups (Lie, analytic or algebraic), denoted as (sgrps). In order to state properly this result, we need to associate in a natural way a supergroup to a SHCP.

Definition D.0.10. Let $(G_0, \mathfrak{g})$ be a SHCP. The sheaf $\mathcal{O}_{G_0}$ of the ordinary group G_0 carries a natural action of $\mathcal{U}(\mathfrak{g}_0)$, since the elements of $\mathcal{U}(\mathfrak{g}_0)$ act on the sections in $\mathcal{O}_{G_0}(U)$ as left invariant differential operators. We define $\mathcal{O}_G(U)$ as:

$$\mathcal{O}_G(U) := \mathrm{Hom}_{\mathcal{U}(\mathfrak{g}_0)}(\mathcal{U}(\mathfrak{g}), \mathcal{O}_{G_0}(U)), \qquad U \subset_{\text{open}} G_0 \,.$$

■

Notice that when $\mathfrak{g} = \mathfrak{g}_0$, a morphism in $\mathrm{Hom}_{\mathcal{U}(\mathfrak{g}_0)}(\mathcal{U}(\mathfrak{g}), \mathcal{O}_{G_0}(U))$ is effectively retrieved by the image of $1_{\mathcal{U}(\mathfrak{g}_0)}$, thus identifying $\mathrm{Hom}_{\mathcal{U}(\mathfrak{g}_0)}(\mathcal{U}(\mathfrak{g}_0), \mathcal{O}_{G_0}(U))$ and $\mathcal{O}_{G_0}(U)$, as one expects.

Proposition D.0.11. *The assignment $U \to \mathcal{O}_G(U)$ is a sheaf of superalgebras on G_0, where the superalgebra structure on $\mathcal{O}_G(U)$ is given by*

$$f_1 \cdot f_2 = m_{\mathcal{O}_{G_0}} \circ (f_1 \otimes f_2) \circ \Delta_{\mathcal{U}(\mathfrak{g})} \,,$$

and the restriction morphisms are

$$\begin{array}{ccc} \mathcal{O}_G(U) & \xrightarrow{\rho_{UV}} & \mathcal{O}_G(V) \\ f & \longrightarrow & \rho_{UV}(f) := \widetilde{\rho}_{UV} \circ f, \end{array}$$

where $\widetilde{\rho}_{UV}$ are the restrictions of the ordinary sheaf $\mathcal{O}_{G_0}$.

Proof. Checking that $f_1 \cdot f_2$ is an associative product is routine, while the sheaf property comes from the fact $\mathcal{O}_{G_0}$ is an ordinary sheaf. ■

The fact that $(G_0, \mathcal{O}_G)$ is a superspace, comes from the fact that it is *globally split*, in other words:

$$\mathcal{O}_G(U) \cong \mathcal{O}_{G_0}(U) \otimes \wedge(\mathfrak{g}_1) \,,$$

as the next theorem shows.

Theorem D.0.12.

(1) The map

$$\begin{array}{ccc} \mathcal{U}(\mathfrak{g}_0) \otimes \wedge(\mathfrak{g}_1) & \xrightarrow{\widehat{\gamma}} & \mathcal{U}(\mathfrak{g}) \\ X \otimes Y & \longrightarrow & X \cdot \gamma(Y) \,, \end{array}$$

is an isomorphism of super left $\mathcal{U}(\mathfrak{g}_0)$-modules, where

$$\begin{array}{ccc} \wedge(\mathfrak{g}_1) & \xrightarrow{\gamma} & \mathcal{U}(\mathfrak{g}) \\ X_1 \wedge \cdots \wedge X_p & \longrightarrow & \frac{1}{p!}\sum_{\tau \in S_p} (-1)^{|\tau|} X_{\tau(1)} \cdots X_{\tau(p)} \end{array}$$

is the symmetrizer map and $|\tau|$ denotes the parity of the permutation τ.

(2) $(G_0, \mathcal{O}_G)$ is globally split i. e., for each open subset $U \subseteq G_0$, there is an isomorphism of superalgebras

$$\mathcal{O}_G(U) \simeq \mathrm{Hom}(\wedge(\mathfrak{g}_1), \mathcal{O}_{G_0}(U)) \simeq \mathcal{O}_{G_0}(U) \otimes \wedge(\mathfrak{g}_1)^* . \tag{D.1}$$

Hence $\mathcal{O}_G$ carries a natural $\mathbb{Z}$-gradation.

(3) $(G_0, \mathcal{O}_G)$ is a superspace.

Proof. For the proof, the reader can consult Ref. [24]. ∎

We now state the main result of the theory of SHCP's, which establishes the equivalence of categories between the SHCP's category and the supergroups one.

Theorem D.0.13. *Let us define the functors $\mathcal{H}$ and $\mathcal{K}$ as*

$$\begin{array}{ccc} (\mathrm{sgrps}) & \xrightarrow{\mathcal{H}} & (\mathrm{shcps}) \\ G & \longrightarrow & (G_0, \mathrm{Lie}(G)) \\ \phi & \longrightarrow & (|\phi|, (\mathrm{d}\phi)_e), \end{array}$$

$$\begin{array}{ccc} (\mathrm{shcps}) & \xrightarrow{\mathcal{K}} & (\mathrm{sgrps}) \\ (G_0, \mathfrak{g}) & \longrightarrow & \left(G_0, \underline{\mathrm{Hom}}_{\mathcal{U}(\mathfrak{g}_0)}(\mathcal{U}(\mathfrak{g}), \mathcal{O}_{G_0})\right) \\ \psi = (\psi_0, \rho^\psi) & \longrightarrow & \psi_0^* \circ f \circ \rho^\psi , \end{array}$$

where G and $(G_0, \mathfrak{g})$ are objects and ϕ, ψ are morphisms of the corresponding categories (in the definition of $\mathcal{H}$, G_0 is the ordinary group underlying G).

Then $\mathcal{H}$ and $\mathcal{K}$ define an equivalence between the categories of supergroups (Lie, analytic or algebraic) and super Harish-Chandra pairs (Lie, analytic or algebraic, respectively). ∎

Appendix E

Quantum Supergroups

In this appendix we want to write some details and calculations that are important and instrumental to the definition of the *quantum general linear supergroup*. We have not included them in the text in Chapter 5, since they are very technical and could disrupt the exposition. We invite the reader to consult [53, 110, 163] for more details and the proofs of our statements, that cannot be included here.

Let k be our ground field and let $k_q = k[q, q^{-1}]$ where q is an indeterminate (for us $k = \mathbb{R}$ or $k = \mathbb{C}$). We start by recalling the definition of the bialgebra of quantum supermatrices discussed in Chapter 5.

Definition E.0.14. We define the superalgebra of *quantum supermatrices* as the quotient

$$k_q[M(m|n)] =_{\text{def}} k_q\langle a_{ij}\rangle / I_M ,$$

where I_M is the two sided homogeneous ideal generated by the relations

$$a_{ij}a_{il} = (-1)^{\pi(a_{ij})\pi(a_{il})} q^{(-1)^{p(i)+1}} a_{il}a_{ij}, \quad j < l ,$$

$$a_{ij}a_{kj} = (-1)^{\pi(a_{ij})\pi(a_{kj})} q^{(-1)^{p(j)+1}} a_{kj}a_{ij}, \quad i < k ,$$

$$a_{ij}a_{kl} = (-1)^{\pi(a_{ij})\pi(a_{kl})} a_{kl}a_{ij}, \quad i < k, j > l \quad \textit{or} \quad i > k, j < l ,$$

$$a_{ij}a_{kl} - (-1)^{\pi(a_{ij})\pi(a_{kl})} a_{kl}a_{ij}$$

$$= (-1)^{\pi(a_{ij})\pi(a_{kj})} (q^{-1} - q) a_{kj}a_{il} \quad i < k, j < l .$$

■

$k_q[M(m|n)]$ is a super bialgebra with comultiplication and counit

$$\Delta(a_{ij}) = \sum a_{ik} \otimes a_{kj}, \qquad \epsilon(a_{ij}) = \delta_{ij} .$$

We consider the following elements inside the superalgebra of quantum matrices:

$$D_1 =_{\text{def}} \sum_{\sigma\in S_m} (-q)^{-\ell(\sigma)} a_{1\sigma(1)} \dots a_{m\sigma(m)},$$

$$D_2 =_{\text{def}} \sum_{\sigma\in S_n} (-q)^{\ell(\sigma)} a_{m+1,m+\sigma(1)} \dots a_{m+n,m+\sigma(n)} \cdot$$

D_1 and D_2 are respectively the quantum determinant of the upper and the lower diagonal block. Notice that in the expression of D_1 we have $(-q)^{-\ell(\sigma)}$, while in the expression of D_2 we have $(-q)^{\ell(\sigma)}$; this is because in the lower diagonal blocks the indeterminates obey the Manin relations with q replaced by q^{-1}.

We now want to establish the commutation relations between the generators of $k_q[M(m|n)]$ and D_1 and D_2.

Proposition E.0.15. *Let the notation be as above. Then D_1 and D_2 commute respectively with a_{ij} with $1 \leq i,j \leq m$ and $m+1 \leq i,j \leq m+n$. As for the other generators we have*

$$D_1 a_{ij} = q^{-1} a_{ij} D_1, \quad D_2 a_{ij} = q^{-1} a_{ij} D_2\,,$$

where $1 \leq i \leq m$, $m+1 \leq j \leq m+n$ or $m+1 \leq i \leq m+n$, $1 \leq j \leq m$.

Proof. The proof amounts to a direct check, which is, however, non trivial, since it makes use of the Laplace expansion of quantum determinants, explained in Ref. [119]. For all of the details the reader can consult Lemma 4.7 in Ref. [53]. ■

We now want to study the comultiplication and counit of D_1 and D_2.

Proposition E.0.16. *Let the notation be as above. Then*

$$\Delta(D_1) = \sum_{i=0}^{m} R_i\,,$$

with

$$R_i = \sum_{\substack{\sigma\in S_m \\ (k_1,\dots,k_m)\in\rho_i}} (-q)^{-\ell(\sigma)} a_{1,k_1} \dots a_{m,k_m} \otimes a_{k_1,\sigma(1)} \dots a_{k_m,\sigma(m)}, \quad \textit{and}$$

$$\rho_i = \{(k_1,\dots,k_m) \mid 1 \leq k_1,\dots,k_m \leq m+n, \\ |\{k_1,\dots,k_m\} \cap \{1\dots m\}| = m-i\}\,.$$

Similarly

$$\Delta(D_2) = \sum_{i=0}^{n} S_i ,$$

with

$$S_i = \sum_{\substack{\theta \in S_n \\ (l_1,\dots,l_n)\in\sigma_i}} (-q)^{\ell(\theta)} a_{1,l_1} \dots a_{n,l_n} \otimes a_{l_1,\theta(1)} \dots a_{l_n,\theta(n)}, \quad \textit{and}$$

$$\sigma_i = \{(l_1, \dots, l_n) \mid 1 \le l_1, \dots, l_n \le m+n, \\ |\{l_1, \dots, l_n\} \cap \{m+1 \dots m+n\}| = n-i\} .$$

Furthermore,

$$\epsilon(D_1) = \epsilon(D_2) = 1 .$$

Proof. In this case the proof amounts to a patient writing of the comultiplication. ■

We are now ready to define the general linear supergroup and to elucidate its bialgebra structure.

Definition E.0.17. We define *quantum general linear supergroup*

$$k_q[\mathrm{GL}(m|n)] =_{\mathrm{def}} k_q[M(m|n)]\langle {D_1}^{-1}, {D_2}^{-1}\rangle ,$$

where ${D_1}^{-1}$, ${D_2}^{-1}$ are even indeterminates such that

$$D_1 D_1^{-1} = 1 = {D_1}^{-1} D_1, \qquad D_2 {D_2}^{-1} = 1 = {D_2}^{-1} D_2 .$$

■

Proposition E.0.18. *Let the notation be as above. Then,* $k_q[\mathrm{GL}(m|n)]$ *is a coalgebra with comultiplication*

$$\Delta(a_{ij}) = \sum a_{ik} \otimes a_{kj},$$

$$\Delta({D_1}^{-1}) = \sum_{i=1}^{2mn+2} (-1)^{i-1} R_0^{-1} [R_0^{-1}(q^{-2} R_1 + \dots + q^{-2m} R_m)]^{i-1},$$

$$\Delta({D_2}^{-1}) = \sum_{i=1}^{2mn+2} (-1)^{i-1} S_0^{-1} [S_0^{-1}(q^{-2} S_1 + \dots + q^{-2n} S_n)]^{i-1} ,$$

and counit

$$\epsilon(a_{ij}) = \delta_{ij}, \qquad \epsilon({D_1}^{-1}) = 1, \qquad \epsilon({D_2}^{-1}) = 1 .$$

Proof. This is a direct although non trivial calculation (see Ref. [53], Proposition 4.12). ■

We now want to give the antipode S^q on $k_q[\mathrm{GL}(m|n)]$, so to prove that $k_q[\mathrm{GL}(m|n)]$ is indeed an Hopf superalgebra. In order to define it properly we need to introduce another piece of notation. Let us call the generators a_{ij}'s of $k_q[\mathrm{GL}(m|n)]$ in a different way so to keep track of their position:

$$\begin{pmatrix} a_{11} & \dots & a_{1m+n} \\ \vdots & & \vdots \\ a_{m+n,1} & \dots & a_{m+n,m+n} \end{pmatrix} = \begin{pmatrix} X_{11} & \Xi_{12} \\ \Xi_{21} & X_{22} \end{pmatrix},$$

where X_{ab} and Ξ_{cd} are matrices of indeterminates.

As we remarked in the text (see Section 5.5), X_{11} and X_{22} are both quantum matrices (that is, their indeterminates generate a quantum matrix bialgebra), however the indeterminates in X_{11} obey the usual Manin relations, while the indeterminates in X_{22} obey the Manin relations with q replaced by q^{-1}. The subalgebra in $k_q[\mathrm{GL}(m|n)]$ generated by the entries in X_{11} and D_1^{-1} is a Hopf algebra, as one can readily check. We call S_{11}^q its antipode. Similarly, the subalgebra in $k_q[\mathrm{GL}(m|n)]$ generated by the entries in X_{22} and D_2^{-1} is a Hopf algebra. We call S_{22}^q its antipode.

In the next theorem we give the antipode in $k_q[\mathrm{GL}(m|n)]$. We send the reader to Refs. [125, 163] for the proof.

Theorem E.0.19. *$k_q[\mathrm{GL}(m|n)]$ is a Hopf algebra with antipode S^q given by*

$$S^q(X_{11}) = S_{11}^q(X_{11}) + S_{11}^q(X_{11})\Xi_{12}S_2^q(X_{22} - \Xi_{21}S_{11}^q(X_{11})\Xi_{12})\Xi_{21}S_{11}^q(X_{11}),$$

$$S^q(\Xi_{12}) = -S_{11}^q(X_{11})\Xi_{12}S_2^q(X_{22} - \Xi_{21}S_{11}^q(X_{11})\Xi_{12}),$$

$$S^q(\Xi_{21}) = -S_2^q(X_{22} - \Xi_{21}S_{11}^q(X_{11})\Xi_{12})\Xi_{21}S_{11}^q(X_{11}),$$

$$S^q(X_{22}) = S_2^q(X_{22} - \Xi_{21}S_{11}^q(X_{11})\Xi_{12}),$$

$$S^q({D_{1\dots m}^{1\dots m}}^{-1}) = D_{m+1\dots m+n}^{m+1\dots m+n}, \qquad S^q({D_{m+1\dots m+n}^{m+1\dots m+n}}^{-1}) = D_{1\dots m}^{1\dots m},$$

where $X_{22} - \Xi_{21}S_{11}^q(X_{11})\Xi_{12}$ is also a quantum matrix (see Section 4 in Ref. [125] or Ref. [163]) and S_2^q denotes its quantum antipode. ■

We end this section with an important result regarding a basis for $k_q[\mathrm{GL}(m|n)]$ (see Ref. [110]). We first need to establish an ordering on the generators a_{ij}. We say that

$$a_{ij} < a_{kl}$$

if either $i > k$ or $i = k$ and $j > l$. We say that a monomial in the a_{ij}'s is *normally ordered* if whenever $a_{ij} < a_{kl}$, then a_{ij} is to the left of a_{kl} and no odd a_{ij} enters the monomial twice.

Theorem E.0.20. *The normally ordered monomials form a basis for* $k_q[\mathrm{GL}(m|n)]$. *In particular* $k_q[\mathrm{GL}(m|n)]$ *is torsion free.* ∎

Bibliography

[1] R. Abraham, J. E. Marsden, *Foundations of mechanics*, Second Edition, Addison Wesley, 1978.

[2] R. Abraham, J. E. Marsden, T. Ratiu, *Manifolds, tensor Analysis and applications*, Springer, 1993.

[3] M. F. Atiyah, R. Bott, A. Shapiro, *Clifford modules* Topology, **3** (Suppl.1), 3-38, 1963.

[4] M. F. Atiyah, I. G. McDonald, *Introduction to commutative algebra*, Perseus books, 1969.

[5] L. Balduzzi, C. Carmeli, G. Cassinelli, *Super G-spaces*, Symmetry in Mathematics and Physics, Contemp. Math., 490, Amer. Math. Soc., 159–176, 2009.

[6] L. Balduzzi, C. Carmeli, R. Fioresi *Quotients in supergeometry*, Symmetry in Mathematics and Physics, Contemp. Math., 490, Amer. Math. Soc., 177–187, 2009.

[7] A. Ballesteros, F.J. Herranz, M.A. del Olmo, M. Santander, *Four-dimensional quantum affine algebras and space-time q-symmetries*. J. Math. Phys, **35**, 4928-4940, 1994; *A new "null–plane" quantum Poincaré algebra*, Phys. Lett. B, **351**, 137-145, 1995.

[8] V. Bargmann, *Irreducible Unitary Representations of the Lorentz Group*, Ann. of Math., **48**, 568-640, 1947.

[9] V. Bargmann, E. P. Wigner, *Group theoretical discussion of relativistic wave equations*, Proc. Natl. Acad. Sci. U.S.A., **34**, No. 5, 211–223, 1948.

[10] J. M. Bardeen, B. Carter, S. W. Hawking, *The four laws of black hole mechanics*, Comm. Math. Phys., **31**, 161-170, 1973.

[11] C. Bartocci, U. Bruzzo, D. Hernandez-Ruiperez, *The Geometry of Supermanifolds*, MAIA 71, Kluwer Academic Publishers, 1991.

[12] F. Bayen, M. Flato, C. Fronsdal, A. Lichnerowicz, D. Sternheimer, *Deformation theory and quantization. I and II*, Ann. Phys., **111**, 1, 61-151, 1978.

[13] J.D. Bekenstein, *Black Holes and Entropy*, Phys. Rev. DRD, **7**, 2333–2346, 1973.

[14] F. A. Berezin, *Introduction to superanalysis*, D. Reidel Publishing Com-

pany, Holland, 1987.

[15] F. A. Berezin, D. A. Leites, *Supermanifolds*, Dokl. Akad. Nauk SSSR, Vol. 224, No. 3, 505–508, 1975.

[16] G. Bergman, *The diamond lemma for ring theory*, Adv. Math., **29**, 178-218, 1978.

[17] Z. Bern, L. J. Dixon, R. Roiban, *Is $N = 8$ Supergravity Ultraviolet Finite?*, Phys. Lett. B, **644**, 265–271, 2007.

[18] A. Borel, *Linear algebraic groups*, Springer Verlag, New York, 1991.

[19] H. Boseck, *Affine Lie Supergroups*, Math. Nachr., **143**, 303–327, 1989.

[20] H. Boseck, *Classical Lie Supergroups*, Math. Nachr., **148**, 81–115, 1990.

[21] N. Bourbaki, *Algebre*, Ch. 9, Hermann, Paris, 1958.

[22] U. Bruzzo, R. Cianci, *Differential equations, Frobenius theorem and local flows on supermanifolds*, J. Phys. A Math. Gen., **18**, 417-423, 1985.

[23] C. G. Callan, S. Coleman, R. Jackiw, *A new improved energy-momentum tensor*, Annals Phys., **59**, 42-73, 1970.

[24] C. Carmeli, R. Fioresi, *Super Distributions, Analytic and Algebraic Super Harish-Chandra pairs*, Pac. J. Math, **263**, No. 1, 29–51, 2013.

[25] C. Carmeli, L. Caston, R. Fioresi, *Mathematical Foundation of Supersymmetry*, with an appendix with I. Dimitrov, EMS Ser. Lect. Math., European Math. Soc., Zurich, 2011.

[26] D. Cervantes, R. Fioresi, M. A. Lledo, *The quantum chiral Minkowski and conformal superspaces*, Adv. Theor. Math. Phys., Vol. 15, No. 2, 565-620, 2011.

[27] D. Cervantes, R. Fioresi, M. A. Lledo, *On chiral quantum superspaces*, Supersymmetry in Mathematics and Physics, Springer, Vol. 2027, 69-99, 2011.

[28] V. Chari, A. Pressley, *A guide to quantum groups*, Cambridge Press, 1994.

[29] C. Chevalley, *The Algebraic Theory of Spinors*, Columbia University Press, 1954.

[30] Y. Choquet-Bruhat, C. DeWitt-Morette, *Analysis, Manifols and Physics. Part II, Applications*, North-Holland, 1989.

[31] N. Ciccoli, R. Fioresi, F. Gavarini, *Quantization of Projective Homogeneous Spaces and Duality Principle*, J. of Noncommut. Geom., 449–496, **2**, 2008.

[32] S. Coleman, R. Jackiw, *Why dilatation generators do not generate dilatations*, Annals Phys., **67**, 552-598, 1971.

[33] S. R. Coleman and J. Mandula, *All possible symmetries of the S matrix*, Phys. Rev., **159**, 1251-1256, 1967.

[34] A. Connes, *Noncommutative geometry*, Academic Press, San Diego, CA, 1994.

[35] R. Coquereaux, *Modulo 8 periodicity of Real clifford Algebras and Particle Physics*, *Phys. Lett.*, **115B**, 389, 1982.

[36] R. D'Auria, S. Ferrara, M. A. Lledo, V. S. Varadarajan, *Spinor algebras*, J. Geom. Phys., Vol. 40, No. 2, 2001.

[37] P. Deligne, J. W. Morgan, Notes on supersymmetry following Bernstein, *Quantum fields and strings; a course for mathematicians, Vol. 1* (Princeton, NJ, 1996/1997), 41–96, Amer. Math. Soc., Providence, RI, 1999.

[38] B. DeWitt, *Supermanifolds*, Cambridge University Press, London, 1984.

[39] M. Demazure, P. Gabriel, *Groupes Algébriques, Tome 1*, Mason&Cie, éditeur, North-Holland Publishing Company, The Netherlands, 1970.

[40] P. A. M. Dirac, *The principles of quantum mechanics* (fourth edition), Oxford University Press, 1958.

[41] J. Dixmier, *Enveloping Algebras*, AMS, 1996.

[42] S. Doplicher, K. Fredenhagen, J. E. Roberts, *The quantum structure of spacetime at the Planck scale and quantum fields*, Comm. Math. Phys., **172**, 187-220, (1995).

[43] V. G. Drinfeld, *Quantum Groups*, Proc. Int. Cong. Math. Berkeley 1986, 798–874, AMS, Providence, RI, 1987.

[44] D. Eisenbud, *Commutative algebra with a view towards algebraic geometry*, GTM Springer Verlag, 1991.

[45] D. Eisenbud, J. Harris, *The Geometry of Schemes*, Springer Verlag, New York, 2000.

[46] L. D Faddeev, N. Y. Reshetikhin, L. A. Takhtajan, *Quantization of Lie groups and Lie algebras*, Algebraic analysis, Vol. I, 129–139, Academic Press, Boston, MA, 1988.

[47] B. V. Fedosov, *A simple geometrical construction of deformation quantization*, J. Diff. Geom., **40**, No. 2, 213-447, 1994.

[48] G. Feinberg, *On the possibility of faster-than-light particles*, Phys. Rev., **159**, 1089-1105, 1967.

[49] S. Ferrara, B. Zumino, *Supergauge invariant Yang-Mills theories*, Nucl. Phys. B, **79**, 413, 1974.

[50] S. Ferrara, Daniel Z. Freedman, P. van Nieuwenhuizen, *Progress toward a theory of supergravity*, Phys. Rev. D, **13**, 3214-3218, 1976.

[51] R. P. Feynman, *Quantum Electrodynamics*, W. A. Benjamin, 1961.

[52] M. Fierz *Über die relativistische Theorie kräftefeier Teilchen mit beliebigem Spin*, Helvetica Physica Acta, **12**, 317, 1939.

[53] R. Fioresi, *On algebraic super-groups and quantum deformations*, J. Algebra Appl., **2**, 403-423, 2003.

[54] R. Fioresi, *Quantizations of flag manifolds and conformal space time*, Rev. Math. Phy., **9**, No. 4, 453-465, 1997.

[55] R. Fioresi, *A deformation of the big cell inside the Grassmannian manifold* $G(r,n)$, Rev. Math. Phy., **11**, 25-40, 1999.

[56] R. Fioresi, *Quantum deformation of the flag variety* Comm. Algebra, **27**, No. 11, 5669–5685, 1999.

[57] R. Fioresi, *Commutation relations among generic quantum minors in* $O_q(M_n(k))$, J. Algebra, **280**, no. 2, 655–682, 2004.

[58] R. Fioresi, *Supergroups, quantum supergroups and their homogeneous spaces*, Modern Physics Letters A, 16, No. 4-6, 269-274, 2001.

[59] R. Fioresi, *On algebraic supergroups and quantum deformation*, J. Algebra Appl., **2**, No. 4, pg 403-423, 2003.

[60] R. Fioresi, *Compact forms of Complex Lie Supergroups*, J. Pure Appl. Algebra 218, No. 2, 228–236, 2014.

[61] R. Fioresi, F. Gavarini, *Chevalley Supergroups*, Mem. Amer. Math. Soc.,

215, No. 1014, 2012.

[62] R. Fioresi, F. Gavarini, *Algebraic supergroups with Lie superalgebras of classical type*, J. Lie Theory, **23**, No. 1, 143–158, 2013.

[63] R. Fioresi, F. Gavarini, *On the construction of Chevalley supergroups*, Supersymmetry in Mathematics and Physics, Lecture Notes in Math., **2027**, 101–123, Springer, Heidelberg, 2011.

[64] R. Fioresi, C. Hacon, *Quantum coinvariant theory for the special linear group and quantum Schubert varieties*, J. Algebra, **242**, No. 2, 433-446, 2001.

[65] R. Fioresi, S. D. Kwok *On SUSY curves*, in "Advances in Lie Superalgebras" M. Gorelik, Papi, P. (Eds.), Springer INdAM Series, **7**, 2014.

[66] R. Fioresi, M. A. Lledo, V. S. Varadarajan *The Minkowski and conformal superspaces*, J. Math. Phys., **48**, 113505, 2007.

[67] P. Di Francesco, P. Mathieu, D. Sénéchal, *Conformal field theory*, Springer Verlag, New York, 1997.

[68] V. Fock, *Konfigurationsraum und zweite Quantelung*, Z. Phys., **75**, 622-647, 1932.

[69] L. Frappat, P. Sorba, *Dictionary on Lie algebras and superalgebras*, Elsevier, 2000.

[70] D. Freedman, A. van Proeyen, *Supergravity*, Cambridge University Press, 2012.

[71] W. Fulton, *Young Tableaux*, Cambridge University Press, 1997.

[72] W. Fulton, J. Harris, *Representation Theory*, Springer GTM, 1991.

[73] H. Georgi, S. Glashow, *Unity of All Elementary-Particle Forces*, Phys. Rev. Lett., **32**, 438, 1974.

[74] R. Giachetti, R. Ricci, **R**-*actions, derivations, and Frobenius theorem on graded manifolds*, Adv. Math., Vol. 62, No. 1, 84–100, (1986 .

[75] R. Godement, *Topologie algebrique et theorie des faisceaux*, Paris, Hermann, 1973.

[76] F. Gürsey, L. A. Radicati, *Spin and Unitary Spin Independence of Strong Interactions*, Phys. Rev. Lett., **13**, 173, 1964.

[77] Yu . A . Gol'fand, E.P. Likhtman, *Extension of the Algebra of Poincare Group Generators and Violation of P invariance*, JETP Lett. 13, 323–326, 1971.

[78] K. Goodearl *Commutation relations for arbitrary quantum minors*, Pac. J. of Math., 228, No. 1, 63-102, 2006.

[79] P. Griffiths, J. Harris, *Principles of Algebraic Geometry*, Wiley Classics Library, 1994.

[80] A. Grothendieck, M. Demazure, M. Artin [et al.]. *Schémas en groupes; séminaire de géométrie algébrique du Bois Marie*, 1962/64, SGA 3. Springer-Verlag, Berlin-New York, 1970.

[81] A. Grothendieck, *Fondaments de la geometrie algebrique*, Secretariat Mathematique, Paris, 1962.

[82] A. Grothendieck, *Résumé de la théorie métrique des produits tensoriels topologiques*, Reprint of Bol. Soc. Mat. São Paulo, **8**, 1–79, 1953.

[83] A. Grothendieck, *Technique de descente et théorèmes d'existence*

en géométrie algébrique, I. Généralités. Descente par morphismes fidèlement plats, Séminaire Bourbaki, Vol. 5, Exp. No. 190, 299–327, Soc. Math. France, Paris, 1995.

[84] C. Gruson, V. Serganova, *Cohomology of generalized supergrassmannians and character formulae for basic classical Lie superalgebras*, Proc. London Math. Soc., 101, 3, 852-892, 2010.

[85] R. Hartshorne, *Algebraic geometry*, Graduate Text in Mathematics, Springer-Verlag, New York, 1977.

[86] S. W. Hawking, *Particle creation by black holes*, Comm. Math. Phys., **43**, 199-220, 1975.

[87] S. Helgason, *Differential Geometry, Lie Groups, and Symmetric Spaces*, AMS, 1978.

[88] N. I. Ivanova, A. L. Onishchik, *Parabolic subalgebras and gradings of reductive Lie superalgebras*, Journal of Mathematical Sciences, **152**, No. 1, 1-60, 2008.

[89] M. Jimbo, *A q-analogue of $U(\mathfrak{gl}(N+l))$-Hecke algebra and the Yang-Baxter equation*, Lett. Math. Phys., **11**, 247-252, 1986.

[90] V. G. Kac, *Lie superalgebras* Adv. Math., **26**, 8–96, 1977.

[91] V. G. Kac, *Representations of classical Lie superalgebras*, Springer LNM, **676**, 597-626, 1978.

[92] V. G. Kac, *A sketch of Lie superalgebras Theory*, Comm. Math. Phys., **53**, 31-64, 1977.

[93] C. Kassel, *Quantum Groups*, GTM, Springer, 1994.

[94] I. Kolář, P. W. Michor, J. Slovák, *Natural operations in differential geometry*, Springer-Verlag, Berlin, 1993.

[95] M. Kontsevich, *Deformation quantization of Poisson manifolds*, Lett. Math. Phys., **66**, Issue 3, 157-216, 2003.

[96] B. Kostant, *Quantization and unitary representations*, Lecture Notes in Mathematics **170**, Springer-Verlag, New York, 1970.

[97] B. Kostant, *Graded manifolds, graded Lie theory, and prequantization. Differential geometrical methods in mathematical physics*, Proc. Sympos., Univ. Bonn, Bonn, 1975 , 177–306, *Lecture Notes in Math.*, Vol. 570, Springer, Berlin, 1977.

[98] J. L. Koszul, *Graded manifolds and graded Lie algebras*, Proceedings of the international meeting on geometry and physics (Florence, 1982), 71–84, Pitagora, Bologna, 1982.

[99] S. Lang, *Algebra*, Addison Wesley, 1984.

[100] S. Lang, *Real and functional analysis*, GTM 142, Springer-Verlag, New York, 1993.

[101] D. A. Leites, *Introduction to the theory of supermanifolds*, Russian Math. Surveys **35**, 1, 1-64, 1980.

[102] D. A. Leites *Quantization and supermanifolds*, Appendix 3 in Berezin F., Shubin M., Schroedinger equation, Kluwer, Dordrecht, 1990.

[103] J. Lukierski, A. Nowicki, H. Ruegg, V. N. Tolstoy, *q-Deformation of Poincaré algebras*, Phys. Lett. B, **268**, 331–338, 1991.

[104] S. Maclane, *Categories for the Working Mathematician*, Graduate Texts in

Mathematics, Springer, 1998.

[105] G.W. Mackey, *Induced representations of locally compact groups, I and II*, Ann. Math., **55**, 101-139, 1952, II, Ann. Math., **58**, 193-221, 1953.

[106] S. Majidn, *Hopf algebras for physics at the Planck scale*, Class. Quant. Grav., **5**, 1587–1607, 1988.

[107] J. Maldacena, *The large N limit of superconformal field theories and supergravity*, Adv. Theor. Math. Phys., **2**, 231-252, 1998.

[108] Y. I. Manin, *Gauge field theory and complex geometry*; translated by N. Koblitz and J.R. King, Springer-Verlag, Berlin-New York, 1988.

[109] Y. I. Manin, *Topics in noncommutative geometry*, M. B. Porter Lectures, Princeton University Press, Princeton, NJ, 1991.

[110] Y. Manin, *Multiparametric quantum deformation of the general linear supergroup*, Comm. Math. Phy., **123**, 163-175, 1989.

[111] H. Matsumura, *Commutative Ring Theory*, Cambridge Studies in Advanced Mathematics, 1989.

[112] P. W. Michor, *Topics in differential geometry*, Graduate Studies in Mathematics 93, American Mathematical Society, 2008.

[113] S. Montgomery, *Hopf Algebras and Their Actions on Rings*, Providence, RI, Amer. Math. Soc., 1993.

[114] D. Mumford, *The red book of varieties and schemes*, Springer-Verlag, Berlin-New York, 1988.

[115] Y. Nakayama, *A lecture note on scale invariance vs conformal invariance*, arXiv:1302.0884 [hep-th].

[116] J. Nestruev, *Smooth manifolds and observables*, GTM 220, Springer-Verlag, Berlin-New York, 2003.

[117] O. Ogievetskii, W. B. Schmidke, J. Wess, B. Zumino, *q-Deformed Poincaré Algebra*, Comm. Math. Phys., **150**, 495-518, 1992.

[118] P. Olver, *Applications of Lie groups to differential equations*, Springer Verlag, New York, 199.

[119] B. Parshall, J. Wang, *Quantum linear groups*, Mem. Amer. Math. Soc., 439, 1991.

[120] W. Pauli, *The connection between spin and statistics*, Phys. Rev., **58**, 716–722, 1940.

[121] I. Penkov, V. Serganova, *Generic irreducible representations of finite-dimensional Lie superalgebras*, Internat. J. Math., **5**, 389-419, 1994.

[122] I. Penkov, *Characters of typical irreducible finite-dimensional $q(n)$-modules*, Functional Anal. Appl., **20**, 30-37, 1986.

[123] I. Penkov, *Borel-Weil-Bott theory for classical Lie supergroups*, Sovr. Probl. Mat., **32**, 71–124, 1988.

[124] R. Penrose, *Twistor algebra*, J. Math. Phys., **252**, 189-258, 1967.

[125] Phung, Ho Hai, *On the structure of the quantum supergroups* $\mathrm{GL}_q(m|n)$, q-alg/9511-23, 1999.

[126] J. Polchinski, *Scale and conformal invariance in quantum field theory*, Nucl. Phys., **B303**, 226-236, 1988.

[127] J. Rabin, P. G. O. Freund, *Supertori are elliptic curves*, Comm. Math. Phys., 114(1), 131–145, 1988.

[128] Riemann, B. *Collected mathematical works, scientific Nachlass and addenda*, R. Narashimhan, ed. Springer, Berlin, 1990 (In German).

[129] W. Rudin, *Functional Analysis*, McGraw and Hill, NY, 1991.

[130] B. Sakita, *Supermultiplets of elementary particles*, Phys. Rev. B, **136**, 1756-1760, 1964.

[131] A. Salam, J. Strathdee, *Super-gauge transformations*, Nucl. Phys. B, **76**, 477-482, 1974.

[132] A. Salam, J. Strathdee, *Supersymmetry and non abelian gauges*, Phys. Lett. B, **51**, 353, 1974.

[133] M. Scheunert, *The Theory of Lie Superalgebras*, Springer-Verlag LNM 716, 1978.

[134] J. Schwinger, *Spin, statistics and the TCP theorem*, **44**, 223–228, 1958.

[135] J-P. Serre, *Faiseaux Algebrique Coherent*, Ann. Math., 61, 197–278, 1955.

[136] Sorieau, *Structure des systémes dynamiques*, Dunod, Paris, 1970.

[137] T. A. Springer, *Linear Algebraic Groups*, Birkhäuser, Boston, 1998.

[138] R. Steinberg, *Lectures on Chevalley groups*, Yale, 1967.

[139] M. H. Stone, *Linear transformations in Hilbert space II: operational methods and group theory*, Proc. Nat. Acad. Sci. USA, **16**, 172-175, 1930.

[140] L. Susskind, *The world as a hologram*, J. Math. Phys., **36**, 6377-6396, 1995.

[141] M. E. Sweedler, *Hopf algebras*, Mathematics Lecture Note Series, W. A. Benjamin, Inc., New York, 1969.

[142] G. 't Hooft, *Dimensional reduction in quantum gravity*, arXiv:gr-qc/9310026.

[143] C. B. Thorn, *Reformulating String Theory with the* $1/N$ *Expansion*, First International A. D. Sakharov Conference on Physics, P. N. Lebedev Institute, Moscow, May 27-31, 1991, published in *Sakharov Memorial Lectures in Physics*, L. V. Keldysh and V. Ya. Feinberg, Eds., Nova Science Publishers Inc. 447-454, 1992.

[144] P. Tipler, R. Llewellyn, *Modern Physics* (4th ed.), W. H. Freeman, 2002.

[145] G. M. Tuynman, *Supermanifolds and supergroups*, MAIA 570, Kluwer Academic Publishers, 2004.

[146] V. S. Varadarajan, *Lie groups, Lie algebras, and their representations*, Graduate Text in Mathematics, Springer-Verlag, New York, 1984.

[147] V. S. Varadarajan. *Supersymmetry for mathematicians: an introduction*, Courant Lecture Notes Series, New York, 2004.

[148] V. S. Varadarajan, *Geometry of the quantum theory*, 2nd edition, Springer, 2007.

[149] M. Vasiliev, *Higher-spin gauge theories in four, three and two dimensions*, Int. J. Mod. Phys., **D5**, 763-797, 1996.

[150] A. Vistoli, *Notes on Grothendieck topologies, fibered categories and descent theory*, preprint `arXiv:math.AG/0412512`, 2008.

[151] D.V. Volkov, V.P. Akulov, *Is the neutrino a Goldstone particle?*, Phys. Lett. B, 46, 109, 1973.

[152] J. von Neumann, *Die Eindeutigkeit der Schrödingerschen Operatoren*, Math. Ann. **104** 1931, 570-578. Reprinted in J. von Neumann, Collected Works, vol. 2, A. H. Taub, ed., Pergamon Press, New York and Oxford,

220-229, 1961.

[153] A. A. Voronov, Yu. I. Manin, *Supercell decompositions of the superspaces of ags*, Sovr. Probl. Mat., **32**, 27–70, 1988.

[154] F. W. Warner, *Foundations of Differentiable Manifolds and Lie Groups*, GTM, Springer, 1983.

[155] W. C. Waterhouse, *Introduction to affine group schemes*, Graduate Texts in Mathematics, 66, Springer-Verlag, New York-Berlin, 1979.

[156] S. Weinberg, *The quantum theory of fields.* Cambridge University press, 1995.

[157] J. Wess, J. Bagger, *Supersymmetry and Supergravity*, Princeton Series in Physics (second edition), USA, 1991.

[158] J. Wess, B. Zumino, *Supergauge transformations in four dimensions*, Nucl. Phys. B, **70**, 39-50, 1974.

[159] J. Wess, B. Zumino, *Supergauge invariant extension of Quantum Electrodynamics*, Nucl. Phys. B, **78**, 1, 1974.

[160] H. Weyl, *The Theory of Groups and Quantum Mechanics*, Hirzel, Leipzig 1931 (in german), Dover Publications, New York, 1950 (english translation).

[161] E. P. Wigner, *Unitary representations of the inhomogeneous Lorentz group*, Ann. Math., **40**, 149-204, 1939.

[162] E. Witten, *Some comments on string dynamics*, In 'Future perspectives in string theory'. Proceedings, Conference, Strings'95, Los Angeles, USA, March 13-18, 1995, I. Bars, P. Bouwknegt, J. Minahan, D. Nemeschansky, K. Pilch, H. Saleur, N.P. Warner (eds.) 1996, Conference: C95-03-13, hep-th/9507121.

[163] H. Zhang, R. B. Zhang, *Dual canonical bases for the quantum general linear supergroup*, J. Algebra 304, 1026–1058, 2006.

Index